HUMAN MACHINE INTERFACE

Concepts and Projects

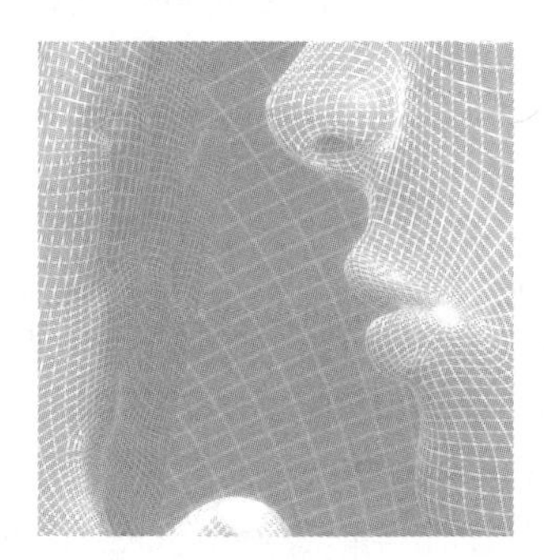

Samuel Guccione
James McKirahan

Industrial Press, Inc.

Industrial Press, Inc.
32 Haviland Street, Suite 3
South Norwalk, Connecticut 06854
Tel: 203-956-5593, Toll-Free: 888-528-7852; Fax: 203-354-9391
E-mail: info@industrialpress.com

Authors: Samuel Guccione and James McKirahan.
Title: *Human Machine Interface: Concepts and Projects, First Edition.*
Library of Congress Control Number: 2016944539

ISBN print: 978-0-8311-3582-9
ISBN ePDF: 978-0-8311-9372-0
ISBN ePUB: 978-0-8311-9373-7
ISBN eMOBI: 978-0-8311-9374-4

Sponsoring Editor: Jim Dodd
Developmental Editor: Robert Weinstein
Interior Text and Cover Designer: Janet Romano-Murray

industrialpress.com
ebooks.industrialpress.com

10 9 8 7 6 5 4 3 2 1

Table of Contents

Preface *ix*

Acknowledgments *xi*

Chapter 1: Human Machine Interface for Automated Control **1**

1.1 Introduction 1

1.2 Overview of PLCs 4

1.3 Introduction to Ladder Diagrams 5

1.4 Ladder Logic Diagrams for PLCs 10

1.5 Programming Ladder Logic Diagrams Using RSLogix 500 Software 11

1.6 Using HMI with PLCs 12

1.7 Additional Ladder Logic Diagram Programming 13

Chapter 2: Types of HMI Displays and HMI Software **15**

2.1 Introduction 15

2.2 Graphic Terminal HMI Display and HMI Software 15

2.3 Rockwell Automation PanelBuilder32 Graphic Terminal HMI Software 17

2.4 Computer-Based HMI Display and Software 18

2.5 RSView32 HMI Software for Computer-Based HMI Displays 19

Chapter 3: Field Devices and Sensors for PLCs **21**

3.1 Introduction 21

3.2 Annunciator Buzzers 21

3.3 Contact Relays 22

3.4 Directional Control Valves with Pneumatic Pistons 22

3.5 Fans 24

3.6 Inductive Sensors 25

3.7 Lights 25

3.8 Microswitches 26

3.9 Rotary Tables 27

3.10 Photo Sensors and Reflectors 27

3.11 Switches 28

3.12 Temperature Sensors 29

TABLE OF CONTENTS

Chapter 4: PanelBuilder32 / PanelView Graphic Terminal Labs		**31**
	Introduction	31
PV Lab 4-1: Basic Motor Starter		31
4.1.1	Initial Setup of the PanelView 550 (PV550) Operator Terminal	32
4.1.2	"CONFIG on power up" Procedure	33
4.1.3	Programming the MicroLogix PLC Using Rockwell RSLogix 500 Software and RSLinx Communication Software	34
4.1.4	Programming the PanelView Using PanelBuilder32	41
4.1.5	Creating and Entering Tags	44
4.1.6	Creating Button and Indicator Icons in the HMI Diagram	45
4.1.7	Loading the HMI Display Screen into the PanelView	49
PV Lab 4-2: Analog and Digital Displays		**51**
4.2.1	Introduction	51
4.2.2	Programming the MicroLogix PLC	51
4.2.3	Creating the HMI Diagram with PanelBuilder32	56
4.2.4	Gauge and Digital Display	56
4.2.5	Creating a Scrolling Message	59
4.2.6	Creating a Message Display	60
4.2.7	Creating a Bar Graph	61
PV Lab 4-3: Pump and Tank		**63**
4.3.1	Introduction	63
4.3.2	Programming the MicroLogix PLC	63
4.3.3	Creating the HMI Pump and Tank Display	64
4.3.4	Adding the Config and Motor Starter Screen Selectors to the HMI Display	70
PV Lab 4-4: Alarms		**73**
4.4.1	Introduction	73
4.4.2	Programming the MicroLogix PLC	73
4.4.3	Begin Creating the HMI Diagram with PanelBuilder32	74
4.4.4	Creating a Second Screen for the Alarm	74
4.4.5	Creating Tags	74
4.4.6	Adding the Bar Graph	75
4.4.7	Setting the Properties and States of the Bar Graph	76
4.4.8	Creating a Print List of Alarms	76
4.4.9	Alarm Banner Screen	78
4.4.10	Confirming that the Print List Button Works Correctly	80

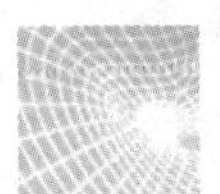

TABLE OF CONTENTS

PV Lab 4-5: Analog Input and Math Operations **83**

4.5.1 Introduction 83

4.5.2 Programming of the MicroLogix 1100 PLC and Subprogramming 83

4.5.3 Programming Math Operations, How PLC Math Works, and Subprogramming 84

4.5.4 Creating the HMI Display with a Temperature Bar Graph 85

4.5.5 Creating the Celsius HMI Display 85

4.5.6 Creating the Fahrenheit HMI Display 86

PV Lab 4-6: Security **89**

4.6.1 Introduction 89

4.6.2 Using Lab #2 to Add a Security Screen 89

Chapter 5: RSView32 Labs **95**

Introduction 95

RSView Lab 5-1: Basic Motor Starter **95**

5.1.1 Creating the PLC Program in RSLogix and Transferring to the PLC 96

5.1.2 Launching and Configuring RSView to Begin the HMI Creation Process 100

5.1.3 Creating Tags for Inputs and Outputs in the Tag Database 105

5.1.4 Creating Graphic Images for the Control Process 109

5.1.5 Button and Motor Creation 110

5.1.6 Adding Date and Time to Display 113

5.1.7 Animating Inputs and Outputs 115

5.1.8 Programming Animation of Inputs and Outputs 116

5.1.9 Adding Invisibility to Animation 117

RSView Lab 5-2: Analog and Digital Displays **121**

5.2.1 Introduction 121

5.2.2 Creating and Downloading the PLC Diagram 121

5.2.3 Creating and Configuring the Timer PLC 123

5.2.4 Creating the PLC Limit Test Symbols 124

5.2.5 Starting RSView in Preparation to Build the HMI 125

5.2.6 Creating Tags in the Tag Database 129

5.2.7 Creating Graphic Images for the HMI 135

5.2.8 Creating Buttons and Motor Graphics 136

5.2.9 Creating Digital Clock Display 139

5.2.10 Animating Input and Output Graphics 141

5.2.11 Adding Visibility and Invisibility to Animation 143
5.2.12 Adding an Analog Gauge and a Digital Meter Graphic 144
5.2.13 Adding a Warning Message Display with Color 146
5.2.14 Running the Project 149

RSView Lab 5-3: Pump and Tank 151
5.3.1 Introduction 151
5.3.2 Creating the PLC Program and Configuring RSView 151
5.3.3 Creating Existing Tags and Two New Tags 152
5.3.4 Creating the Pump and Tank Graphics 152
5.3.5 Animating the Pump and Tank and Displays 153
5.3.6 Creating Warning Messages and Display Features 155
5.3.7 Creating Pump Animation and Display Messages 157

RSView Lab 5-4: Alarms 159
5.4.1 Introduction 159
5.4.2 Configuring RSLogix, RSLinx, and RSView 159
5.4.3 Creating Tags in the Tags Database 160
5.4.4 Creating Graphics for the Alarms 162
5.4.5 Creating the Bar Graph 166
5.4.6 Creating an Alarm Button and Alarm String 167
5.4.7 Creating the Alarm Display Screen 171
5.4.8 Creating the Alarm Close Button 174

RSView Lab 5-5: Math Operations, Subprograms, and Alarms 179
5.5.1 Introduction 179
5.5.2 Programming the PLC 179
5.5.3 Configuring RSView 181
5.5.4 Creating the Tag Database 181
5.5.5 Creating the Graphic Screens 183
5.5.6 Creating a Digital Temperature Display 183
5.5.7 Creating an Analog Temperature Display 184
5.5.8 Creating the Celsius Button 184
5.5.9 Creating the Celsius Screen 185
5.5.10 Creating an Analog Display 186
5.5.11 Creating the Activate Button 186
5.5.12 Return to Main Screen Button 187
5.5.13 Creating the Alarm Button and Alarm Screen 188
5.5.14 Running the Project 189

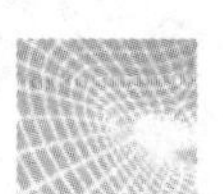

TABLE OF CONTENTS

RSView Lab 5-6: Security **191**

5.6.1 Introduction 191
5.6.2 Configuring RSView 191
5.6.3 Creating the Tag Database 192
5.6.4 Creating the Graphic Screens 195
5.6.5 Buttons 196
5.6.6 Entering Security Settings 197
5.6.7 Entering Security Codes 198
5.6.8 Entering User Accounts 199
5.6.9 Running and Testing the Security Program 199

RSView Lab 5-7: TrendX Charting and Data Logging **201**

5.7.1 Introduction 201
5.7.2 Configuring RSView 202
5.7.3 Setting Up the Tag Database 202
5.7.4 Creating the Main Screen 202
5.7.5 Creating Data Logging, Trends, and Activating the Data Logging 202
5.7.6 Creating TrendX Display to Show Logged Data 205
5.7.7 Configuring Project Startup Properties 211
5.7.8 Testing the Trend Graphics 211
5.7.9 Creating a Trend Graph 212

RSView Lab 5-8: Moving Animation **215**

5.8.1 Introduction 215

Appendix I: Hardware and Software Required for Chapters 4 and 5 Lab Activities ***221***

Appendix II: PLC and HMI Manuals for Software and Hardware Used in Labs ***222***

Appendix III: Constructing Digital Temperature Sensors ***223***

Index ***225***

About the Authors ***231***

Preface

Nowadays, students as young as fifth graders are working with robots by building, programming, and testing them. Robots do not operate in a vacuum in industrial processes: they are connected to Programmable Logic Controllers (PLCs) and Human Machine Interfaces (HMIs).

Although the concepts of PLCs and HMI are in the realm of engineering, this textbook was created primarily for beginners in the field of industrial automation. The automation presented consists of examples of actual industrial processes that exist in manufacturing. The large number of images and text comments were created to help students and industrial workers follow the complex processes of HMI and its interaction with PLCs.

Human Machine Interface: Concepts and Projects is based on notes and lab activities created and collected over several years of teaching PLCs with HMI at both the collegiate level and at industrial workshops. It will benefit students from high school through four-year colleges who may need exposure to PLCs and HMI. Faculty who are considering using this book in their curriculum will find that the first three chapters cover the fundamentals of HMI for automated control, types of HMI displays, and Rockwell Automation HMI software.

Starting in Chapter 4 are introductory labs that use beginning-level PLC hardware and HMI. The first few labs of this text can demonstrate how PLCs and HMI are connected to a small, inexpensive moveable arm robot and the communications between PLCs / HMI and the robot. Several advanced labs, presented in both Chapters 4 and 5, involve higher levels of programming of the PLC and HMI.

All labs are written for Allen-Bradley PLCs and Rockwell Automation HMI software and hardware. However, users of other brands of PLCs and HMI will still find that this text serves as a valuable guide to using PLC hardware and HMI programming.

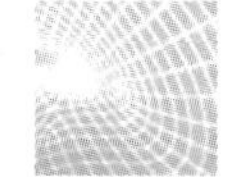

Acknowledgments

In our years of work preparing this textbook, others have participated in its creation. In particular, students at Eastern Illinois University (EIU) were the audience that helped us to develop the contents in this text.

However, one student, Mr. Uros Marjanovic, who was Dr. Guccione's graduate assistant, helped in the automation laboratory at EIU for several years. Uros was instrumental to the construction of our Industrial Automated Lab. This lab has enabled students and others to learn automation without any prior knowledge of PLCs and HMI. He helped these students understand the complexities of Industrial Automation. Thank you, Uros.

HUMAN MACHINE INTERFACE

Concepts and Projects

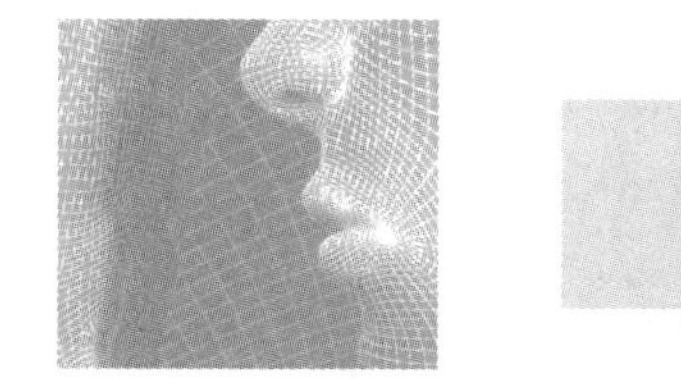

CHAPTER ONE

Human Machine Interface for Automated Control

1.1 Introduction

Human Machine Interface (HMI), once known as MMI (Man Machine Interface), is the instant remote control of machines in industrial processes, whether that control is automatic or by humans. Information that is both computerized and immediately available facilitates that instant remote control. It is key to HMI use.

The feedback information from the process machines allows employees to make immediate operational control decisions for overriding automatic control. To facilitate automatic and human monitoring of machines and processes, HMI requires use of Programmable Logic Controllers (PLCs).

HMI came into existence due to problems with simple electrical circuit control of industrial machines processes. Prior to HMI, automated control of industrial process machines typically involved the use of electromechanical devices mounted on or inside machinery enclosures. Human interface for control purposes between the external input and output devices of the machine was through these panels. The enclosures, then and now, are known as the Operator Interface or the Operator Panel.

The most common type of input and output control devices used in these panels were momentary pushbuttons (PB), pilot lights, and relays. PBs are electrically operated mechanical devices for controlling electricity flow to other devices.

Pilot lights (PL), which are basically small light bulbs in control circuits, are used to signal machine operation. The PLs are often controlled using pushbuttons in a circuit and signal the turning on or turning off of devices depending on operation configurations.

The relays used in circuits to control equipment are electrically-activated mechanical switches. These electrically-activated mechanical switches control electrical flow to devices such as motors. Other electromechanical input and output devices, such as solenoids and photo-switches, are used as well.

A big problem within Operator Panels was the failure of devices like the pilot lights and pushbuttons. Relays failed rather frequently too. In addition, pilot lights were deficient too for signaling purposes. They providing limited information to an operator monitoring and controlling a machine or process. Often when these devices failed entire systems stopped working.

For example, pilot lights are used frequently in industrial systems to signal an operator whether a machine is powered on or off. Situations would occur where a pilot light would electrically or mechanically fail. The operator would have to determine whether the pilot light had failed or whether the pilot light was actually signaling something about the process. The operator would ask, "Did the light go out because the machine actually had shut down or did the light go out because the light bulb in the pilot light burned out?"

It was not until the introduction of electronics and computers that situations like the one described could be easily resolved. With the introduction of electronics and computers, graphic electronic symbols for pushbuttons, lights, electronic relays, and other devices were developed. These were used to represent operations of industrial processes. The electro-mechanical Operator Panels were replaceable too.

These improvements eliminated problems once experienced using electro-mechanical devices since the electronic devices very rarely failed. Because automated industrial processes are becoming more and more complex, it is much more critical now to use graphic symbols of electronic devices provided through HMI to communicate component, operation, and systems failures.

HMI as presented in this text, consists of two types of digital graphic display. One type is a digital graphic display terminal that shows machine or process controls.The second type of digital graphic display uses the computer display screen to show not just a machine or process control but also multiple machine operations, multiple processes, or other types of devices that could be a part of an industrial manufacturing process.

The digital graphic display terminal is programmed by using specialized HMI software that uses graphical symbols to assemble the operations on the screen. Each type of HMI may require different software for this programming.

Typically, a display can include monitoring, controls, display of animated activities, alarms, and other tasks of the automated machine.

HMI is a primary part of providing an active display of what the automated machines or processes are doing. Techniques such as using color changes of graphics, icons of devices, animated graphics, etc., can show the flow of the production operation and many other operations.

In addition, HMI can become a part of a SCADA (supervisory control and data acquisition) system that can be in a large-scale process. An example of this type of process is the automotive manufacturing production line that uses several kinds of machines to produce assembled parts for automobiles or other types of vehicles.

HMI can show the reactions of outputs from a machine in response to inputs to the machine from various kinds of control devices. Furthermore, HMI can capture monitoring data collected in

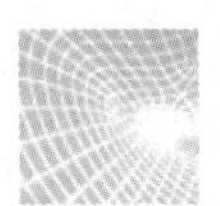

a specialized industrial automated control computer called a known as a programmable logic controller (PLC). This data is accessed and processed in the HMI software for display as a graph. Spreadsheets and other types of business and engineering presentation media are used today.

Through its programming, the HMI graphic terminal can display more information, for example, a specific condition that took place in a machine. The HMI graphic terminal display can provide information such as alarm activations due to malfunctions and, with directions and images, how to deal with the malfunctions.

HMI systems were specifically developed to interface to PLCs. Besides displaying input and output animated reactions, the HMI can be set up to display data from the PLC in various kinds of graphics, including charts, histograms, X-Y charts, and bar graphs.

Today, more critical uses of computer-based HMI involves acquiring data stored in the PLC's memory that is captured as part of the monitoring process. The HMI-captured data is processed by the computer for storage in a database.

Due to the creation of business office and industrial products, more powerful capabilities have been developed so that they can interface with today's HMI software. Business office products such as spreadsheets and databases are now in standard use with HMI.

Therefore, data captured in a PLC can be ported by the HMI system to a spreadsheet for use by upper management to make decisions about process operations. In addition to the use of spreadsheets, the ability to store data in a database is now a very powerful tool for managing processes.

In addition, industrial products, such as Rockwell Automation's Factory Talk Transaction Manager software, also are interfaced to HMI. Factory Talk Transaction Manager is an industrial transaction software that shares the shop floor data collected from PLCs and other devices (machines/industrial processes) with the top floor enterprise systems (spreadsheets/managerial level) where this industrial data is stored in the Transaction Manager's internal database. Industrial transactions are activities such as automating the industrial data logging, controlling the plant floor business rules, monitoring and controlling quality, and managing recipes used in PLC-controlled processes. Integration with ERP software is also possible.

Moreover, the animated display of the machines process itself can be configured so that the machine operator can clearly view what the process is doing at any time, including the ability to easily change the process operations through programmed commands to change the process.

The control process can be displayed on the HMI display device in an animated form by using moving or fixed objects with color choices and color changes to represent changing settings, blinking simulated lights, making objects invisible under certain conditions, and many other graphic conditions. Introductory level laboratories provide experiences in programming and using Rockwell Automation human machine interface software as well as Allen-Bradley HMI and PLC hardware. We assume that readers have little or no experience with automated control devices such as PLCs, HMI hardware, and HMI software.

Because the PLC is the device that HMI interacts with and controls, the next section provides an overview of PLCs. This discussion will help when working with the PLCs used in the HMI labs in Chapters 4 and 5.

1.2 Overview of PLCs

Prior to the mid-to-late 1960s, industrial processes used electromechanical relays in automatic control of manufacturing processes and machines. A relay is an electromechanical switch that consists of an electromagnet, called a solenoid, with contacts attached to an arm that can move, and open and close fixed contacts. When the electromagnet is electrically activated, the moveable contacts move together and make a path for electricity to flow. When the electromagnet is deactivated, the moveable contacts move apart. This interrupts the path for current flow. Although relays perform well in some situations, like all mechanical devices they are plagued with high failure rates, especially when applied in high speed manufacturing processes.

In late 1968, a key event took place in the automation arena. The General Motors Company sent a request for proposals for an electronic device to replace electromechanical relays in their automotive manufacturing lines.[1] The winning proposal was from Bedford and Associates of Bedford, Massachusetts, which created the programmable logic controller[1].One employee who worked on the project was Dick Morley, who is considered to be the "father" of the PLC.[2, 3]

Figure 1-1a: Allen-Bradley MicroLogix 1200 PLC (Courtesy of Rockwell Automation, Inc.)

Figure 1-1b: Allen-Bradley CompactLogix PLC (Courtesy of Rockwell Automation, Inc.)

Figure 1-1 shows an up-to-date version of a PLC. As an industrial control computer, a PLC performs special computer activities, but does not look like the computer or tablet in daily use. It is not Windows-based and does not have a keyboard, mouse, DVD drive, etc. It must be programmed with a traditional computer using special programming software that uses a graphical computer language.

A single PLC, such as the one shown in Figure 1-1a, is known as a fixed I/O PLC; it typically controls a single machine or possibly a small machine process and is limited by its own inputs and outputs. Multiple PLCs or a modular PLC, as shown in Figure 1-1b, have additional, readily avail-

[1] Hayden, E., Assante, M. & Conway, T. (2014, August). *An abbreviated history of automation and ICS cybersecurity.* Retrieved online from https://ics.sans.org/media/An-Abbreviated-History-of-Automation-and-ICS-Cybersecurity.pdf

[2] Moody, Patricia E. & Morley, Richard E. (2007). *The Technology Machine: How Manufacturing Will Work in the Year 2020.* NY: The Free Press.

[3] Dunn, Allison. (2008). The father of invention: Dick Morley looks back on the 40th anniversary of the PLC. *Automation Magazine.* Retrieved online at http://www.automationmag.com/features/the-father-of-invention-dick-morley-looks-back-on-the-40th-anniversary-of-the-plc.html.

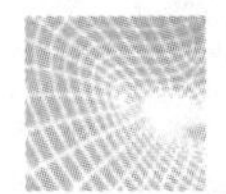

able inputs and outputs that can control multiple machines in a factory-wide or enterprise-wide operation; they may be connected together. In addition, PLCs can interface to other sophisticated automation devices, especially robots, in addition to HMI.

Since the 1970s, improvements in computer electronics and programming as well as computer communication through networking, have changed how computers interact with each other and later how PLCs and HMI were included in this interaction.

In the 1990s, specialized human machine interface (HMI) hardware and software became widely available, enabling PLCs to interact that have Graphic Terminal Displays with electronic (non-mechanical) switching devices. Furthermore, data values collected within the PLC were able to be exported into spreadsheets and databases using computerized HMI software.

The development of the Ethernet and the Internet completely changed the communication process in distributed control systems (DCS), including PLCs, HMIs, and computer systems such as database servers that could be located anywhere in the world.

Additional sophisticated specialized software, such as Rockwell Automation Industrial Transaction Manager software, established the ability for data retrieved from a PLC to be processed through HMI in order to share data and information between enterprise applications such as corporate-wide databases and plant floor equipment and automated systems.

The sophistication of automation systems through the inclusion of HMI increases the speed and efficiency; it also makes crucial information about a control process available to all who need this data.

1.3 Introduction to Ladder Diagrams

Before briefly describing how a PLC works, this section introduces the Ladder Diagram, which is used for programming PLCs. The Ladder Diagram in Figure 1-2 is called a Relay Logic Ladder Diagram. This diagram provides a schematic format of the original form displayed by a machine's automated controls. It is called a ladder diagram because the two vertical lines on the left and right sides, combined with the horizontal lines (called rungs) that connect the two vertical lines, resemble a ladder. Electrical power is applied to the rungs, with electricity flowing from hot to neutral.

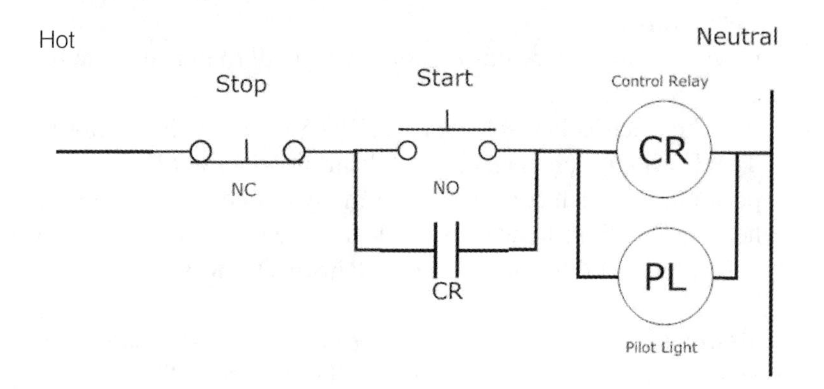

Figure 1-2: Relay Ladder Logic Diagram

Figures 1-3: Momentary STOP pushbutton (red) right, and 1-4: Momentary START pushbutton (green) left

Figure 1-5: Momentary pushbutton electrical contacts

This Relay Ladder Logic Diagram shows the devices used to start and stop a motor. It contains the four most common electromechanical devices used in many automated control circuits, momentary open or closed pushbuttons, control relays, and pilot lamps. The STOP and START pushbuttons are input devices that are connected to input terminals in the input module. The CONTROL RELAY and PILOT LIGHT are output devices that are connected to the output terminals of the output module of the PLC.

The CONTROL RELAY CONTACT (CR)-| |- provides a special condition that will be described below.

Figure 1-3 above shows the momentary STOP pushbutton, which is red, and Figure 1-4 shows the momentary START pushbutton, which is green. Pushbuttons are mechanical; they are operated by pressing, then releasing them, which is the momentary action. Pushing on the button causes it to "activate"; releasing then causes it to "deactivate".

Figure 1-5 provides a bottom view of a momentary pushbutton. Two small round gold beads hold the contacts of the pushbutton.

Figure 1-6 shows the operation of the electrical switch contacts. The STOP switch is a closed circuit (electricity flows) and the START switch is an open circuit (electricity does not flow).

In Figure 1-6a left, the STOP pushbutton switch contacts are normally closed (NC) as shown. This means electricity will flow when the STOP pushbutton is "NOT PRESSED". In Figure 1-6b left, electricity will cease to flow when the STOP pushbutton is "PRESSED", which opens the switch contacts.

In Figure 1-6a right, the START pushbutton switch contacts are normally open (NO), as shown. This means electricity will not flow when the START pushbutton is "NOT PRESSED". In Figure

Figure 1-6:
Switch contacts for
STOP (NC) and START (NO)

Normally Closed (NC) Pushbutton

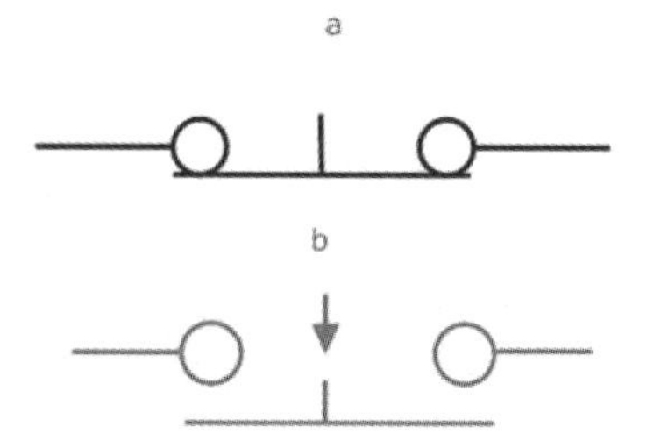

Normally Open (NO) Pushbutton

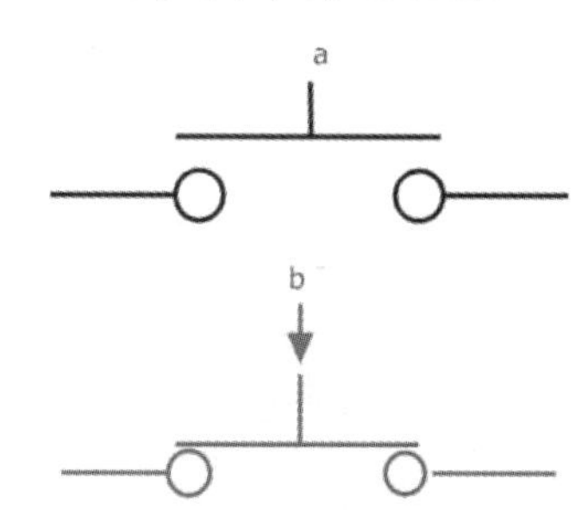

1-6b right, electricity will flow when the START pushbutton is "PRESSED", which closes the switch contacts.

Figure 1-7 shows a Pilot Light, which is an output. A small light bulb inside the Pilot Light will turn on when electricity flows through it. A red light might indicate a stopped condition. Lights of other colors are used as well for other conditions.

Figure 1-7: Pilot light

Figure 1-8 below shows a Contact Relay. It has an electromagnet and sets of contacts attached to an arm.

Figures 1-9a and 1-9b shows the schematic diagram and operations diagram for a Contact Relay operation. Note: the solenoid is the electromagnet. In Figure 9a, the contact relay has two sets of contacts called Poles. Each Pole has a Normally Closed (NC) state and a Normally Open (NO) state. If this Contact Relay was used in Figure 1-2, the solenoid connections would be wired to the START button, then to NEUTRAL. The NO CONTACT in Pole 1would be the Contact Relay CR wired across the START button.

Returning to the Relay Ladder Logic Diagram (Figure 1-2), when the circuit is powered up, the Control Relay and Pilot Light will be off.

Figure 1-8: Contact relay

Pressing the START button closes the START NO contacts. Electricity can now flow from the HOT side to the NEUTRAL side through the STOP button NC contacts, through the START closed contacts, and finally through both the Control Relay solenoid and the Pilot Light bulb. Because the Control Relay solenoid is an electromagnet attached to a mechanical linkage, it causes the Control Relay (CR) Contacts (Pole 1) to close, thereby establishing an alternate path for electricity to flow around the START button.

These contacts maintain a path for current flow around the START button, which keeps it flowing through the Control Relay solenoid and the Pilot Light bulb. It is customary to say that the automated control circuit is now "running". The portion of the Relay Ladder Logic Diagram circuit with the

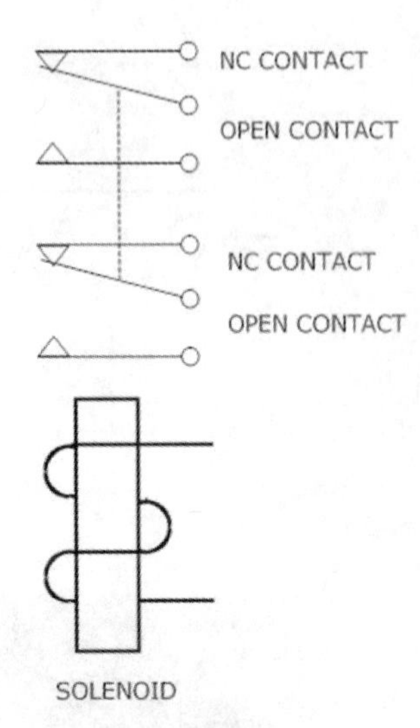

Figure 1-9a

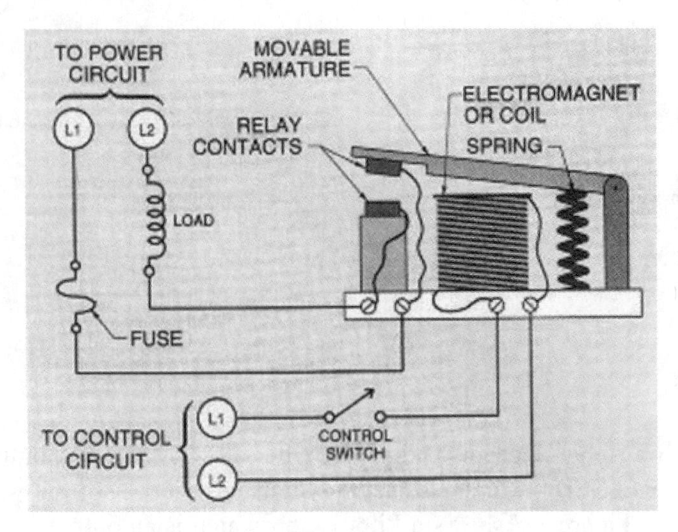

Figure 1-9b

Control Relay (CR) contacts, where the relay maintains power to itself through its own contacts, is often referred to as a *memory circuit* or a *latching circuit.*

Although not shown in this ladder diagram, an additional Control Relay contact (Pole 2) is used toward actual manual control of a motor or other device.

To stop the circuit from running requires only that the STOP pushbutton be pressed momentarily, then released. This action interrupts the flow of electricity, causing the Control Relay solenoid and Pilot Light to deactivate. When the Control Relay deactivates, it will also deactivate both Poles. The circuit has now "stopped".

All of the devices depicted in the Ladder Diagram have a fairly high failure rate because they are electromechanical. The PLC and HMI can replace these electromechanical devices with electronic devices that have a very high reliability as well as additional capabilities.

Because some industrial operations need only the electromechanical devices described above, a PLC can sometimes be used without HMI. For example, Figure 1-10 shows a PLC trainer board with pushbutton switches, lights, and micro-switches on the lower left side permanently wired to the PLC.

More details about the connections of the switches and lights to the PLC follow. PLCs have both input and output modules. Switches, sensors, and other input devices

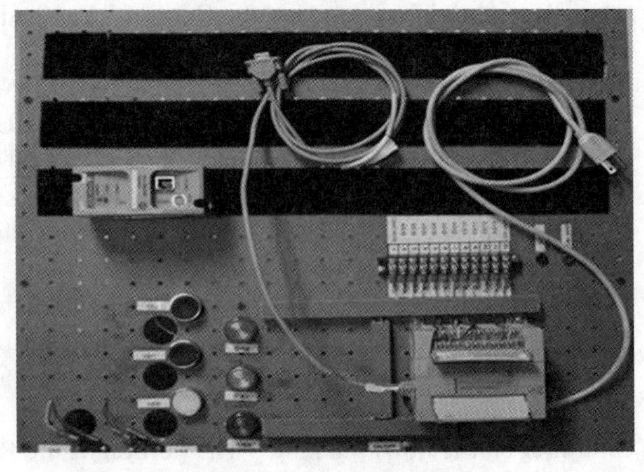

Figure 1-10: PLC trainer board

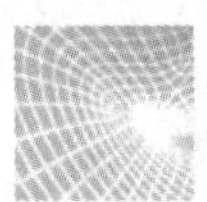

are wired to the input module. Lights, contact relays, and other output devices are wired to the output module.

In the PLC shown in Figure 1-11, the input module is the top two rows of screw connections whereas the output module is the bottom two rows of screw connections.

Input and output devices can be permanently wired to the PLC as on the trainer board. The devices can be placed in the ladder logic diagram as needed without the necessity of physically swapping or rewiring them. In the traditional Relay Ladder Logic Diagram, a new wiring of devices would need to occur when a change was necessary.

The trainer board also has a black terminal strip to accommodate additional devices. Therefore, you can construct numerous types of complex automated control circuits to easily test circuit operation.

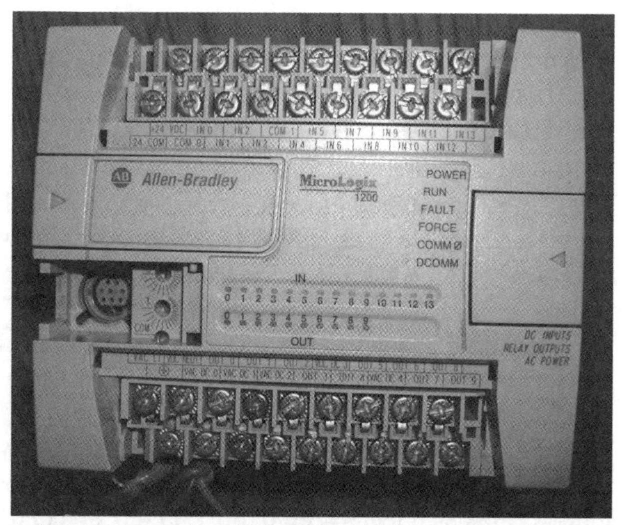

Figure 1-11: Input modules (top) and output modules (bottom) (Courtesy of Rockwell Automation, Inc.)

PLCs typically have input and output modules that operate as digital devices; thus they have only two states of operation: open or closed, on or off, etc. The input module responds to an on or off electrical signal because the device connected to the input has a source of electricity, typically VDC (Voltage Direct Current). When the device is activated, the VDC source sends the electrical signal to the input. The PLC has a source of internal power for input devices.

The output module sends an on or off electrical signal to an output device. The internal power source is not used for outputs. A separate power source must be provided to connect to the output device.

However, some PLCs may also have special input and output modules that operate in an analog state. Here is an example of an analog operation. If a PLC is monitoring the temperature of a process using a temperature sensor, it would be connected to an analog input. The temperature sensor sends a continuous stream of electrical signals (voltages) to the analog input. This electrical signal is

Figure 1-12: Ladder Logic Diagram for PLC (Courtesy of Rockwell Automation, Inc.)

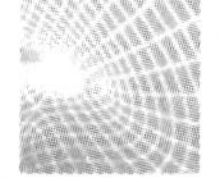

processed in the PLC and the HMI. An HMI would be used here to display the temperature. Chapter 4 (Lab 5) and Chapter 5 (Lab 5) demonstrate this concept.

1.4 Ladder Logic Diagrams for PLCs

The PLC programming process involves translating the Relay Ladder Logic Diagram to a Ladder Logic Diagram. This process involves using icons and specialized software. With practice, it is common to create a Ladder Logic Diagram directly without needing to create a Relay Ladder Diagram first.

Figure 1-12 shows an example of a Ladder Logic Diagram for Figure 1-2 (Relay Ladder Logic Diagram). Note the similarity in the circuit construction between the two diagrams. Also note the names above the icon.

The Ladder Logic Diagram is the computer program that needs to be created and subsequently entered into the PLC so that it provides the desired control actions. The alphanumerical values are addresses for each icon — I:0/0, I:0/1, O:0/0, and O:0/1 — which will be part of the PLC and HMI programming.

The addresses for the PLCs used in this text, called single line addresses, consist of three parts that provide the data needed by the PLC internally. The first part is a letter that identifies the part of the PLC that is being used. For example, the letter I is the Input module and the letter O is the Output module.

The second part of the address identifies the slot number, which is 0 for the PLCs used here. The PLCs used here have a single slot that has both Input and Output modules.

The third part of the address identifies the channel number being used in the Input or Output module. Channels 0 and 1 are used in the Ladder Logic Diagram shown in Figure 1-12. The PLC in Figure 1-11 can have 14 input channels numbered 0 to 13 and 10 output channels numbered 0 to 9.

For the Ladder Logic Diagram to do its job once it is loaded into the PLC, physical input and output devices need to be wired to the PLC. In the logic diagram, there are two inputs that need physical momentary pushbutton switches and two outputs that need a control relay and a pilot light.

For the STOP icon in the logic diagram to work, a physical STOP (NC) pushbutton needs to be wired to channel 0 of the 0 slot Input Module and connected to the internal power source of the PLC. Also, for the START icon, a physical START (NO) switch needs to be wired to channel 1 of the 0 slot Input module and connected to the internal power source.

Similarly, a physical Control Relay needs to be wired to channel 0 of the 0 slot Output module and connected to an external power source. Also, a physical Pilot Light needs to be wired to channel 1 of the 0 slot Output module and connected to an external power source.

The circuit works by momentarily pressing and releasing the START button. Releasing the button activates the Control Relay and makes the Pilot Light turn on and stay on. Momentarily pressing and releasing the STOP button will deactivate the Control Relay and turn off the Pilot Light.

This PLC has monitor lights for the I/O modules, as shown in Figure 1-13. The STOP button Input light 0 is on because of the NC contacts. Pressing the START button will cause Input light 1 to turn on and Output lights 0 and 1 to light up. These lights are used to be sure the program is work-

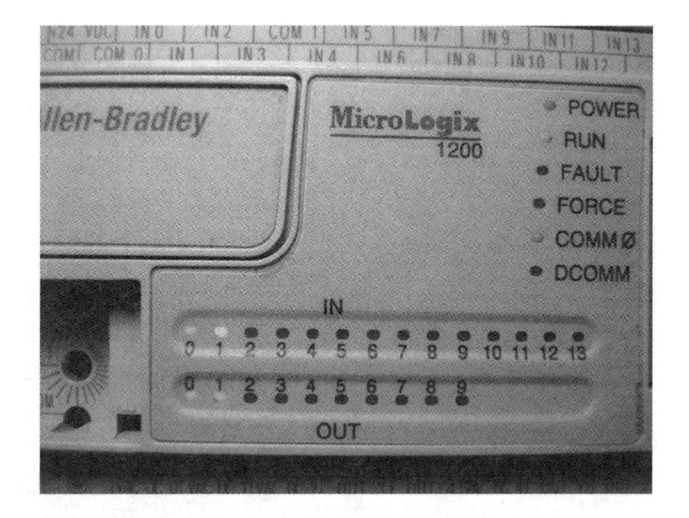

Figure 1-13: Monitoring lights on PLC panel (Courtesy of Rockwell Automation, Inc.)

ing, but also in troubleshooting to be sure the Input and Output modules are functioning correctly as well as the program

Note the six lights in the upper right corner of Figure 1-13. The COMM 0 light means the PLC is connected to the computer and both are communicating. The RUN light shows that the PLC is in RUN Mode and, of course, the POWER light shows that the PLC is power up. The unlit FORCE light identifies when the PLC causes an input change of state from the RSLogic 500 control panel only. When it is on, the FAULT light shows that a hardware error has occurred in the PLC. The DCOMM light is a special communication setting.

The internal operation of the PLC is based on binary electrical signals, not only for its internal processing, but also for the operation of the inputs and outputs. The inputs receive binary signal voltages that are applied to electronic circuits in the PLC; the outputs send binary signal voltages to output devices. The input and output modules for the PLCs in this text are connected for 24 volts DC. This level is for safety reasons because beginners often may make mistakes. The devices wired to the PLCs in this text were purposely selected to be at this low voltage level for both costs and safety.

This completes the introduction to the PLC operation and Ladder Logic Diagramming.

1.5 Programming Ladder Logic Diagrams Using RSLogix 500 Software

Creating and entering the Ladder Logic Diagram into the PLC requires specialized software. Rockwell Automation RSLogix 500 and RSLinx are the software packages commonly used to program Allen-Bradley PLCs.

The first step is to install the RSLinx communication software, then install the RSLogix 500 programming software. Follow the installation procedures included with these software packages. Also, ensure the PLC is connected to the computer that has RSLogix installed.

It is assumed that, when working on the labs in Chapters 4 and 5, the software packages listed in Appendix I have been installed.

Because each lab in Chapter 4 and 5 has complete standalone step-by-step instructions on how to assemble the Ladder Logic Diagram and enter it into the PLC, the programming process for a PLC will not be described here.

1.6 Using HMI with PLCs

Figure 1-14 shows an example of a Ladder Logic Diagram created for the PLC when using an HMI. This diagram is an edited version of the Relay Ladder Logic Diagram shown in Figure 1-2. Note the similarity of the circuit construction and use of graphic icons between the Figure 1-2 Ladder Diagram and the Figure 1-12 Ladder Logic Diagram for PLC.

The names above the icons and the alphanumerical values — B3:0/0, B3:0/1, and B3:0/10 — become part of the programming process in the PLC and especially in the HMI programming.

The following example of a PLC address is similar to the I:x/x and O:x/x addresses used in the PLC binary memory values stored in the PLC memory and in the HMI display memory.

The B3s are internal These values are described below.

The B3:0/0 icon -|/|- is called an "Internal Binary Input". The B3 identification means binary. That is, it can control a memory location identified as B3:0/0 where a binary 1 or 0 (on/off) is entered. In addition, the B3:0/0 icon is equivalent to the operation of an electromechanical Normally Closed (NC) STOP switch. It also has all of the same properties described above for the I:0/0 address with the following exception: B3 addresses cannot control any external physical device or be controlled by any external physical devices.

Similarly, the B3:0/1 icon -| |- is an Internal Binary Input. It has the same properties as the electromechanical Normally Open (NO) START switch described above for the I:0/1 address.

Note the two B3:0/10 icons labeled MTR_LATCH in Figure 1-14.The left icon is an Internal Binary Input and the right is an Internal Binary Output. Both of these icons control B3 memory locations in the PLC memory and in the HMI device memory.

All B3 icons provide an internal operation that can be controlled by the PLC Ladder Diagram, which in turn can control an HMI icon on a Graphic Terminal Display HMI or a Computerized Dis-

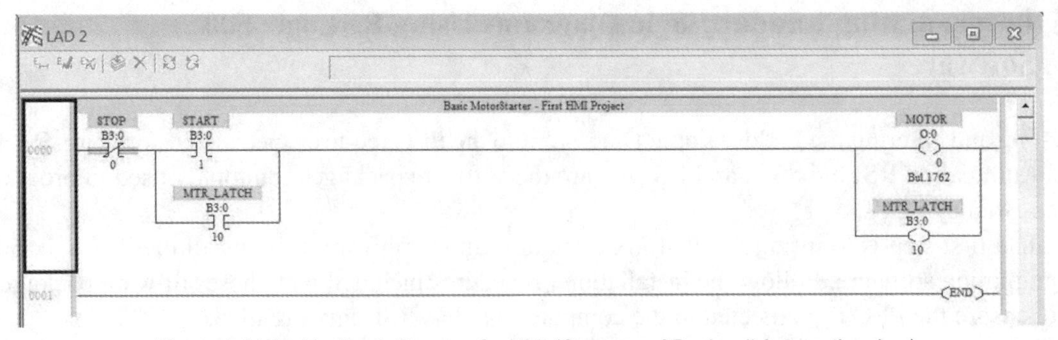

Figure 1-14: Ladder Logic Diagram for HMI (Courtesy of Rockwell Automation, Inc.)

play HMI. The switch icons in the HMI diagram use the memory addresses B3:0/0 and B3:0/1 to turn inputs on or off by touching those switch icons on the HMI device screen or using a mouse to click an icon. The B3:0/0 and B3:0/1 turn on the B3:0/10 and B3:0/10 outputs as well as the O:0/0 output.

Here is where the HMI interacts with the PLC. The HMI display device is programmed with a diagram of the automated controls, as shown in Chapter 3, Figure 3-2, for graphic terminal display and Figure 3-3 for computer-based display.

The PLC ladder logic diagram in Figure 1-14 shows the electronic icons that implement the Figure 1-2 and 1-3 controls.

1.7 Additional Ladder Logic Diagram Programming

Automated control circuits in an industrial environment can be very large. Single ladder diagram control circuits are used in this text, but single diagrams are not common in industry. Large industrial control circuit processes are typically broken down into several smaller circuits called subprograms. Subsequently, most industrial Ladder Logic Diagrams consist of a large number of subprograms. The industrial diagram has a Main program at the beginning that interacts with the subprograms that make up the remainder of the diagram.

To implement these large diagrams and other industrial circuits, there are numerous other types of icons used to implement control devices. Figure 1-15 shows the graphic icon area of the PLCs programming software RSLogix 500.

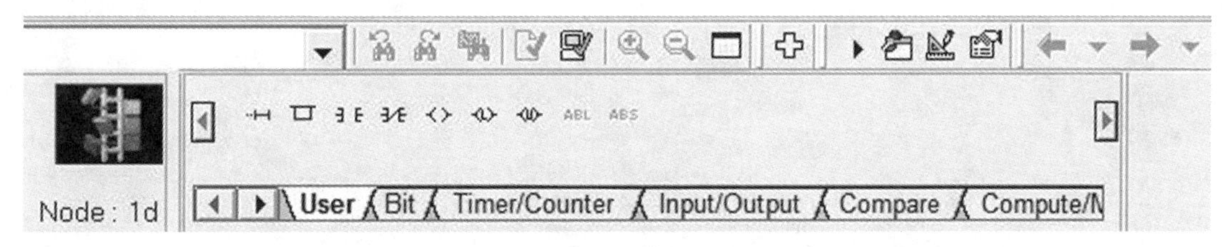

Figure 1-15: RSLogix 500 logic icon panel (Courtesy of Rockwell Automation, Inc.)

Beside the standard logic symbols for inputs, outputs, and others, as shown in Figure 1-15, note the other tab choices. These include logic symbols such as Compare called LIM (Limit Test), CTU (Count Up), Logical Operators called LEQ (Less Than or Equal) and GEQ (Greater Than or Equal), Math Operation called MUL (Multiply) and DIV (Divide), Move Operation called MOV (Move), RET (Return from Subroutine), Subprogram called JSR (Jump to Subroutine), and Timers called TON (Timer On-Delay). All of these icons are used in the Chapter 4 and 5 labs.

In PLC and HMI hardware and software, these devices appear as rectangles with the needed data values entered in fields located inside the rectangle. The Help selection in the top command line gives additional information about each icon; the Labs have step-by-step instructions of what to do for creating the Ladder Logic Diagram (Figure 1-16).

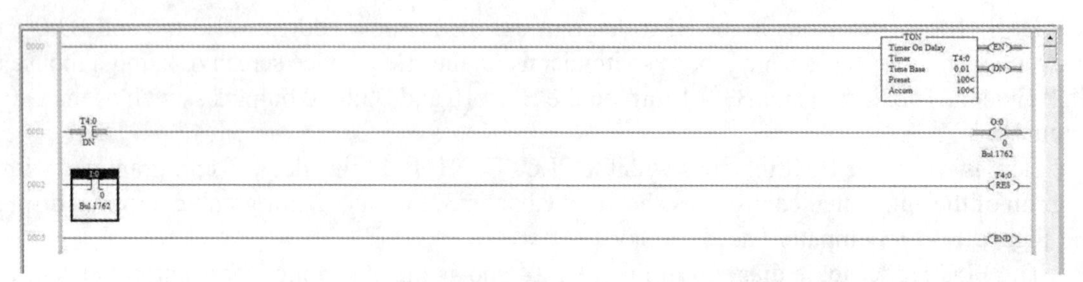

Figure 1-16: TON Ladder Logic Symbol Diagram (Courtesy of Rockwell Automation, Inc.)

The PLC programming process requires specialized software because the creation of Ladder Logic Diagrams involves a graphic-based programming technique. Each brand of PLC tends to use proprietary programming software program due to the internal operation of the PLC and the types of icons used in PLC ladder diagrams.

Assume that some type of trainer similar to the PLC trainer shown in Figure 1-10 is used to implement the labs. Also, assume that Rockwell Automation RSLinx communication software and RSLogix 500 PLC programming software are already installed on the programming computer for the labs.

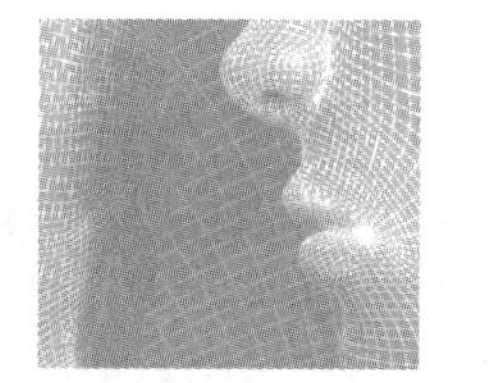

CHAPTER TWO

Types of HMI Displays and HMI Software

2.1 Introduction

This chapter briefly describes the types of HMI displays and HMI software programs available from Allen-Bradley/Rockwell Automation, as used in this text. Types of HMI can be characterized by the display device. There are two types: a Graphic Terminal HMI Display and a computer-based HMI screen.

Each HMI type has a specific use that is described below.

2.2 Graphic Terminal HMI Display and HMI Software

The Graphic Terminal HMI Display is a standalone video display terminal with a computer and memory. This type of HMI display is often called *operator interface* or *operator panel*.

Graphic Terminal HMI Displays have many types of features such as touch screen, serial connection, Ethernet connection, keyboard, monochrome display, color display, and various screen sizes and resolutions.

Figure 2-1 shows two PanelView (PV) models; both are Allen-Bradley graphic displays. The PanelView model on the left has an integrated keypad located below the screen. The PanelView on the right has a touch screen type display. The terminals are monochromatic. Each panel has a model number. The panel on the left, a PanelView 300, is referred to as PV300. The panel on the right, a PanelView 550, is referred to as PV550. A PV550 is used in the Chapter 4 Labs. Note the different screen size of these displays.

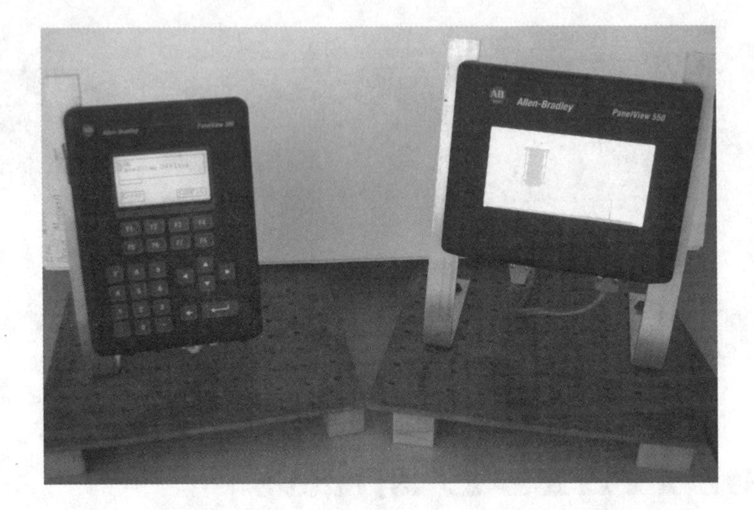

Figure 2-1:
Allen-Bradley PanelView 300 (left) and PanelView 550 Operator Terminals (right) (Courtesy of Rockwell Automation, Inc.)

Because Allen-Bradley HMI display devices are the focus of this text, exploring the Allen Bradley website, to review the different types of Graphic Terminal Displays available to a user is highly encouraged. See:

http://ab.rockwellautomation.com/Computers

This website provides beneficial information regarding types and features currently available for use as well as future terminal selection and application in automated processes.

The Graphic Display HMI Terminal is typically connected to a single PLC controlling a single machine or a machine with one or more supporting devices. Graphic images are created and transferred to the terminal display screen, which represents an animated version of the machine control process logic.

Physical control devices, such as switches in hardwired industrial ladder circuits, are replaced by electronically created graphics in the machine process control layout with PLCs and on the HMI display screen. Some output devices, such as pilot lights, can also have an electronic equivalent using PLCs and HMI. The HMI display shown in Figure 2-2 is programmed using a separate computer with specialized HMI programming software called *PanelBuilder32*.

Figure 2-2 shows an example of an automated control layout showing some basic graphics. Remember that the Ladder Logic Diagram that is programmed into the PLC essentially becomes the control diagram that is implemented in the HMI display. This diagram was created based on the basic Ladder Logic Diagram shown in Chapter 1, Figure 1-2.

In Figure 2-2, note the STOP and START buttons and the output motor graphic. The graphical icons in HMI — like the STOPPED motor icon — can be animated. For example, the buttons can be animated to show when the motor is running or not. Because the PV550 has a touch screen, touching the START or STOP button controls the motor.

The Goto Config Screen graphic in the lower right corner of the display is used to set controls for the terminal display, such as brightness, contrast, etc.

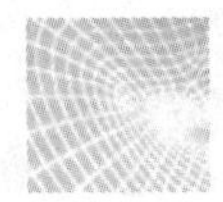

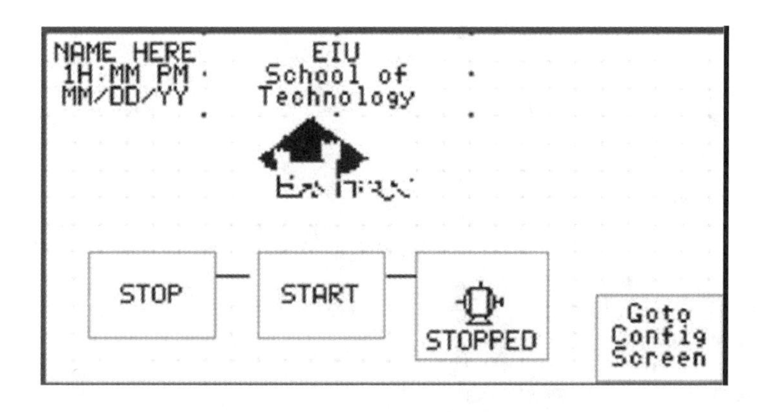

Figure 2-2:
Automated control layout on PV550 Graphic Display HMI Terminal (Courtesy of Rockwell Automation, Inc.)

2.3 Rockwell Automation PanelBuilder32 Graphic Terminal HMI Software

There are two software packages for the two Allen-Bradley HMI display devices. The software for the Graphic Display HMI Terminal is called *Rockwell Automation PanelBuilder32*. The second software, *RSLinx32*, provides communications drivers shared between programs. *AB Utilities* is support software that is a part of, and common to, both PanelBuilder32 and RSLinx. Allen-Bradley *PanelView* graphic terminals are programmed with the PanelBuilder32 software, as used in the figures shown here. Although the figures are monochrome, color display PanelView terminals are also available. See:

http://support.elmark.com.pl/rockwell/Literatura/2711-60.pdf

Figure 2-3 shows a modified version of Figure 2-2 on the Graphic Terminal HMI Display. Besides the starting and stopping graphics, the motor graphic can be blinking (not shown) when the motor is running and not blinking when the motor is stopped. In addition, PanelBuilder32 has the capability of creating analog dial graphs and showing digital numerical displays so the operator can monitor the speed of the motor to be sure it is running at the desired speed.

The PanelBuilder32 software contains a library of graphic images for creating process graphic displays. However, additional images can be uploaded to the library. Importing a company logo is a common addition to the image database. Low resolution pictures can also be imported, depending upon the screen resolution. The screen size of the terminal does limit the quality of the image when displayed on the screen.

Figure 2-4 shows how the process can be displayed on the terminal using graphic representations, such as a pump and tank, which give more visual information to the machine operator.

Text, animated images, histograms, digital analog meters, and alarms are among the capabilities available for inclusion in automated control diagrams. If a color-based terminal is available, diagrams can be colorized to indicate critical conditions, for example, in red. Multiple screen displays are also used in HMI devices because the relatively small display area sometimes require multiple screens to be created in order to produce the desired screen display information.

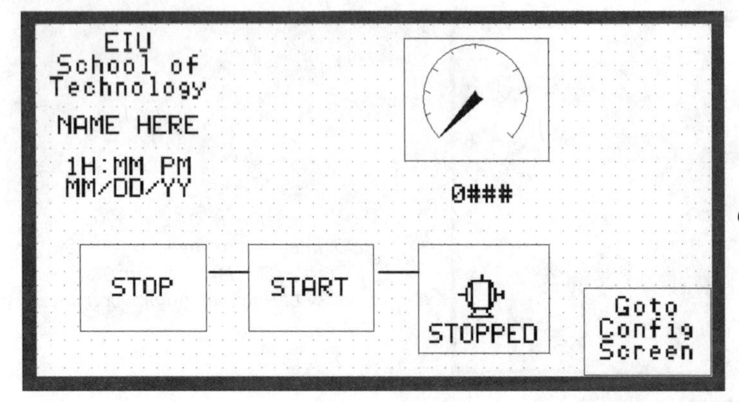

Figure 2-3:
Operator terminal graphic process diagram
(Courtesy of Rockwell Automation, Inc.)

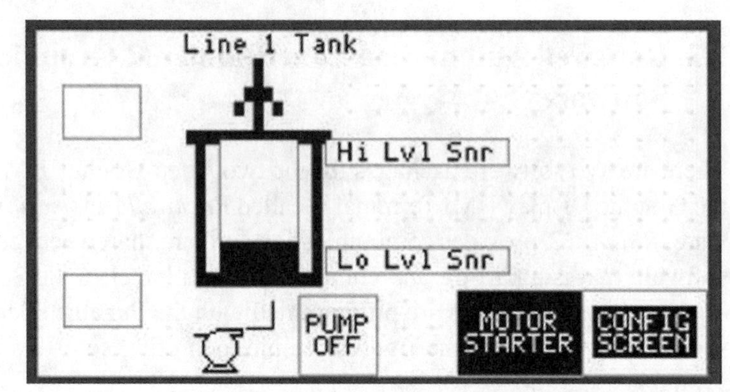

Figure 2-4:
Graphic terminal pump and tank
(Courtesy of Rockwell Automation, Inc.)

The PanelBuilder32/PanelView (PV) labs in Chapter 4 show a majority of basic capabilities of the PanelBuilder32 type of HMI software.

The labs in Chapter 4 provide step-by-step directions about which icons could be used, how they are selected, and why the icons are arranged on the HMI display device in a manner that represents the control process layout. The HMI software is also used to create special items called tags and determine their data values that control the data transferred to and from the HMI display.

Note that text information — including date, time, organization name, logo, and scrolling text at the bottom of the screen — can be placed on the screen.

The Graphic Display HMI Terminal can also be programmed to monitor and display various data in the PLC memory about the process on the display screen that could be of use, such as motor speed, how long the motor ran before stopping, etc.

2.4 Computer-Based HMI Display and Software

The *Computer-Based HMI Display* device is the computer's video screen with specialized computer-based HMI programming software installed. The video screen displays high resolution graphics that show the process and its operations. This feature can be used for small production lines and plant-wide operations, and even enterprise-wide operations.

The computer-based HMI software can be used in a monitoring mode to capture and display PLC data values related to the process being controlled. In addition, this software can be interfaced to spreadsheets, databases, and other specialized industrial software typically used today.

Data from the process can be used to create real-time screen displays; this data can be automatically entered on a spreadsheet which, in turn, could be used to help make decisions about the process. The ability to store the data in a database provides the capability of reviewing processes for efficiency, quality, and other factors.

Computer-based HMI is typically used with multiple PLCs that could be a part of a larger automated control system that is itself part of a plant-size or enterprise-wide system. Both types of HMI — graphic terminal and computer-based HMIs — are typically connected to each other in large control processes, especially processes that are factory- or enterprise-wide.

Control items such as switches, lights, and others are electronically assembled on the video screen of the computer using a library of symbols as part of the HMI programming software.

Computer-based HMI software has even more advanced capability. It can interface with multiple Graphic Display HMI Terminals, multiple PLCs, and other automated machines such as robots. It can also interact with other software programs for the management of control processes.

Figure 2-5 shows a computer-based HMI control diagram for a control process. Figure 2-5 is essentially equivalent to the Graphic Terminal Display type of HMI shown already in Figure 2-2. However, the graphic icons in this version have greater detail and use color.

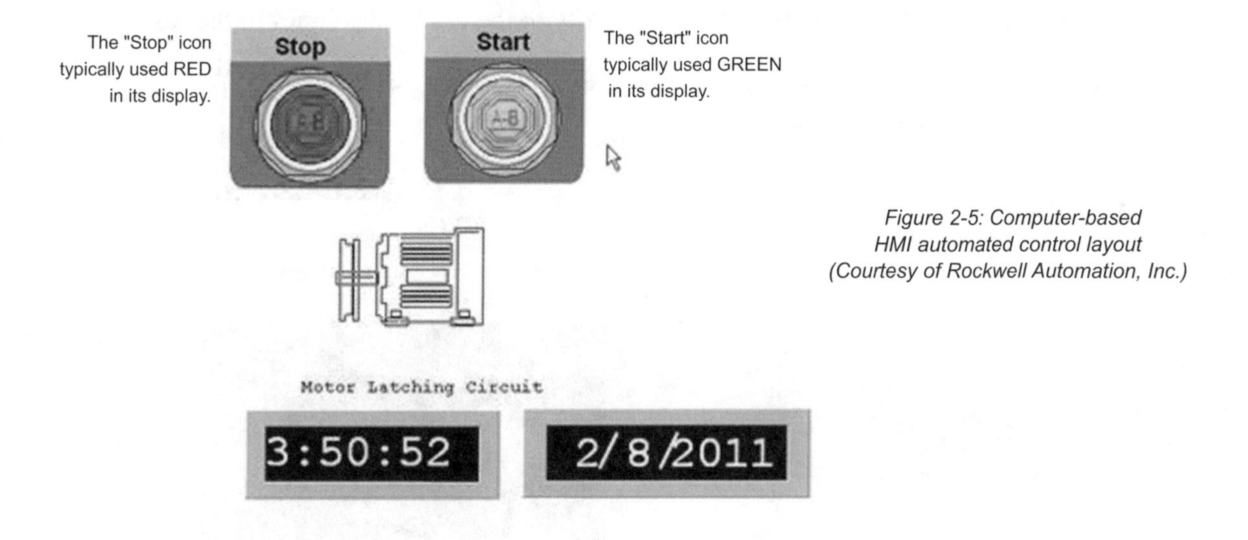

Figure 2-5: Computer-based HMI automated control layout (Courtesy of Rockwell Automation, Inc.)

2.5 RSView32 HMI Software for Computer-Based HMI Displays

The *RSView32 HMI* software is used to create animated displays of control processes from a single work cell up to an entire enterprise campus. It is also used to monitor and control the processes and other tasks. This type of HMI software must be run on a computer that has its video display terminal functioning as the HMI display in addition to fulfilling all of the computer's other needs.

The RSView32 software has a number of features that support the management of larger operations, such as enterprise manufacturing and plant-wide operations. Automated control processes in RSView32 can be extremely complex with high resolution images as well as pictures of devices. Color is used significantly in RSView32. The library of images is extensive, with icons of physical devices that can be animated in numerous ways.

An extensive set of commands are available to animate images. Special images are available, including X–Y graphs, histograms, analog meters, digital numerical displays, and many more. Real time display of trends in a process can be displayed as well as captured in a database.

Also, RSView32 can be programmed with Visual Basic so that additional sophisticated capabilities can be included as part of the automated control. For example, a specialized Excel spreadsheet with specialized data process can be generated and populated from an RSView32 using Visual Basic coding.

Figure 2-6 shows an example of the HMI screen display of an RSView32-created screen. This diagram displays real time trending of a process. Located below the trend graph are the STOP and START buttons and a graphic of the motor that is being measured. Also, note the Close Screen button, which means this control diagram has multiple screens. By closing out this screen, any other screens can be opened that show other parts of the process.

RSView32 can be used to program and display trending data from a database or an Excel spreadsheet. Historical data can be captured in a database for playback and analysis. The advanced capabilities of RSView32 are numerous.

The RSView32 labs in Chapter 5 show a majority of the basic capabilities of the RSView32 software.

HMI software requires installation along with RSLinx and RSLogix 500. Follow the installation procedures included with the software packages.

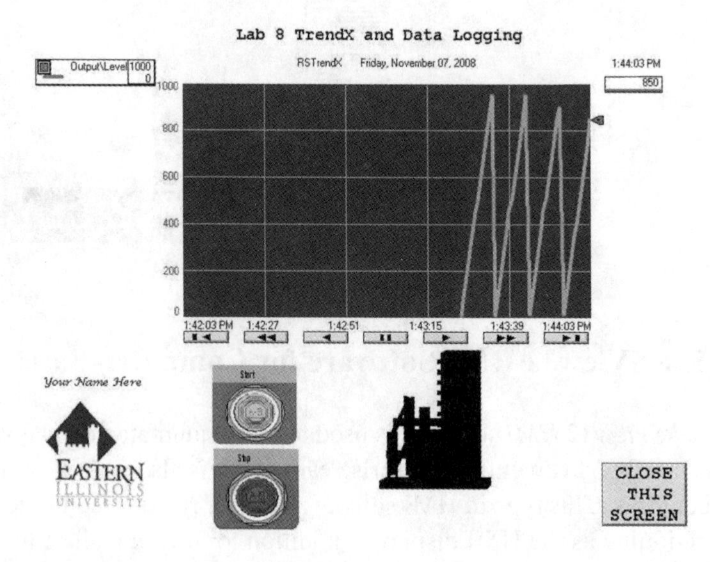

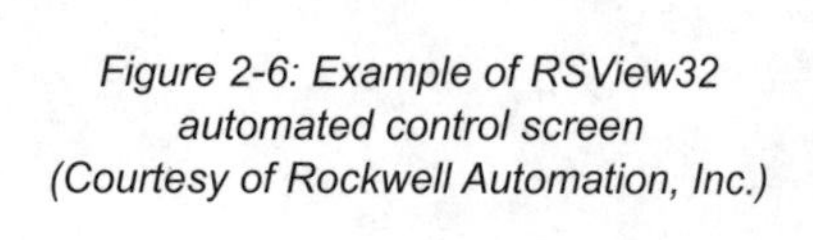

Figure 2-6: Example of RSView32 automated control screen (Courtesy of Rockwell Automation, Inc.)

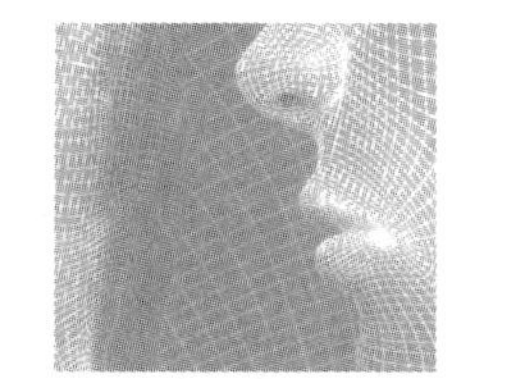

CHAPTER THREE

Field Devices and Sensors for PLCs

3.1 Introduction

This chapter describes a variety of the field devices commonly used in industrial, business, and home settings. They range from fairly simple mechanical/electrical devices to electronic field devices called sensors.

A sensor detects condition changes. These include changes to energy levels, for example from a light beam, an electrical field, or a magnetic field. A sensor might detect a change in state of a physical property, for example, force or pressure. The presence or absence of a product moving through a production line often involves the use of various sensors.

Some of the field devices and sensors discussed are industrial whereas a few were constructed from electrical and electronic parts readily available at electronic stores. See the following web site for pictures and information about field devices and sensors that are used in industrial processes:

http://wwww.automationdirect.com

3.2 Annunciator Buzzers

An annunciator buzzer is a device that produces a sound used to alert machine operators and others that something critical is happening in the automated process. Figure 3-1 shows an example of a small, inexpensive 12–24 volts DC annunciator that can be used to create and program PLCs and HMIs. This annunciator is called a buzzer because it produces a very shrill buzzing tone that catches the attention of any person nearby. Wiring the buzzer to a PLC simply requires connecting the red and black wires to a PLC output connection and to the source of the DC voltage needed to power the buzzer.

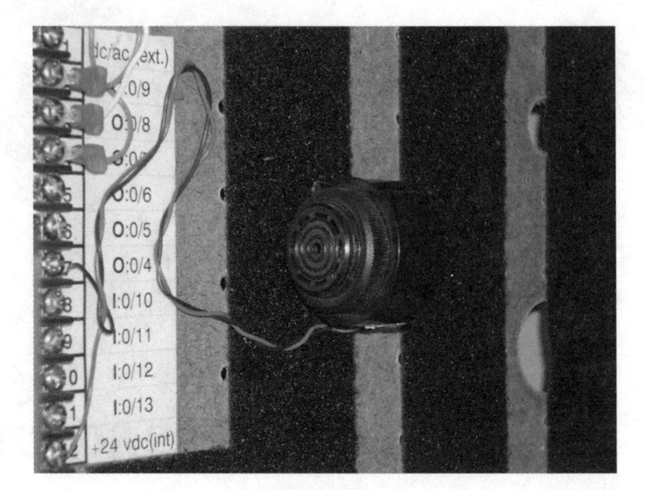

Figure 3-1: 12 VDC Buzzer

There are many other types of annunciators in use in industrial automated control systems.

3.3 Contact Relays

Figure 3-2 shows a contact relay. It has an electromagnet and sets of contacts attached to an arm. When the electromagnet is powered up, the electromagnet pulls the arm down, which causes contacts to move down. This action creates a Normally Closed State, then a Normally open state of the contacts.

Figure 3-3 on the following page, shows the schematic diagram for a contact relay. Note the solenoid is the electromagnet. This contact relay has two groups of contacts, called poles; they allow you to control separate circuits. Each set of contacts has a Normally Closed (NC) state (electricity can flow) and a Normally Open (NO) state (electricity cannot flow). When the solenoid receives an electrical signal from the PLC output to which it is connected, the magnetic field generated pulls the moveable arm down to the bottom, then holds the arm. This action opens the NC contacts and closes the NO contacts.

When the PLC output turns the electrical flow off to the solenoid, the magnetic field disappears, allowing the spring return to move the arm to its resting state of being up.

3.4 Directional Control Valves with Pneumatic Pistons

A Directional Control Valve (DCV) is a pneumatic equivalent of an electrical switch. The word pneumatic means *being driven by air*. In Figure 3-4, small diameter tubing is used to make pneumatic connections to ports on the DCV similar to wires used to connect electrical circuits.

The green and yellow electrical wires with banana plugs shown in Figure 3-4 are connected to an output on the PLC. If the DCV is in a deactivated state (not activated), air from a pneumatic source flows through the purple/red tubing on the left side and is directed by the DCV to exit into the green plastic tubing on the bottom. If the DCV is activated, the air entering the DCV from the left side will be directed to exit into the red tubing on the top right.

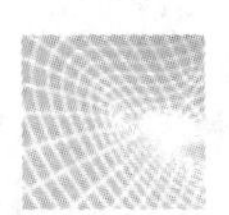

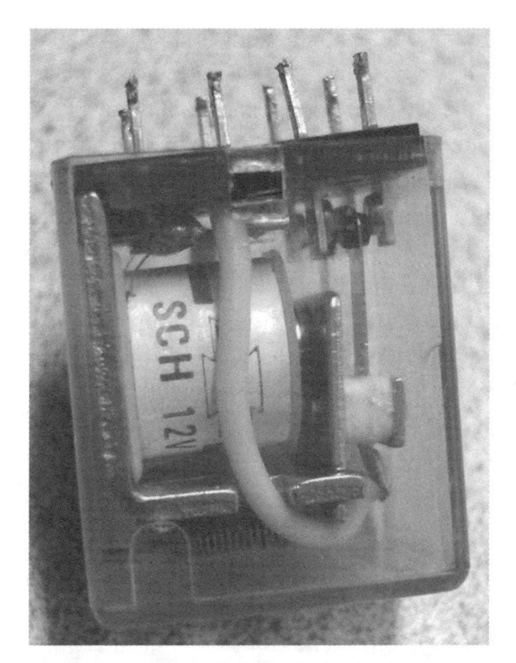

Figure 3-2: Contact relay

This is exactly what happens in a two-way electrical switch. When the switch is set to one position, electricity flows from the input through the first set of contacts. When the switch is set to the other position, electricity flows from the input through the second set of contacts.

DCVs are commonly used to control a pneumatic piston in industrial processes. A pneumatic piston is a device that, when air is applied to it, it will use the air to push a piston rod out or in.

Figure 3-5 below shows a pneumatic piston with a ball on the end of its piston rod. When air is allowed to flow through the red tubing attached to the bottom of the piston, and the green tubing is open, the piston rod is forced to move forward quickly and exhaust the air already in the piston through the green tubing.

Similarly, if air is allowed to flow through the green tubing at the top, and the red tubing is open, the piston rod will be forced to move back quickly and exhaust the air already in the piston through the red tubing.

NC CONTACT
OPEN CONTACT
NC CONTACT
OPEN CONTACT
SOLENOID

Figure 3-3: Contact relay schematic diagram

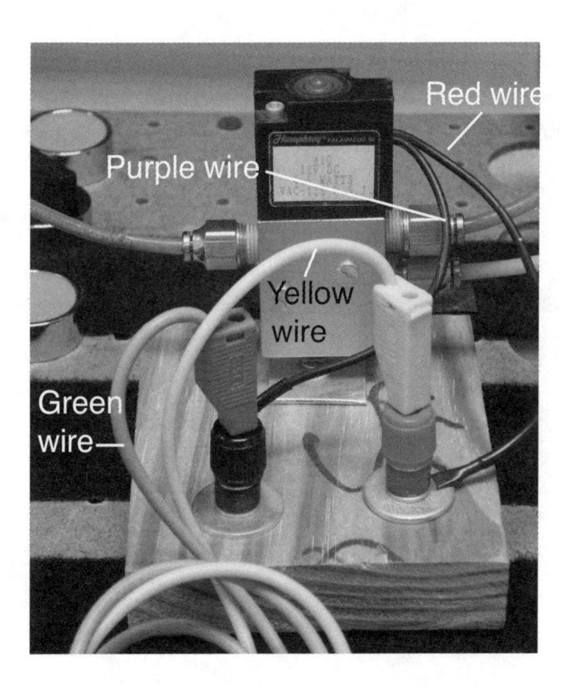

Figure 3-4: Directional Control Valve (DCV)

Figure 3-5: Pneumatic piston

This pneumatic piston action can be automated by using a DCV. In Figure 3-4, the yellow and green wires from the DCV are connected to an output on the PLC. The piston can now be controlled using electrical on and off signals sent to the DCV from the PLC output module. In turn, the DCV controls the air entering and leaving the piston so the piston can move forward and backward.

The device on the right side of Figure 3-5 is a microswitch, positioned so that the ball on the piston will activate the microswitch if the ball is pushed forward by the piston. See the description of the microswitch below.

3.5 Fans

The fan is used often in cooling situations for automated control systems. The mini DC fan in Figure 3-6 is representative of a fan that could be used in an industrial, automated control system. Small fans like this one are often DC-powered. This fan is a 12-VDC fan, which is sufficient for the learning labs in Chapters 4 and 5.

The fan is an output device with a red and black wire connecting to a PLC output. When the PLC output turns on, electricity flows into the fan, causing the fan blades to rotate and move air that could be used to cool a product. When the PLC output is turned off, the electricity flow stops and the fan blades stop rotating.

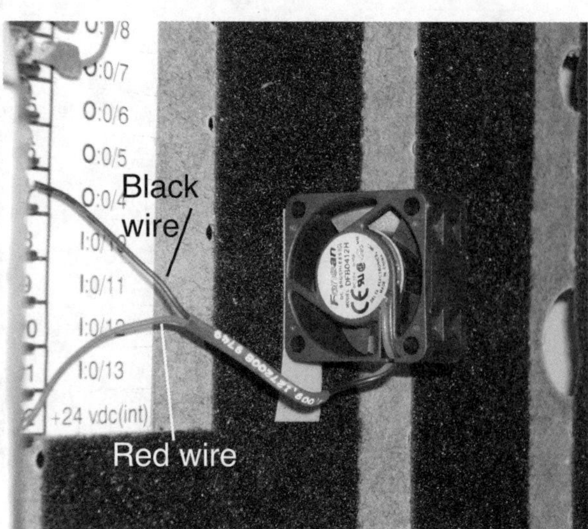

Figure 3-6: Mini DC fan

3.6 Inductive Sensors

Figure 3-7: Inductive sensor

There are two common electronic sensors. One, based on magnetic fields, is called an inductive sensor; the other, based on electrical fields, is called a capacitive sensor.

Figure 3-7 is an example of an inductive sensor based on magnetic fields. It uses an inductor, which is a coil of wire. The bottom part of the sensor contains the inductive devices and the electronic circuitry designed to detect a change in the magnetic field if a metallic object moves close to it, causing an on or off action. Note the red and black wires connected to the PLC.

The sensing range for inductive sensors is very short, less than 1 inch or smaller, and they can usually only detect magnetic metals such as steel.

A reason to use an inductive sensor rather than a photo cell is that the photo cell is susceptible to dust or dirty environments that would cover the light source, the photo sensor, or both. The inductive sensor is completely sealed; if dust or other similar materials are coating it, they will be ignored.

The other sensor that uses electric fields is called a capacitive sensor. It looks similar to an inductive sensor, but has higher sensitivity for detecting many types of objects. The capacitive sensor is based on the electrical device called a capacitor, which is two pieces of metal plate separated with insulation. An electric field is generated when electricity is applied to both plates. When objects get really close to the electric field, they cause a disturbance detected by electronic circuitry as on or off.

3.7 Lights

a. Pilot Lights

Figure 3-8: Pilot light

Figure 3-8 shows a pilot light, which is an example of an output. It has two screw connections for attachment to the PLC output. A small light bulb inside the pilot light is turned on when the PLC output turns on the flow of electricity to it. The light goes out when the electrical flow is turned off. A red light could indicate a possible stopped condition. Lights of other colors are used as well for other conditions.

The silver ring in Figure 3-8 is a nut that, when turned, will unloosen and can be removed. The plastic dome can then be removed so that the small light bulb can be replaced. When reassembling the light, make sure the good light bulb is inserted and tightened correctly, then replace the dome. Finally, screw the nut back in place.

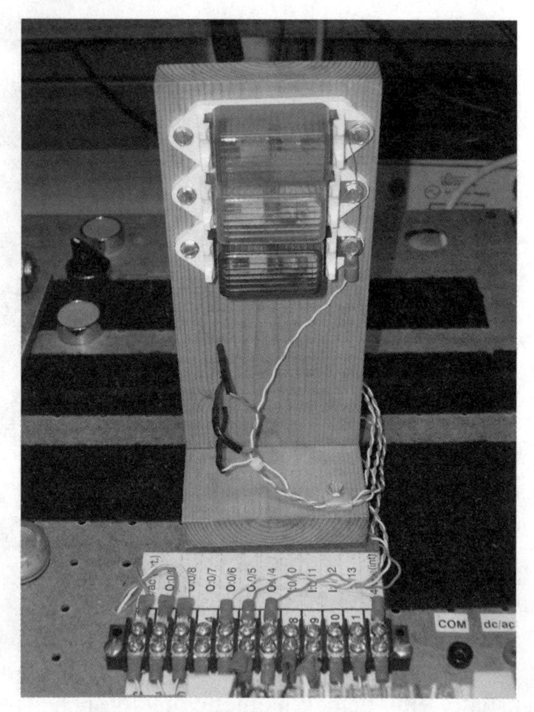

Figure 3-9: Handmade stack light unit

b. Stack Light Units

A stack light unit consists of several lights, typically two to four, stacked one above the other. Each light typically has a different light color. They are used as light annunciators in a production process where a certain color means some event is happening.

Suppose a stack unit has a green light with a yellow light stacked above it and a red light stacked above the yellow light. This stack could represent the operation of a process where the green light means the process is running and red means it has stopped. The yellow light could indicate a pause mode.

Stack lights are very popular in industrial processes.

Figure 3-9 shows a handmade stack light unit using a wooden mounting system with small DC lights covered with colored rectangular plastic covers. The three lights here from top to bottom are red, orange, and blue. The three lights have a common connection on one end of each of the three lights. The other side of each light then has an individual wire that would be connected to three different PLC output module connections.

3.8 Microswitches

Microswitches are another commonly used field device. A microswitch consists of a push-type switch with a roller arm. The microswitch in Figure 3-10 is a two-way switch, meaning that electricity can be switched in two different directions.

Figure 3-10: Microswitch

The microswitch works as shown in Figure 3-11. The switch presently has a path such that electricity can flow from an input connection through one of the two output connections. If the piston ball in the upper left side of Figure 3-5 moves forward, it will travel over the roller. This movement makes the arm flex to the right, which in turn pushes a small button, That button switches internally, causing the action of the electricity being directed to switch to the other output connection.

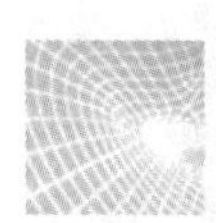

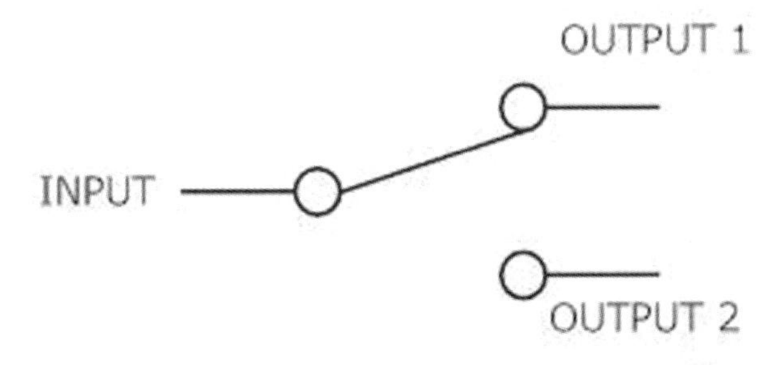

Figure 3-11: Schematic of a microswitch

The activated microswitch could in turn activate some other type of device such as a fan for example.

3.9 Rotary Tables

Figure 3-12: Rotary table

Figure 3-12 shows a small rotary table that could represent a manufacturing process that moves a product from one direction to another or similar types of tasks.

Because this example is a handmade representation of real rotary tables, it provides a learning opportunity with a different type of field device. The table consists of a DC electric motor with a disk attached to its rotational shaft. Rotary tables typically rotate in partial turns, such as 90-degrees of rotation, to move a product going one direction to a direction that is right angle to the original direction. The rotary table is an output field device in that it is connected to an output on the PLC.

3.10 Photo Sensors and Reflectors

A photo sensor uses a special semi-conductor device called a photo cell with some electronic circuitry to detect the presence or absence of an object through the photo cell's ability to detect the presence or absence of light. It becomes essentially a switch by its internal resistance changing to a very low value (switch ON) when light shines on it and its internal resistance changing to a very high value (switch OFF) when no light shines on it.

By combining a photo sensor with a small light source and a reflecting device, it becomes a very sophisticated sensor field device called a photo sensor, as shown in Figure 3-13. Essentially

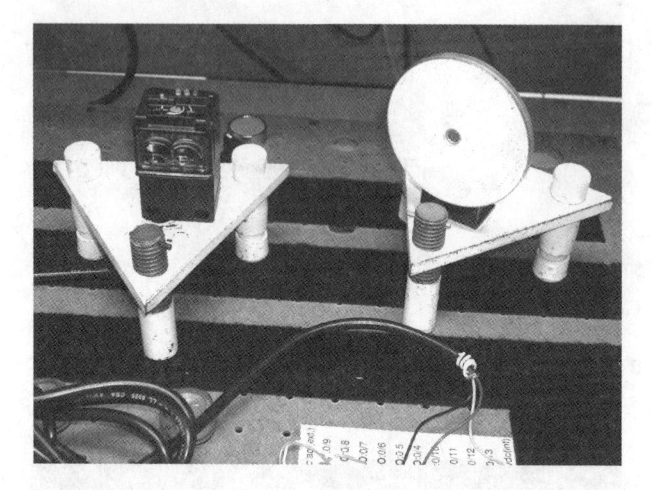

Figure 3-13: Photo sensor and reflector

the photo cell uses a small light source that sends a beam of light in a specific direction toward a light reflecting device shown on the right side of the figure. The light reflecting device is positioned so that the incoming light beam will be directed back toward the photo cell inside the photo sensor.

Here is an example of how a photo sensor is used in a production line. The photo sensor is set to make sure that a product is not missing a special part. The special part is moving on an assembly line; the photo sensor reflector detects if the part is on the line. If the part blocks the light beam, this means the part is present at that point. The sensor can then signal that information to the PLC. If no part is present, the light beam is not blocked. In that case, the sensor can signal the PLC that there is no part.

The wiring of the photo sensor is a little more complicated than that of other field devices. It requires a source of DC voltage and is connected to the PLC as an input device.

The typical use for a photo cell is to detect whether an object is present or not. The sensing range of a photo cell is from inches to feet.

3.11 Switches

a. Momentary Pushbuttons

Figure 3-14 shows a momentary pushbutton. Pushbuttons are mechanical and are operated by pressing and releasing them. Push the button to "activate" and release to "deactivate". Red pushbuttons are used to stop a process; green pushbuttons are used to start a process.

Figure 3-15 shows the internal operation of the pushbuttons. A set of electrical switch contacts are inside the button. They can be an open circuit (electricity will not flow) or a closed circuit (electricity will flow).

Figure 3-14: Momentary pushbutton

Normally Closed (NC) Pushbutton

a

b

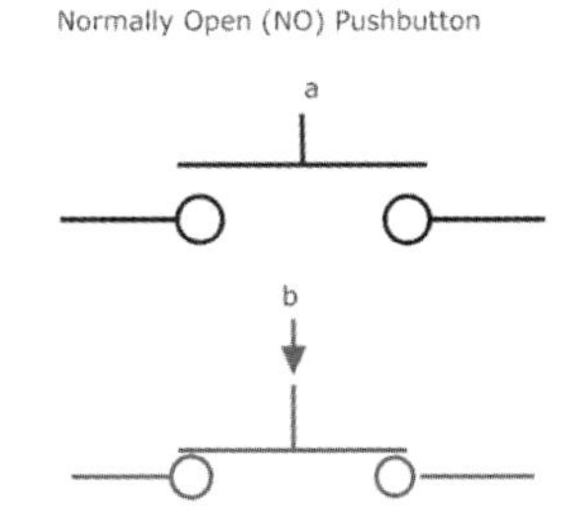

Figure 3-15: Switch contacts for STOP (NC) and START (NO)

In Figure 3-15a left, the STOP pushbutton switch contacts are normally closed (NC), as shown. This means electricity will flow when the STOP pushbutton is "Not Pressed". In Figure 3-15b left, electricity will cease to flow when the STOP pushbutton is "Pressed", which opens the switch contacts. This switch can control two paths of current flow together.

In Figure 3-15a right, the START pushbutton switch contacts are normally open (NO), as shown. This means electricity will not flow when the START pushbutton is "Not Pressed". In Figure 3-15b right, electricity will flow when the START pushbutton is "Pressed", which closes the switch contacts.

3.12 Temperature Sensors

Sensing temperature is a very old technology. With today's solid state devices, digital temperature sensors can be created. Figure 3-16 shows a handmade temperature sensor using a digital chip that can measure Fahrenheit temperature values. A Celsius digital chip is available as well.

The chip becomes the sensor and processing circuit all in one to translate the temperature into a digital value.

Chapter 5, RSView32 Lab 5, requires a digital temperature sensor.

Contact the authors for more information about this sensor.

There are many other field devices available; they are usually specific to a select automated control situation. The ones identified here represent a sample of the devices that are typically used in automated control environments.

This chapter showed some of the more common field devices used in automated control. They are the ones that are included in the lab experiments in Chapters 4 and 5.

Figure 3-16: Digital temperature sensor

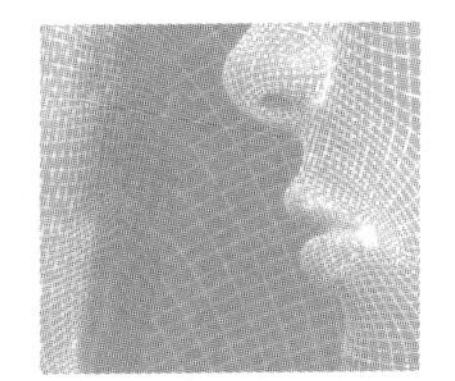

CHAPTER FOUR

PanelBuilder32 / PanelView Graphic Terminal Labs

Introduction

This chapter contains step-by-step procedures for the six laboratories listed below. These labs demonstrate how the graphic terminal HMI device is programmed. They also demonstrate the appearance of this type of HMI display and how interaction occurs as though a machine operator is using the display.

- ***Lab 4-1: Basic Motor Starter***
- ***Lab 4-2: Analog and Digital Displays***
- ***Lab 4-3: Pump and Tank***
- ***Lab 4-4: Alarms***
- ***Lab 4-5: Analog Input and Math Operations***
- ***Lab 4-6: Security***

PV Lab 4-1: Basic Motor Starter

This basic lab demonstrates how Rockwell Automation PanelBuilder32 software is used to program the Allen-Bradley PanelView series graphic terminal displays. A PanelView 550 (PV550) touch screen terminal is used here. The PV550 has a viewing screen area of 4 in x 6 in.

The Allen-Bradley Micrologix series of PLCs is used here for simplicity. Rockwell Automation RSLogix 500 software is needed to program the Micrologix series.

RS232 communications protocol was chosen for programming the PLC and PV550 and for control communications between the PV550 and the PLC.

4.1.1 Initial Setup of the PanelView 550 (PV550) Operator Terminal

1. Figure 4-1-1 shows the back view of the PV550. This model has a printer port (called ***RS-232***) that can be set up as a programming port. The ***RS232 DF1*** Port is the serial communications port that connects to the PLC communications cable.

Figure 4-1-1: View of PV550 ports and power connections (Courtesy of Rockwell Automation, Inc.)

2. Figure 4-1-2 shows the three cables that connect to the PV550. The left cable is a user-supplied power cable that is connected to a 24-volt DC power supply. Be careful in correctly connecting these wires to the PV550 and to the DC supply. The + and – connections are shown on the left side of the terminal in Figure 4-1-1.

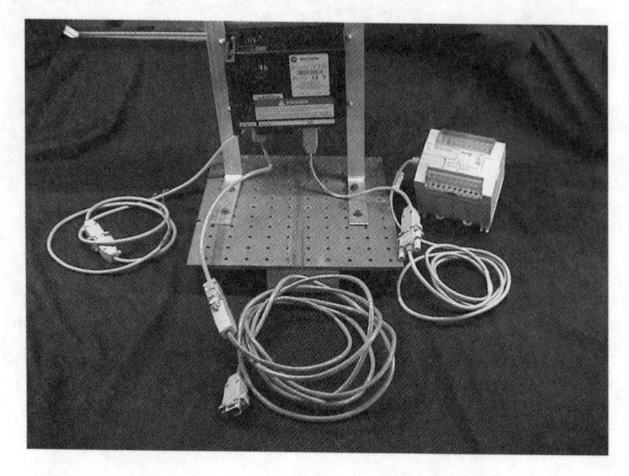

Figure 4-1-2: PV550 with cables attached (Courtesy of Rockwell Automation, Inc.)

3. In Figure 4-1-2, the middle cable is the ***Allen Bradley 2711-NC13***, which is plugged into the ***RS-232 connector*** (called the printer port). It connects the computer with ***PanelBuilder32*** software to the graphic terminal.

4. The cable on the right side in Figure 4-1-2 is the standard ***Allen Bradley 1761-CBL-PM02*** communication/programming cable for Micrologix PLCs. Also shown is a homemade null adapter cable attached to the 1761 cable.

5. Once all of the cables are connected, power up the PLC first; then power up the ***PV550*** graphic terminal.

6. The ***PV550*** display should go through a series of opening screens.

7. When completely powered up, the ***PV550*** should display a ***GOTO CONFIG*** button. If the ***GO TO CONFIG*** button appears, touch it and proceed to Step 10 below.

8. If the ***GOTO CONFIG*** button does not appear on the screen, do a restart and use the "***CONFIG on power up***" procedure described in Section 4.2.2.

4.1.2 "CONFIG on power up" Procedure

9. The ***CONFIG on power up*** procedure begins with powering up the ***PV550*** and watching for a small blinking white square in the lower right side of the screen as shown circled in the lower right of Figure 4-1-3. Quickly place your finger over the small blinking white box for about 3 to 6 seconds. Then remove your finger and the box should have disappeared.
10. The configuration screen shown in Figure 4-1-4 should appear after a short period of time.

11. Set the configuration as described here. Select a choice by using the up or down arrow keys shown by the circle to move the highlight bar.

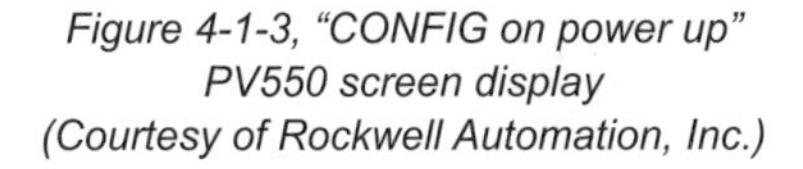

Figure 4-1-3, "CONFIG on power up" PV550 screen display (Courtesy of Rockwell Automation, Inc.)

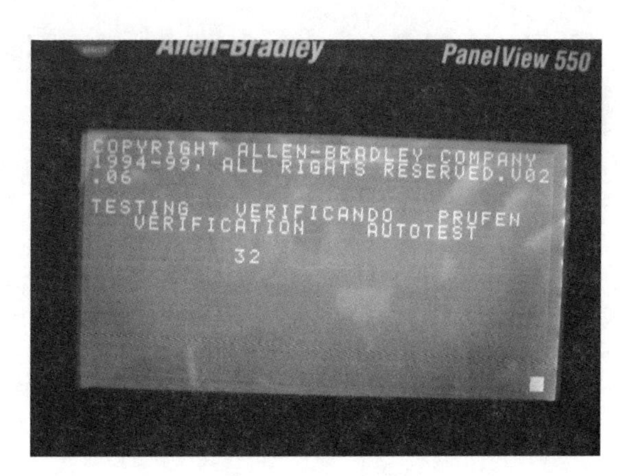

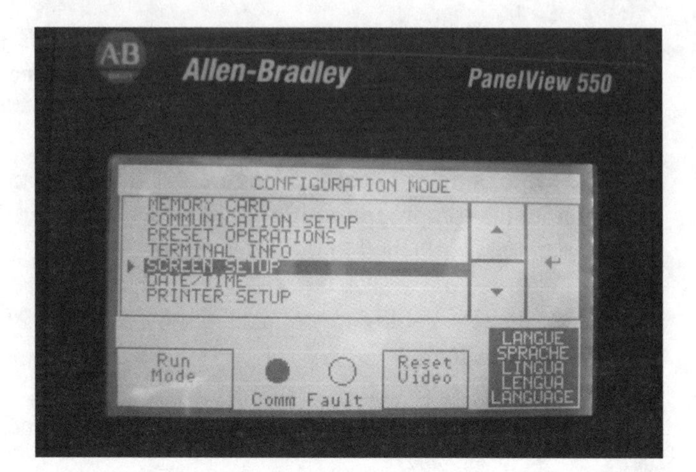

Figure 4-1-4: Configuration mode display (Courtesy of Rockwell Automation, Inc.)

12. Move the highlight bar to ***COMMUNICATION SETUP***; then touch the ***ENTER*** key (to the right side) to select the communication menu.

13. By touching the individual boxes, set the following communication parameters: 1) Baud to 19200, 2) No Parity, 3) 8 Data Bits, and 4) 1 Stop Bit. Also, set Node to 2 and Error Detection to CRC with Handshaking to Off. Press ***EXIT*** to save the values.

14. Under ***SCREEN SETUP***, set the Contrast to 6 or 7, Video Mode to Normal, Backlight to ON and Backlight Timeout to 5 min. Remember to press ***EXIT***.

15. Set the date and time under ***DATE/TIME*** by pressing each box as needed to establish the year, month, day, hour, minute, and second. Press ***EXIT*** to save settings.

16. Finally, under ***PRINTER,*** set Port Mode to ***PRINTING DISABLE*** so it can be used as the programming port. Press ***EXIT***.

17. This concludes the setup configuration of the PV terminal.

4.1.3 Programming the MicroLogix PLC Using Rockwell RSLogix 500 Software and RSLinx Communication Software

18. The ***RSLogix 500*** software is used to program the PLC through RS232 serial communications. ***RSLinx Professional*** is the serial communication software that provides the communication from computer to the PLC.

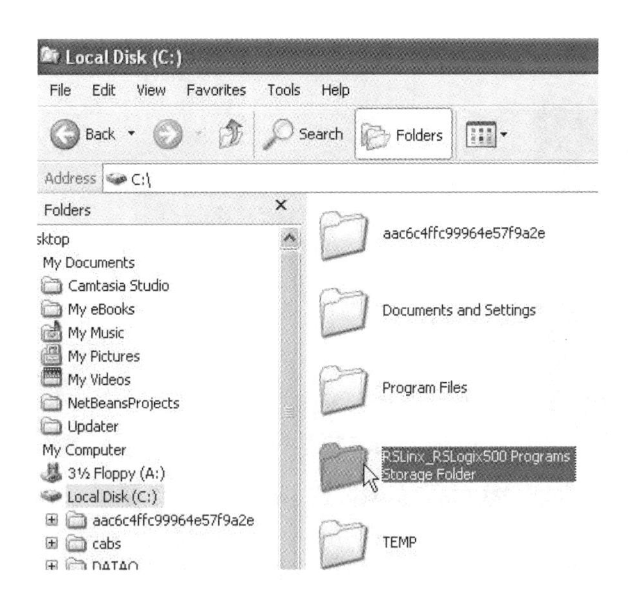

Figure 4-1-5:
Local disk programs storage folder
(Courtesy of Rockwell Automation, Inc.)

19. To establish the correct communication setup between these devices, **the PLC must be connected to the computer and powered on!** This is most critical.

20. If it has not already been done, create a folder on ***Local Disk (C:)*** drive entitled ***RSLinx_RSLogix500 Programs Storage Folder,*** as shown in Figure 4-1-5. Store all of the related files for this laboratory in this folder. Create a folder and save it with a name such as ***Lab_1_XXX*** (XXX = your initials).

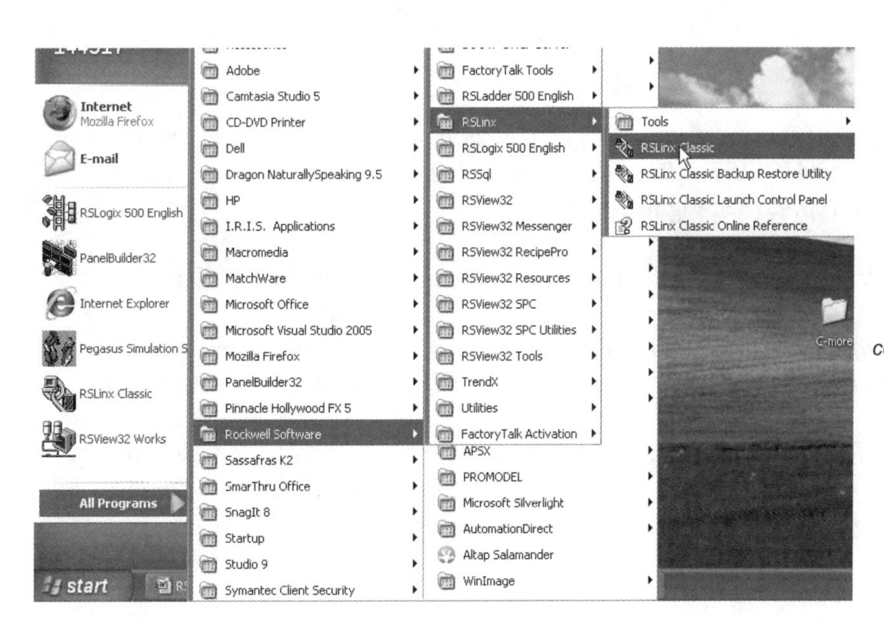

Figure 4-1-6: Rockwell communications RSLinx software (Courtesy of Rockwell Automation, Inc.)

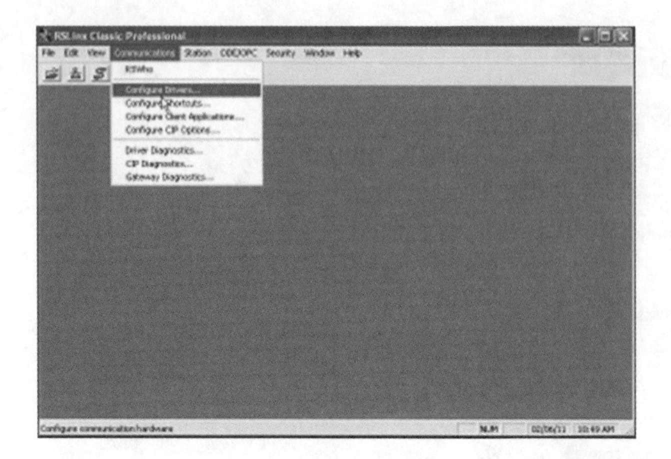

Figure 4-1-7: Communications menu for configuring RS232
(Courtesy of Rockwell Automation, Inc.)

21. Next, open the Rockwell communication software ***RSLinx***, as shown in Figure 4-1-6.

22. Notice the drop down menu that appears, as shown in Figure 4-1-7. Configure ***RS232*** communications as follows:

    a. Click ***Communications***; then select ***Configure Drivers***.

    b. A pop-up screen will appear similar to the one shown in Figure 4-1-8. Under ***Configured Drivers: Name and Description***, click ***AB_DF1-1 DH-485*** selection; then click the ***Con-figure…*** button (on the right) to configure this RSLinx driver. Click ***OK*** and **Close**. Skip to **Step 23**.

    c. If there is no ***AB_DF1-1 DH-485*** selection to select (as shown highlighted in Figure 4-1-8), proceed to Step d.

    d. If under ***Configured Drivers: Name and Description,*** no drivers appear or another driver selection that is not ***AB_DF1-1 DH-485*** appears, click the pull down arrow next to ***Available Driver Types***. (See Figure 4-1-9.)

    e. Select ***RS232 DF1*** devices; then click the ***Add New*** button (immediately to the right).

    f. Next, click ***Configure***. Then if ***AB_DF1-1*** appears and ***RS232 DF-1*** devices appear within the ***Available Driver Types:*** window, click ***OK*** and ***Close.*** Proceed to **Step 23.**

23. Now configure the RS-232 communications network. Doing so will allow ***RSLinx*** to very rapidly create the communication link to the PLC via RSLogix and RSView. When the graphic shown in Figure 4-1-10 appears, click the ***Auto-Configure*** button.

a. A message appears, as shown in Figure 4-1-11, indicating that auto configuration was successful.

b. Click ***OK*** and then click ***Close*** at the ***Configure Drivers*** pop-up screen.

c. If slow response toward successful auto-configuration occurs, either the **PLC was not plugged into the computer** or the **PLC was not turned on**. Remedy the situation by going back through this step. If errors still occur, the ***RS232 DF-1 driver*** will need to be reinstalled.

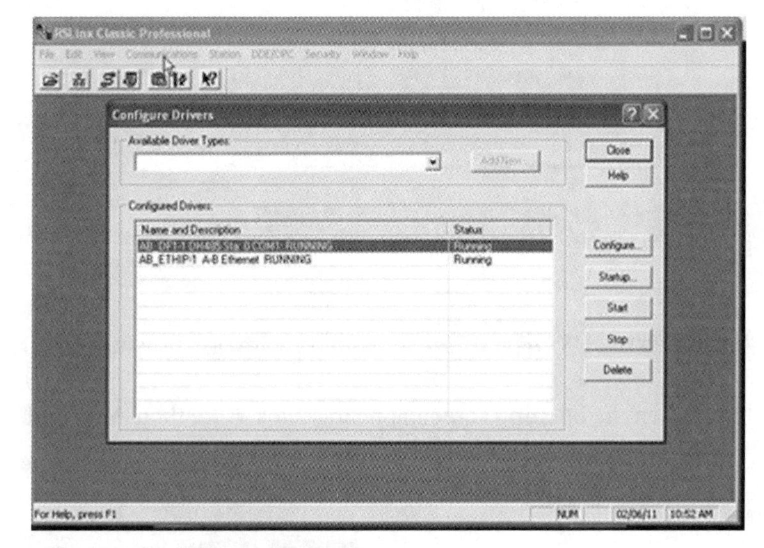

Figure 4-1-8:
Pop-up configure drivers menu screen
(Courtesy of Rockwell Automation, Inc.)

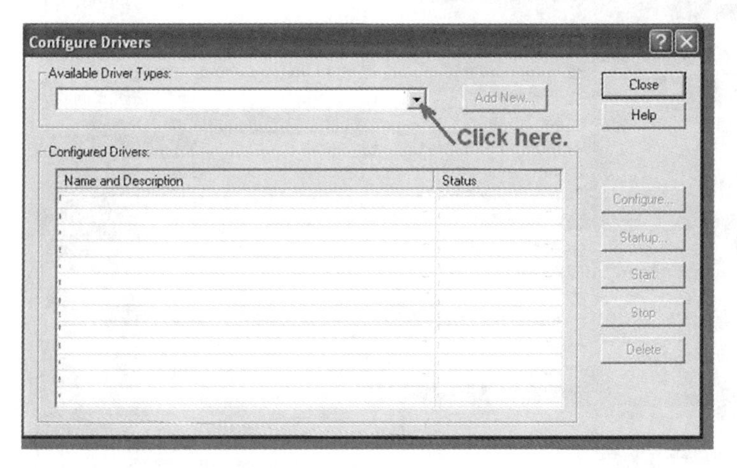

Figure 4-1-9: Available Driver Types screen
(Courtesy of Rockwell Automation, Inc.)

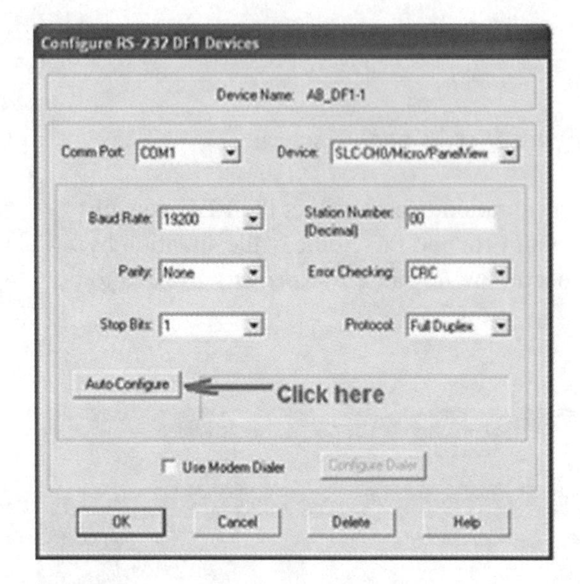

Figure 4-1-10: Configure RS232 DF-1 Devices screen (Courtesy of Rockwell Automation, Inc.)

Figure 4-1-11: Auto Configuration Successful (Courtesy of Rockwell Automation, Inc.)

24. Open the ***RSLogix 500 English*** software, as shown in Figure 4-1-12.

25. When the opening screen appears, click ***File***, then ***New*** and select the ***Processor Type*** for the PLC being used, as shown in Figure 4-1-13.

Figure 4-1-12: RSLogix 500 software window (Courtesy of Rockwell Automation, Inc.)

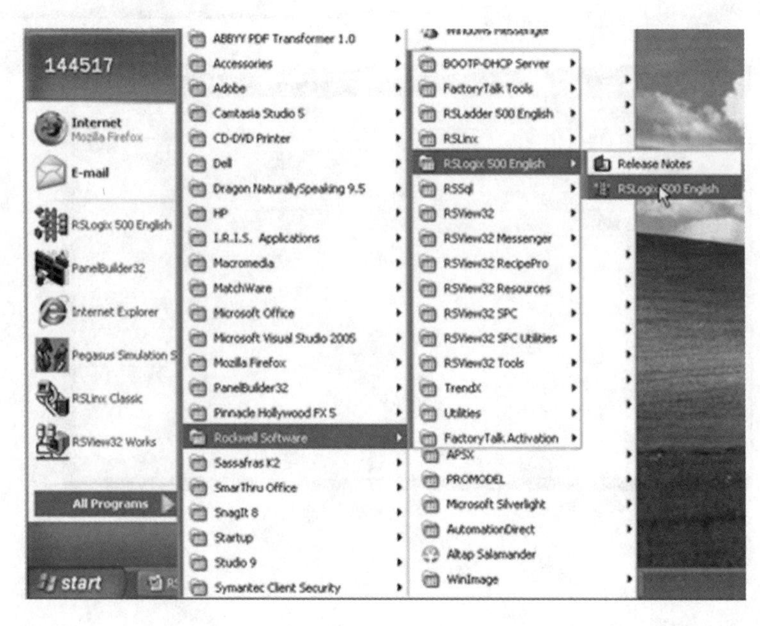

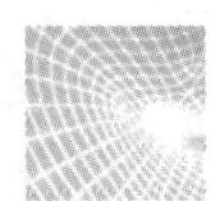

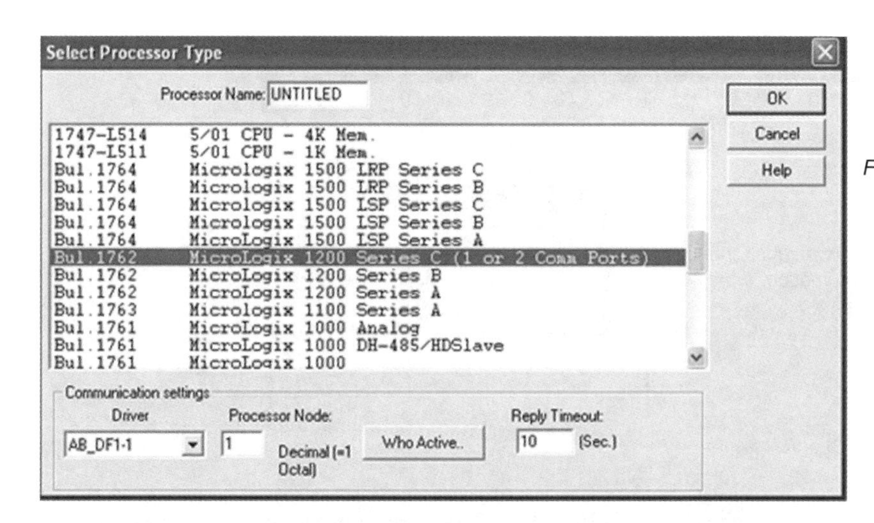

Figure 4-1-13: Select Processor Type screen (Courtesy of Rockwell Automation, Inc.)

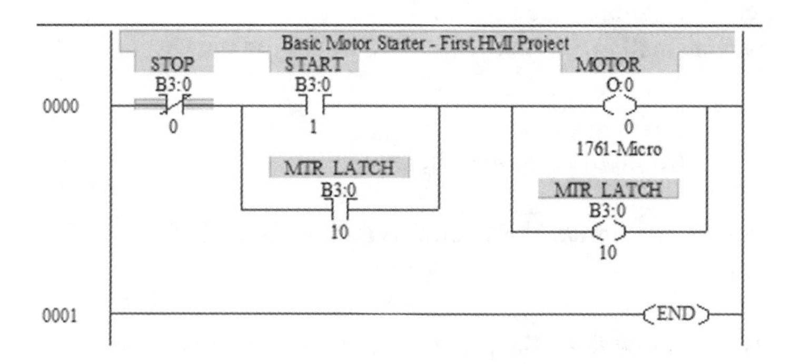

Figure 4-1-14: Basic Motor Starter PLC program (Courtesy of Rockwell Automation, Inc.)

26. Create in ***RSLogix 500*** the PLC program shown in Figure 4-1-14.

27. An animated HMI consists of specific animations and selected icons that control the inputs and outputs of the PLC. Binary B3 address must be used for this type of HMI (Figure 4-1-15).

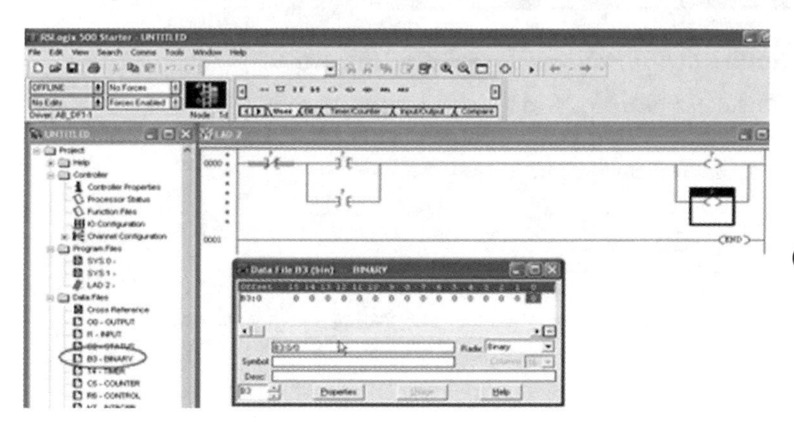

Figure 4-1-15: RSLogix 500 Starter screen (Courtesy of Rockwell Automation, Inc.)

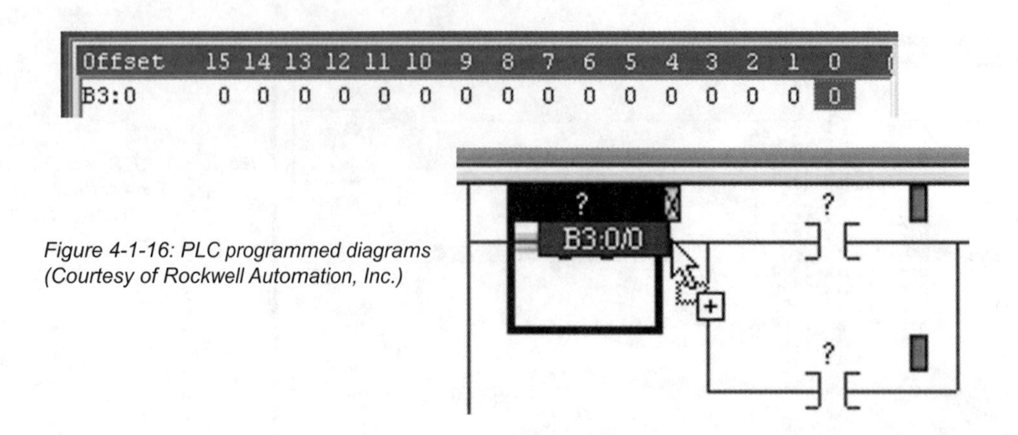

Figure 4-1-16: PLC programmed diagrams (Courtesy of Rockwell Automation, Inc.)

28. In the ***Data Files*** folder in the left window, click the B3 Binary icon. Then **click** and **drag** the binary address ***B3:0/0*** (which appears in a small blue box) to the switch icon (Figure 4-1-16). The screen will have a small green box with an X where the binary address can be inserted.

29. Repeat this procedure for the other two B3 input icons and the two outputs.

30. Save this program with the file name ***Motor Latching Circuit.rss*** (saving to local C: drive is best) and then minimize ***RSLogix 500.***

31. Load the PLC program in ***RSLogix 500***. Download the program into the PLC by clicking ***Comms,*** then ***Download,*** and click ***OK*** at the ***Revision Note*** pop-up screen. When the ***RSLogix 500*** screen appears, as shown in Figure 4-1-17, click ***Yes*** to finish downloading the program and going online.

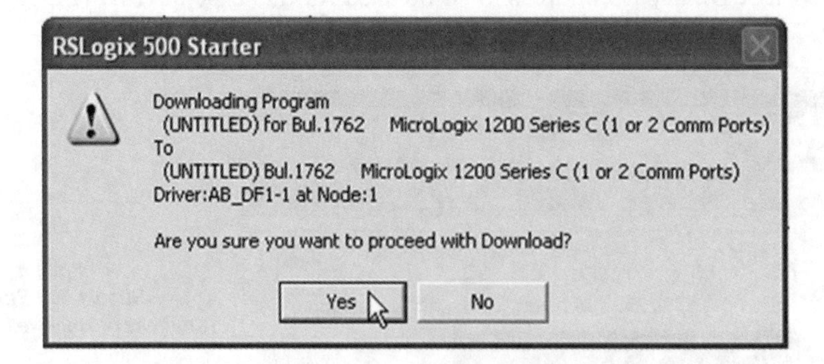

Figure 4-1-17: Pop-up RSLogix 500 Starter Program (Courtesy of Rockwell Automation, Inc.)

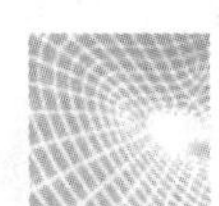

Figure 4-1-18: Select Online RUN mode
(Courtesy of Rockwell Automation, Inc.)

32. Make sure the PLC is in ***RUN*** mode (Figure 4-1-18).

4.1.4 Programming the PanelView Using PanelBuilder32

33. Launch ***PanelBuilder32*** and look for the screen shown in Figure 4-1-19. Make sure that ***Create New Application*** button is selected as shown. Click ***OK***.

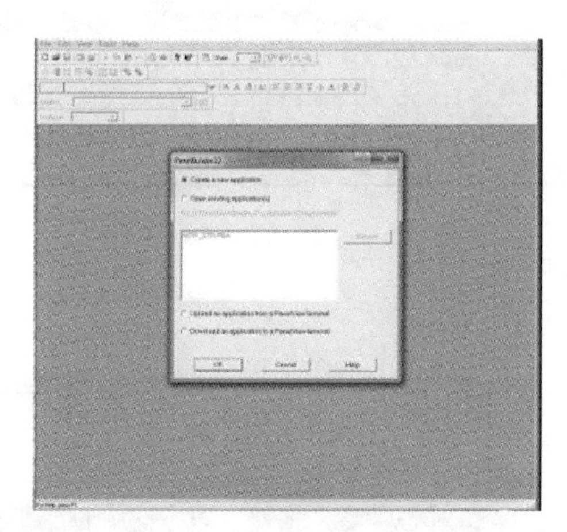

Figure 4-1-19: PanelBuilder32 opening screen
(Courtesy of Rockwell Automation, Inc.)

34. The ***Create New Application*** window will appear, as shown in Figure 4-1-20. Enter ***MtrStarter*** in the ***Application Name box***. Select ***PV550*** as the PanelView Type and select ***DF01*** as the Protocol to use. Select ***2711-T5A16 PV550 Touch, FRN 4.10-4 xx*** in the Catalog and Version box.

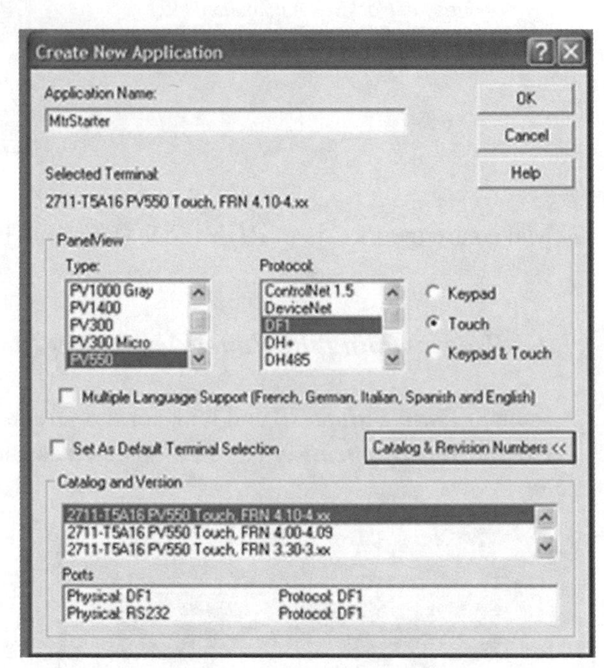

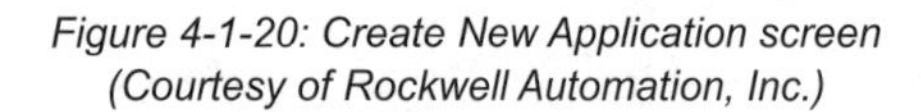
Figure 4-1-20: Create New Application screen (Courtesy of Rockwell Automation, Inc.)

35. Click ***OK***. The screen in Figure 4-1-21 will appear.

36. This screen is where the HMI process diagram will be created. **Expand the headings** in the left window by clicking the ***plus (+) sign***. Also, expand the right window to fill the available space by clicking the rectangle next to the red X close box that appears in the upper right corner of the screen.

37. In the left window of Figure 4-1-21, click ***Communications Setup*** under ***Application Settings***.

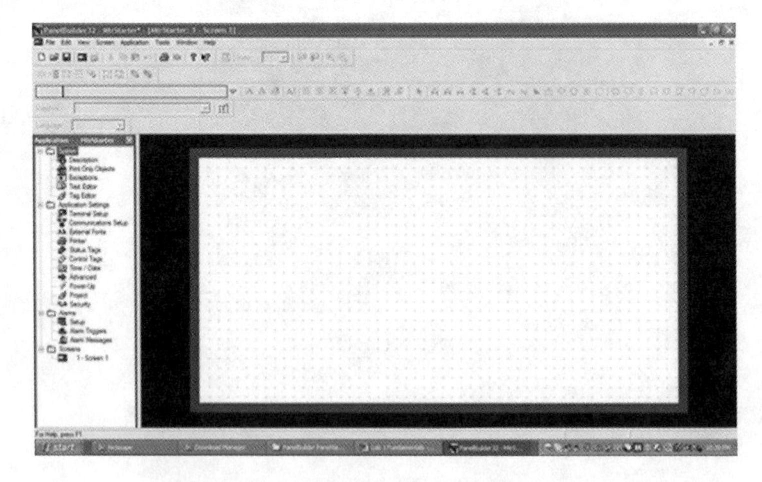
Figure 4-1-21: Create Another Application screen (Courtesy of Rockwell Automation, Inc.)

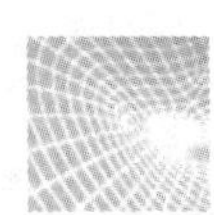

38. The ***Communications Setup*** window will appear as shown in Figure 4-1-22. Change the ***Baud Rate*** to ***19200***. In ***Network Nodes***, right click the line under the *asterisk (*)* and select ***Insert Node Ins.*** Enter a node name such as ***ML1000*** under Node ***Name.*** Enter ***1*** under ***Node Address***. ***Micrologix*** will appear under ***Node Type***. Click ***OK***.

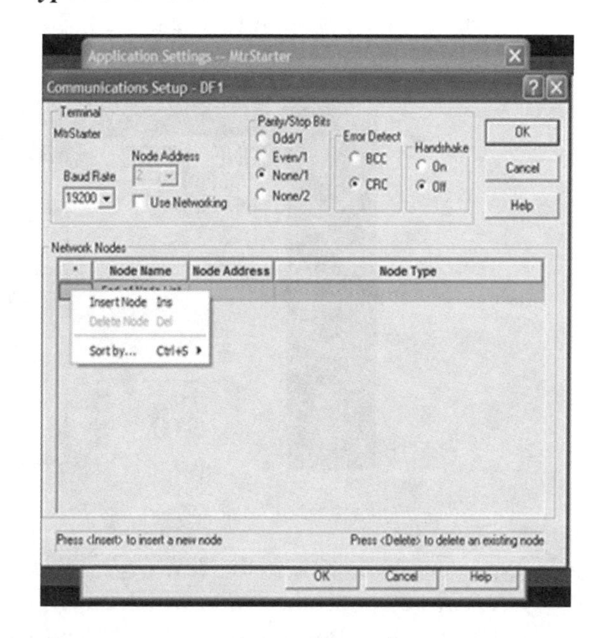

Figure 4-1-22: Application Settings — MtrStarter (Courtesy of Rockwell Automation, Inc.)

39. The final screen for this HMI should look something like that shown in Figure 4-1-23. Use it as a guide for the programming steps that follow.

40. Most of the icons on the screen will be created using the ***Objects*** drop down menu shown circled in Figure 4-1-23.

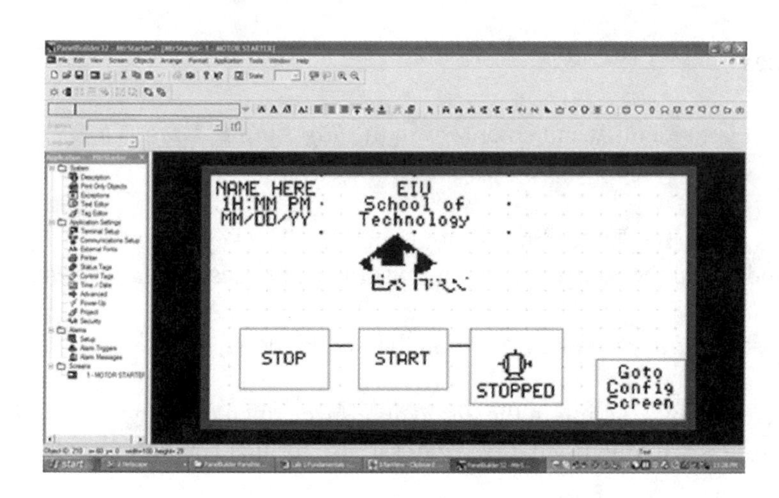

Figure 4-1-23: Example of Final Screen for HMI (Courtesy of Rockwell Automation, Inc.)

41. The initial startup of PanelBuilder will have a blank right window. Click anywhere in the window to activate the task bar menu.

42. Click ***Objects***, then ***Text*** to create the text area ***NAME HERE***. A dotted box will appear, as shown on the left side of Figure 4-1-24. Size this box as needed by dragging it. Then double click in the box to get the ***Text Object*** window, as shown.

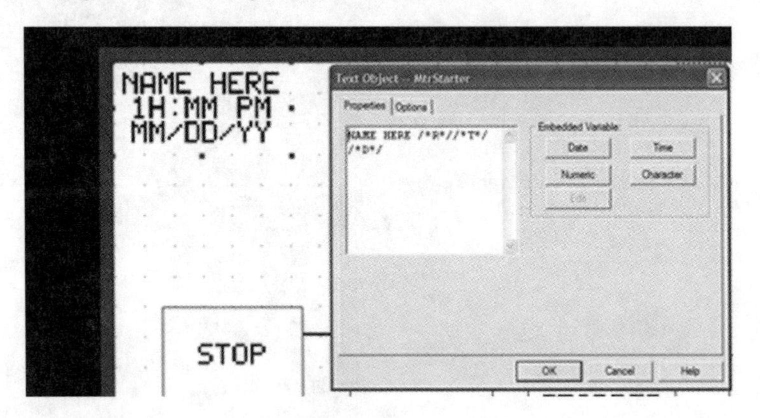

Figure 4-1-24: Create Objects and Texts screen (Courtesy of Rockwell Automation, Inc.)

43. Type your own name in the ***Text Object*** window in place of **Name Here**. Press the keyboard ***ENTER*** key to get the */*R*/,* which means create a new text line. Click the ***Time*** button to get */*T*/* and click ***Date*** button for */*D*/*. These two special commands will display the time and date of PV550's internal clock. Click ***OK***.

44. Follow the step above to create the name of your school, organization, company, division, plant, etc. A logo could be included. **Note: Copy and paste can be used as needed.**

4.1.5 Creating and Entering Tags

45. The next step to creating an HMI is to identify and enter ***tags*** in the ***Tag Editor***. A ***tag*** is a name that is assigned to anyone of the objects in the PLC ladder diagram.

46. The order of entering tag names in the Tag Editor lines is not important. The key thing to keep in mind is that the PLC address in the ***Address*** column must match the desired ***Tag Name***.

47. As shown in Figure 4-1-25, click ***Tag Editor*** under ***System*** in the left window.

48. Each subsequent line must be created by clicking in the ***Tag Name*** space, then clicking the ***INSERT*** icon (circled in Figure 4-1-25). Suggested ***Tag Names*** to use for this HMI are listed in the figure.

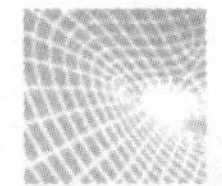

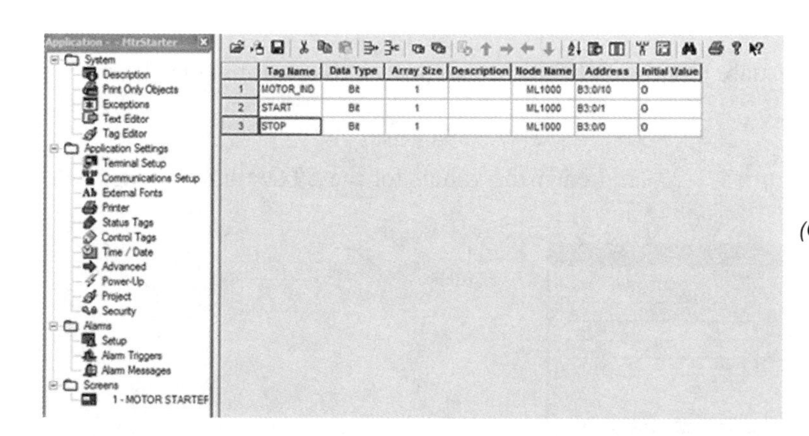

Figure 4-1-25: Tag Editor (Courtesy of Rockwell Automation, Inc.)

49. Once a line appears, type the tag name in the ***Tag Name*** column. Clicking the ***Data Type***, ***Array Size***, ***Node Name*** columns will cause a drop down list of values to appear. Select the values shown in the table. **REMEMBER**, if the input/output address in the PLC ladder diagram changes, then change them here as well.

50. Once all of the tags are entered, be sure to click the ***SAVE*** icon, as shown circled in Figure 4-1-25, to save the tag data in the tag database.

4.1.6 Creating Button and Indicator Icons in the HMI Diagram

51. Create the ***STOP*** and ***START*** pushbuttons by clicking ***Objects***, then ***Pushbutton***, then ***Momentary***. Size the box as needed; then double click inside the box to see the screen shown in Figure 4-1-26.

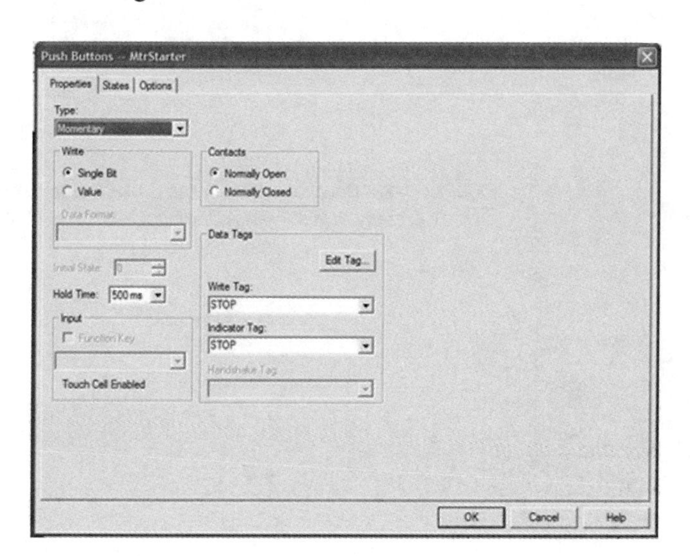

Figure 4-1-26: Creating pushbuttons in the HMI (Courtesy of Rockwell Automation, Inc.)

52. Enter values in the **Properties** tab. Use the right side down arrows to choose values from the drop down menus.

53. Next, click the ***States*** tab (Figure 4-1-27) and enter the values for the ***STOP*** button. Click ***OK***.

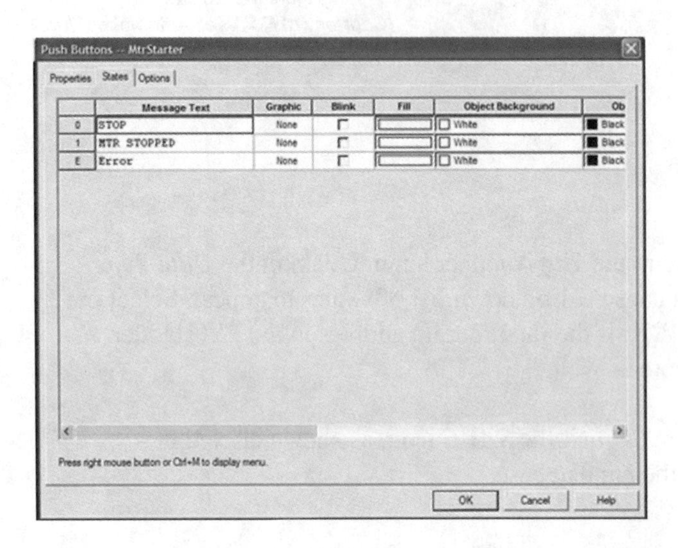

Figure 4-1-27: Creating states of pushbuttons (Courtesy of Rockwell Automation, Inc.)

54. Repeat the above steps for the ***START*** button, as shown in Figures 4-1-28 and 4-1-29. (By the way, the STOP button can be copied and pasted again. Then double click it and change the values as needed.)

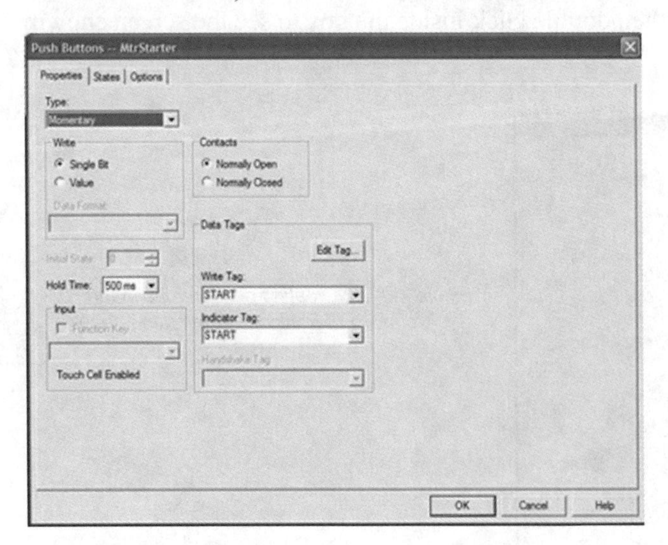

Figure 4-1-28: Set properties for the START button (Courtesy of Rockwell Automation, Inc.)

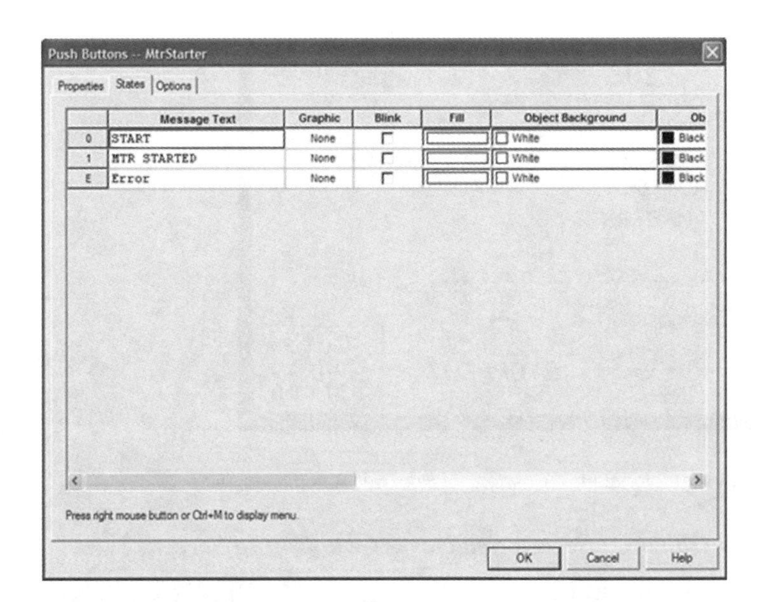

Figure 4-1-29: Create states for the START button (Courtesy of Rockwell Automation, Inc.)

55. Now create the indicator/display box by clicking ***Objects*** > ***Indicators*** > ***Multistate*** (Figure 4-1-30). Size the box, as shown in Step 42. Double click the box; then click ***States***.

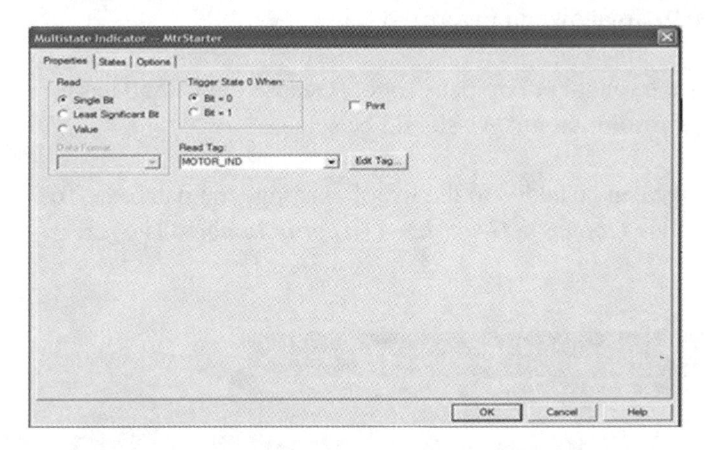

Figure 4-1-30: Multistate Indicator screen (Courtesy of Rockwell Automation, Inc.)

56. The **states 2 and 3 must be deleted to make a single bit indicator**. Under the ***Graphic*** column, right click the cell and a drop down menu will appear. Click ***ISA – Motor***. Repeat for the next line. Note by clicking the ***Blink*** column box that the motor graphic will blink when the motor is running. This is just one example of possible graphic capabilities of the terminal. Do not forget to click ***OK***.

57. One last control icon needs to be added. The ***GOTO CONFIG*** screen is an example of a screen selector (Figure 4-1-31). ***This icon screen selector MUST ALWAYS BE in the display and must be on the first screen!!!!***

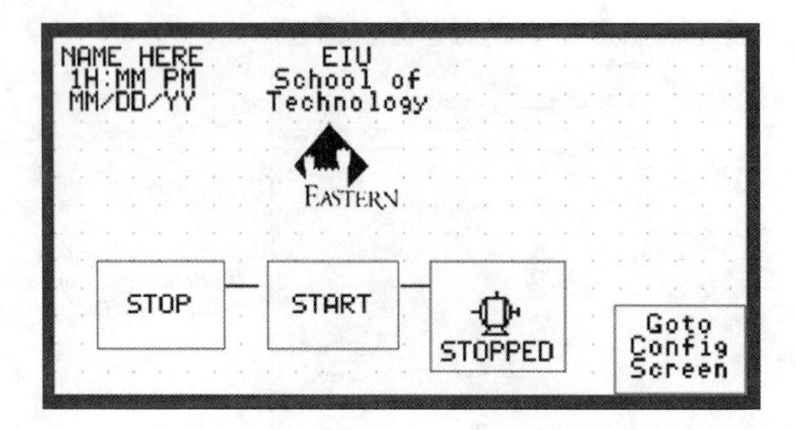

Figure 4-1-31: Final HMI display for motor starter (Courtesy of Rockwell Automation, Inc.)

58. To create this screen selector, click ***Objects*** > ***Screen Selector*** > ***Goto Config Screen***. Size the icon as needed.

59. Don't forget to make the connecting lines between the ***STOP*** and ***START*** buttons. Clicking ***Objects*** > ***Graphics*** > ***Line***, note the connectiong line was constructed from ***START*** button to the ***Motor Indicator box*** (***STOPPED),*** as shown in Figure 4-1-31.

60. The last icon that could be added is the school or company logo. Because the PV550 has a relatively low resolution screen, low resolution images should be selected.

61. By the way, almost any graphic image can be added to the list of existing icon database. To add an image to the icon database, click ***Objects*** > ***Graphics*** > ***Graphic Images***. The screen in Figure 4-1-32 will appear

Figure 4-1-32: Adding objects to icon database (Courtesy of Rockwell Automation, Inc.)

62. Click the pull-down menu circled in Figure 4-1-32 and search through the list of graphics on the computer to find your school or company icon name.

63. Select the icon name, place it on the screen, and size it.

64. Don't forget to save this screen at this point by clicking ***File*** > ***Save As*** to the desired folder on the computer's drive.

4.1.7 Loading the HMI Display Screen into the PanelView

65. To download the HMI into the PV550, Click ***File*** > ***Download***. The screen shown in Figure 4-1-33 should appear.

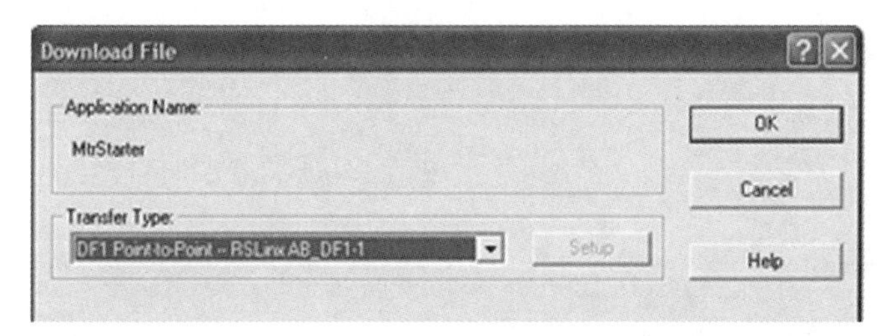

Figure 4-1-33: File download screen (Courtesy of Rockwell Automation, Inc.)

66. Click ***OK*** and the transfer between the computer and PanelView will begin, as shown by Figure 4-1-34.

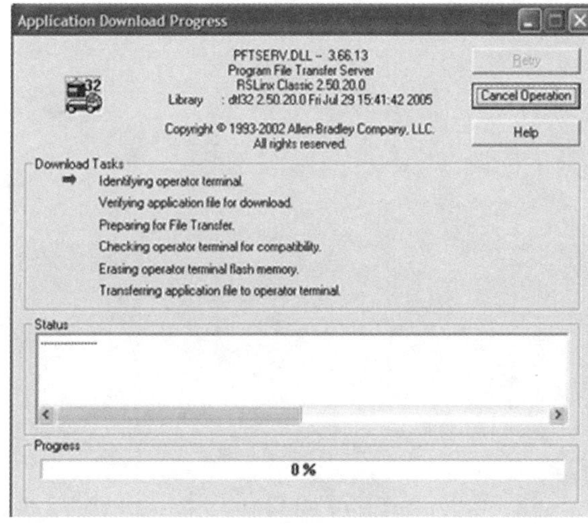

Figure 4-1-34: Application Progress Download screen (Courtesy of Rockwell Automation, Inc.)

67. If the communication parameters are set up correctly, six tasks will appear and checked off in order, and the Progress bar will begin moving. This step will take several minutes to complete.

68. When the transfer is 100% complete, the PV will restart itself, which can take several more minutes to complete. The HMI screen created should appear on the PV display.

69. **If the PLC is not connected to the HMI terminal or is not powered up,** an offline error will appear on the PanelView screen. Connect the PLC and verify that the screen works by touching the buttons. Verify that the output ***O:0/0*** is lit up (if output was used).

70. Also, touch the ***Goto Config Screen*** selector and explore the terminal setup. For example, change the display to ***Reverse Video*** to see how this might give a better display.

71. This completes PV Lab 4-1.

PV Lab 4-2: Analog and Digital Displays

4.2.1 Introduction

This lab illustrates more complex graphic capabilities and how multiple screens are used in operator display terminals (especially small terminal sizes). It will build upon Lab 4-1.

4.2.2 Programming the MicroLogix PLC

1. Copy the PLC program from Lab 4-1 and save it as Lab 4-2. Then modify this activity. Alternatively, follow the steps below to program the PLC.

2. Enter the PLC program shown in Figure 4-2-1 into ***RSLogix 500 (English)***. Notice that ***binary addressing, Limit Tests,*** and ***Time On-Delay (TON)*** timer HMI icons are used.

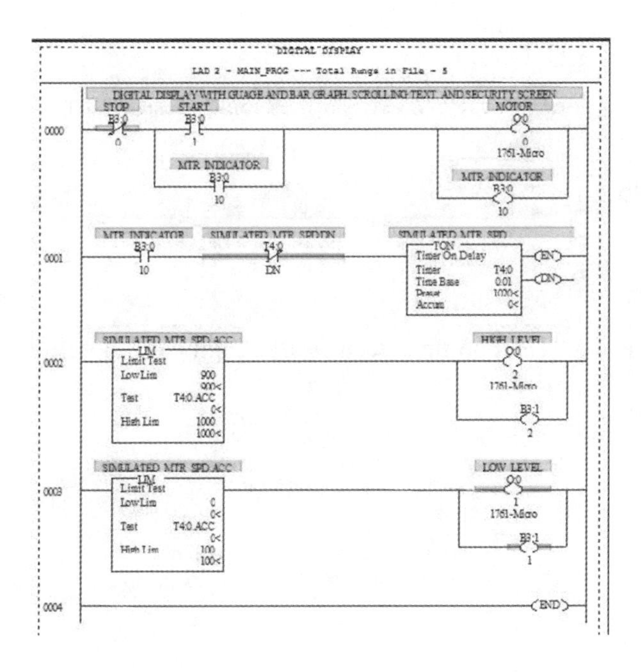

Figure 4-2-1:
Modified Lab 4-1 PLC program for Lab 4-2
(Courtesy of Rockwell Automation, Inc.)

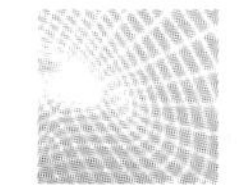

3. It may happen that the RSLogix 500 software needs to be configured if only a limited number of binary addresses are available. In the ***Data File B3*** pop-up menu, as shown in Figure 4-2-2, only addresses from ***B3:0/0 to B3:0/15*** are offered.

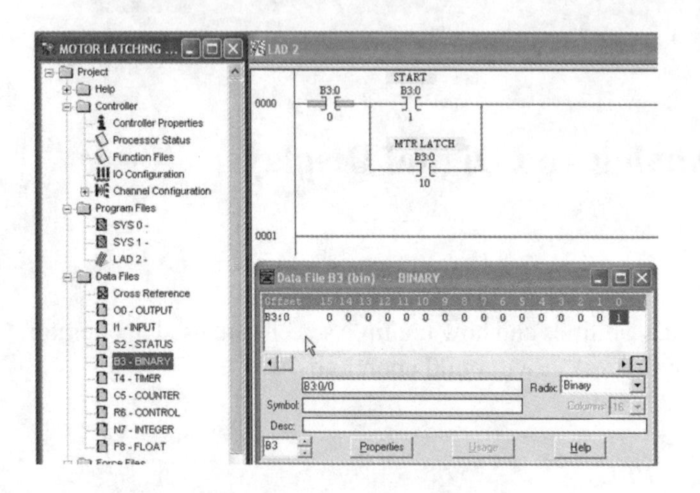

Figure 4-2-2: Using Data File B3
(Courtesy of Rockwell Automation, Inc.)

4. To increase the number of address selections, click the **Properties** button at the bottom of the **Data File B3** pop-up screen (Figure 4-2-3).

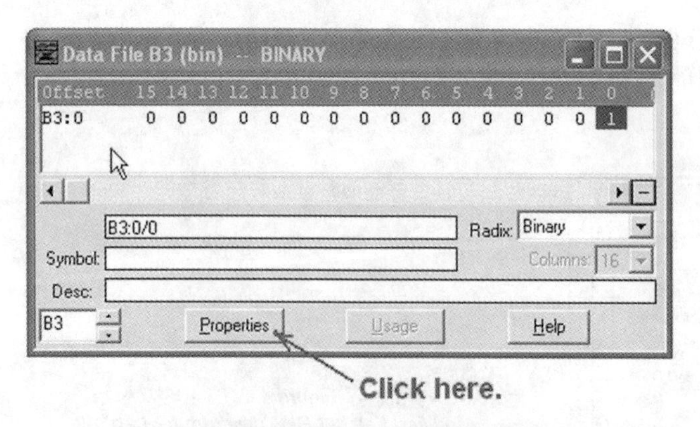

Figure 4-2-3: Properties of Data File B3
(Courtesy of Rockwell Automation, Inc.)

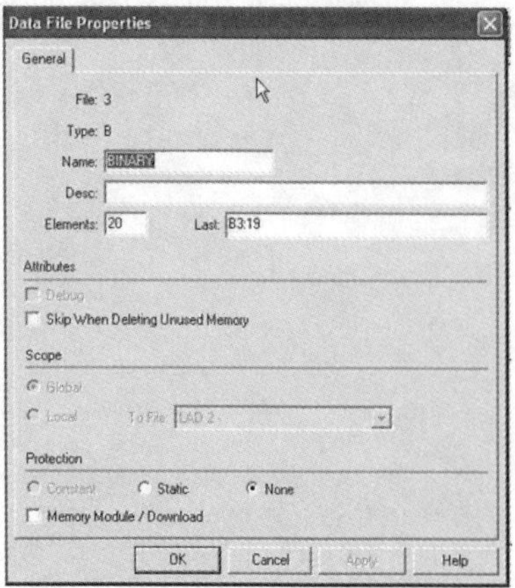

5. The pop-up screen seen in Figure 4-2-4 will appear.

Figure 4-2-4: Data File B3 pop-up display
(Courtesy of Rockwell Automation, Inc.)

6. In the **Elements** dialog box, type in the value of **20**. Click **Apply** and then **OK**.

7. **RSLogix 500** will now provide 20 x 15 binary addresses for input and output labeling purposes (Figure 4-2-5).

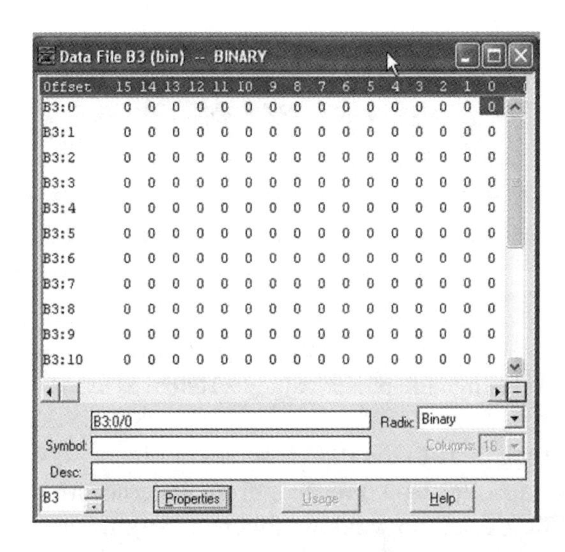

Figure 4-2-5:, Adding Elements to Data File B3 (Courtesy of Rockwell Automation, Inc.)

8. To label the **Time On-Delay (TON)** timer, double click the **T4-Timer** icon underneath the **Data Files** Folder in RSLogix 500 (Figure 4-2-6). A **Data File T4** pop-up screen will appear.

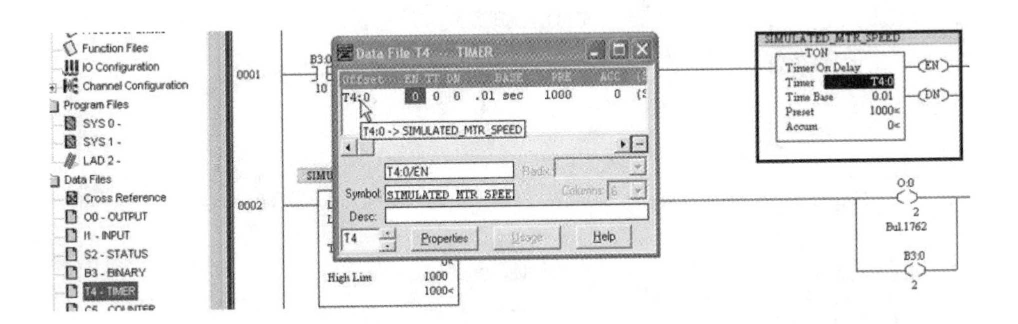

Figure 4-2-6: Time On-Delay (TON) to T4-Timer (Courtesy of Rockwell Automation, Inc.)

9. Click the **T4:0** in the pop-up menu and then drag it over to the Timer labeling area (shown blackened in Figure 4-2-6).

10. On the **TON** icon in the PLC program, click the current **Time Base** value. Change the value to **0.01** (seconds), as shown in Figure 4-2-7. Then click somewhere off of the icon to store that value.

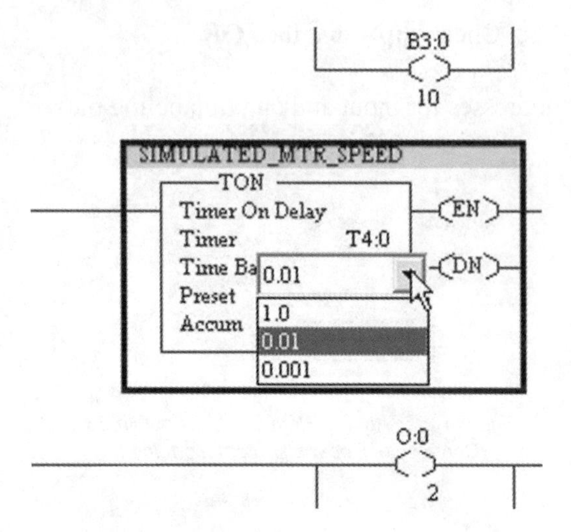

Figure 4-2-7: TON icon (Courtesy of Rockwell Automation, Inc.)

11. Double click the **Preset** value on the **TON** icon to change the preset value to **1000**.

12. To label the normally closed input (as shown in rung 0001 of the PLC program in Figure 4-2-8) so that it is actuated/linked to the **T4:0 timer**, click the **T4-Timer** icon from underneath **Data Files**. Next, click the **DN** from the **Data File T4** pop-up window (Figure 4-2-9). Drag the **DN** symbol to label the input. Place it on top of the normally closed input.

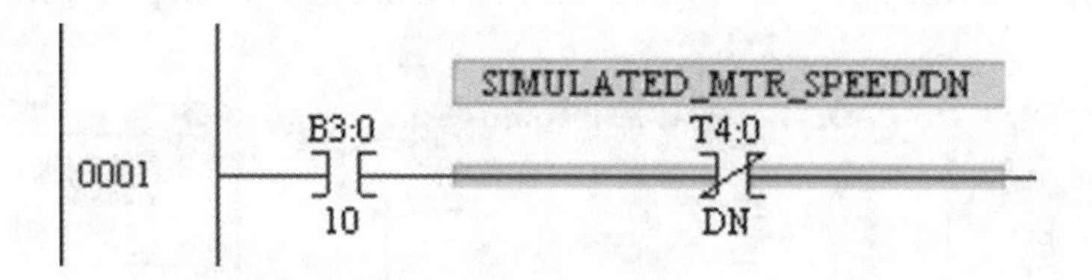

Figure 4-2-8: Setting the T4 timer (Courtesy of Rockwell Automation, Inc.)

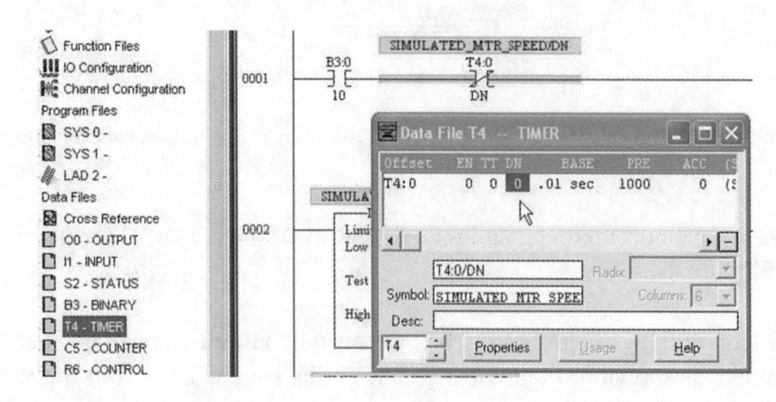

Figure 4-2-9: DN symbol for T4 timer Data File (Courtesy of Rockwell Automation, Inc.)

13. To label a **Limit Test**, click the question mark (**?**) located next to the **Low Lim** symbol on the icon. A dialog box appears, as shown in Figure 4-2-10.

    a. Type in the count value needed for the Low Lim test. Repeat this for the **Test** value and for the **High Lim** value.

    b. To label the **Test** portion of the **Limit Test** icon, note the graphic provided in Figure 4-2-11. Click the **T4 Timer** icon from underneath **Data Files**. Next, click **ACC** from the **Data File T4** pop-up window. Drag the **ACC** symbol so it labels the icon by placing it on top of the **Test** field inside the icon.

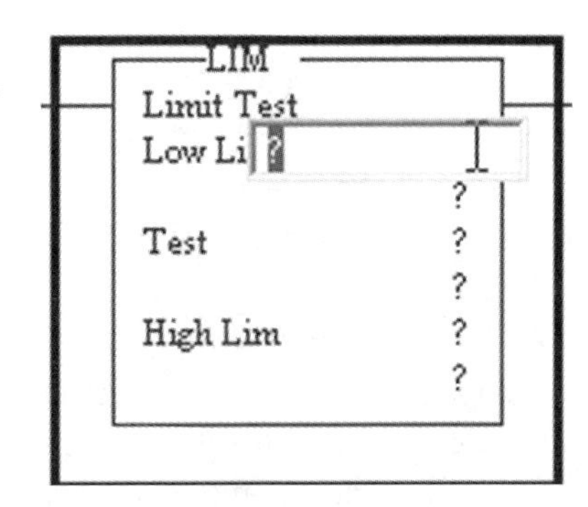

Figure 4-2-10: Limit Test Data File (Courtesy of Rockwell Automation, Inc.)

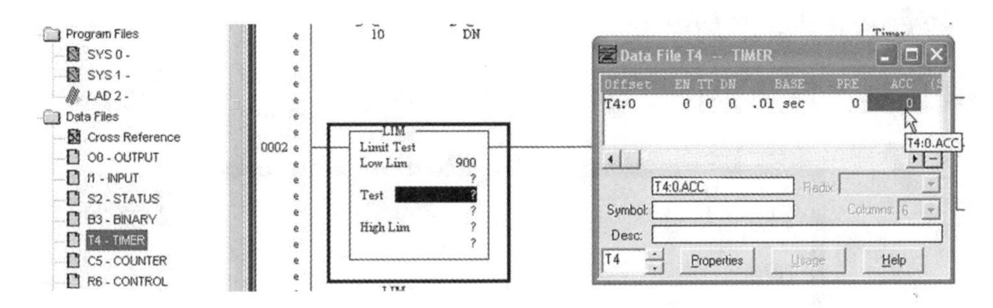

Figure 4-2-11: Limit Test connected to T4 Timer (Courtesy of Rockwell Automation, Inc.)

    c. Click the program somewhere else beside a program icon to save the changes to the **Limit Test** icon.

14. **Download** the ladder program into the PLC. Be sure to place the program in the **RUN** mode to continue.

15. Click **File** and then **New**. In the **Create New Project** pop-up screen that appears, be sure that **Local Disk (C:)** is selected in the **Look in:** dialog box (Figure 4-2-12). Select the **RSLinx_RSLogix500 Programs Storage Folder** and create an animated project using RSLinx.

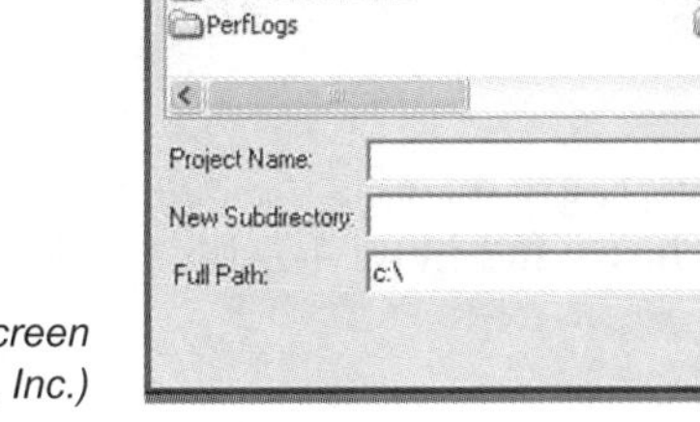

Figure 4-2-12: Create New Project pop-up screen (Courtesy of Rockwell Automation, Inc.)

16. In the **Project Name:** portion of this pop-up screen, type in something distinctive, for example, **Lab_2_XXX** (XXX = your initials), and then click **Open**.

4.2.3 Creating the HMI Diagram with PanelBuilder32

17. Launch ***PanelBuilder32*** and create a new HMI program called ***Digital_Display***. Use techniques from Lab 4-1 to create the Digital_Display program. Don't forget to change the baud rate to 19,200 by clicking ***Communications Setup***. Also, remember to create the Node Name, Node Address, and Node Type as was done in Lab 4-1.

18. The following series of displays illustrate the capabilities of the Panel View hardware and software.

4.2.4 Gauge and Digital Display

Figure 4-2-13 illustrates the gauge and digital type displays.

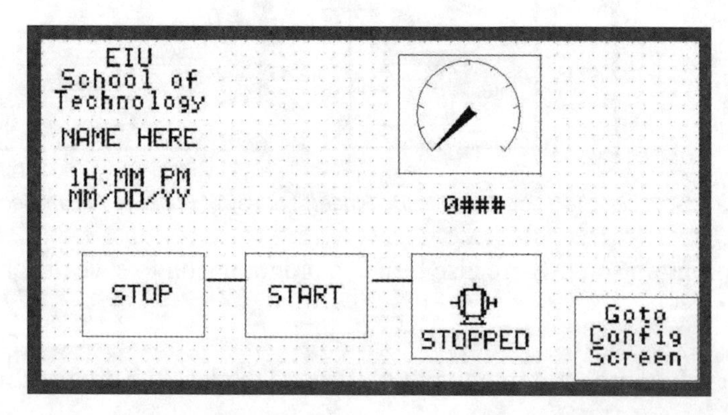

Figure 4-2-13: Gauge and Digital Display (Courtesy of Rockwell Automation, Inc.)

19. With the exception of the digital and gauge displays, the screen in Figure 4-2-13 is very similar to the screen in Lab 4-1. Just load that screen and save it as Lab 4-2. Change the screen name by highlighting the screen name in the left window and then right clicking the highlighted name and select ***Properties***.

20. In the ***Tag Editor***, use the existing tags if possible or modify them as needed. Create new tags as needed as shown in Figure 4-2-14.

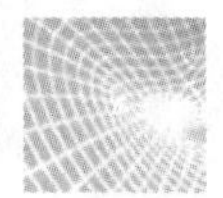

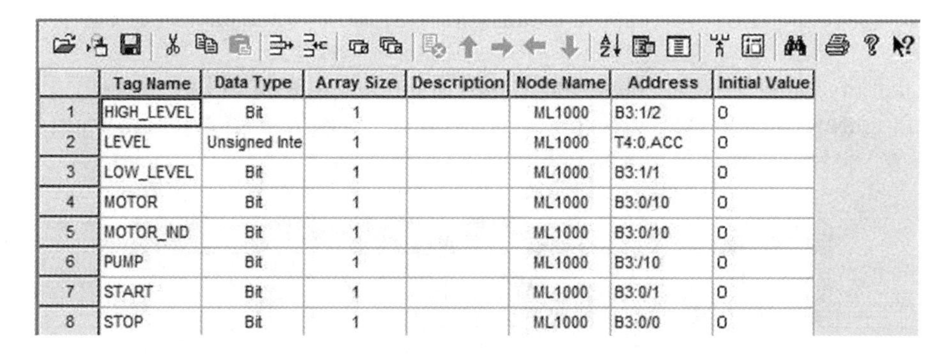

	Tag Name	Data Type	Array Size	Description	Node Name	Address	Initial Value
1	HIGH_LEVEL	Bit	1		ML1000	B3:1/2	0
2	LEVEL	Unsigned Inte	1		ML1000	T4:0.ACC	0
3	LOW_LEVEL	Bit	1		ML1000	B3:1/1	0
4	MOTOR	Bit	1		ML1000	B3:0/10	0
5	MOTOR_IND	Bit	1		ML1000	B3:0/10	0
6	PUMP	Bit	1		ML1000	B3:/10	0
7	START	Bit	1		ML1000	B3:0/1	0
8	STOP	Bit	1		ML1000	B3:0/0	0

Figure 4-2-14: Tag editor (Courtesy of Rockwell Automation, Inc.)

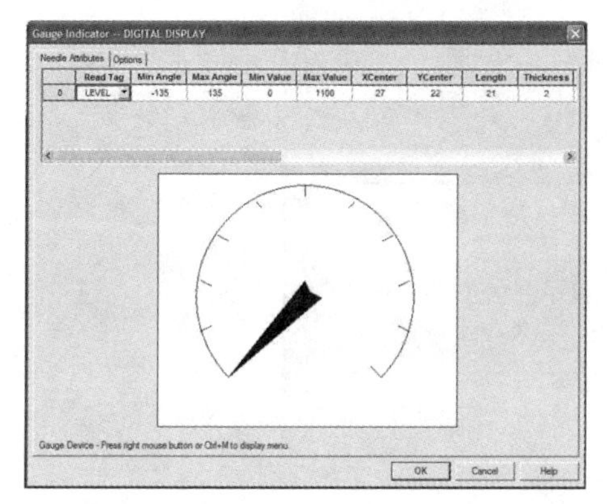

Figure 4-2-15: Graphic Gauge Indicator (Courtesy of Rockwell Automation, Inc.)

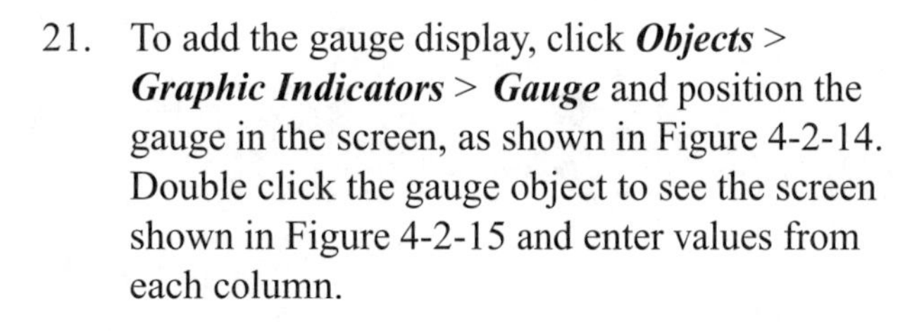

21. To add the gauge display, click ***Objects*** > ***Graphic Indicators*** > ***Gauge*** and position the gauge in the screen, as shown in Figure 4-2-14. Double click the gauge object to see the screen shown in Figure 4-2-15 and enter values from each column.

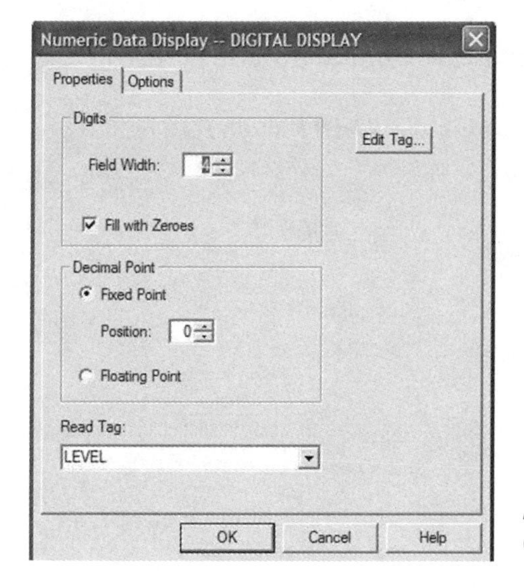

22. Now add the digital display by clicking ***Objects*** > ***Numeric Data Display***; position this display under the Needle Attributes, as shown in Figure 4-2-15. Double click the digital display to see the screen in Figure 4-2-16 and enter the values as shown.

Figure 4-2-16: Numeric Digital Data Display (Courtesy of Rockwell Automation, Inc.)

23. Be sure to save the HMI project by clicking ***File*** > ***Save As*** to the chosen folder.

24. Check that the Panel View is connected to the computer and PLC before downloading the the HMI in Figure 4-2-16 to the PanelView. Click ***File*** > ***Download*** and click through the screens to see the transfer begin, as in Figure 4-2-17.

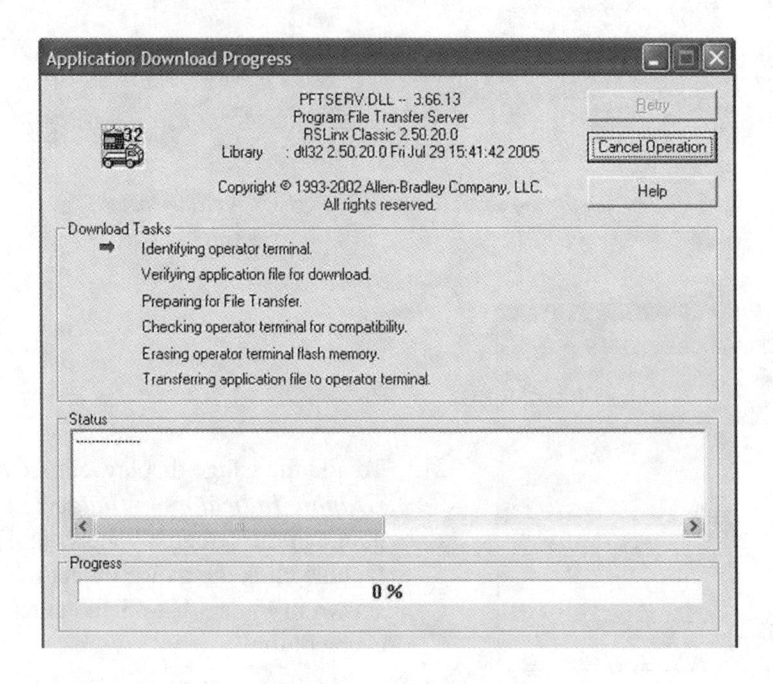

Figure 4-2-17: Application Download Progress display (Courtesy of Rockwell Automation, Inc.)

25. Remember to check that the PLC is connected; otherwise, an offline error will occur.

26. Make sure the gauge and digital display work by activating the ***START*** and ***STOP*** buttons. Also make sure the **Goto** screen works.

4.2.5 Creating a Scrolling Message

27. Now add a scrolling message to the bottom of the screen, as shown in Figure 4-2-18.

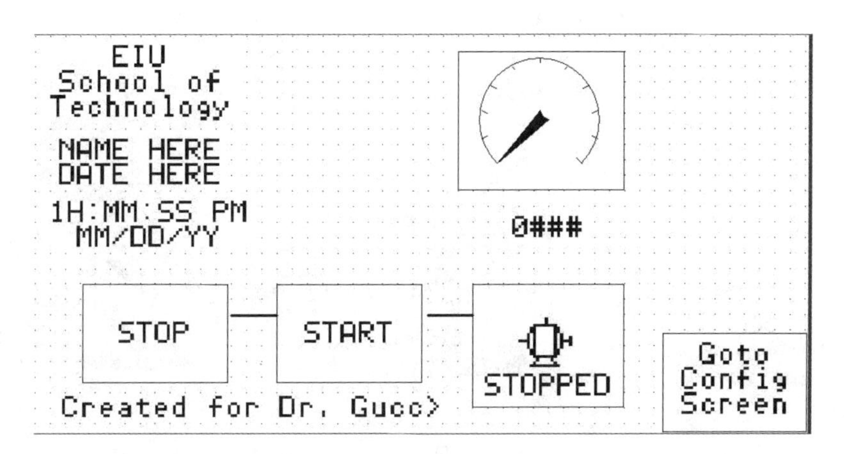

Figure 4-2-18: HMI with Scrolling Message at bottom of screen (Courtesy of Rockwell Automation, Inc.)

28. Click ***Object*** > ***Scrolling Text*** and position the text message box at the bottom left side of the screen shown in Figure 4-2-19. Double click the box and enter a text message such as the one shown in the figure.

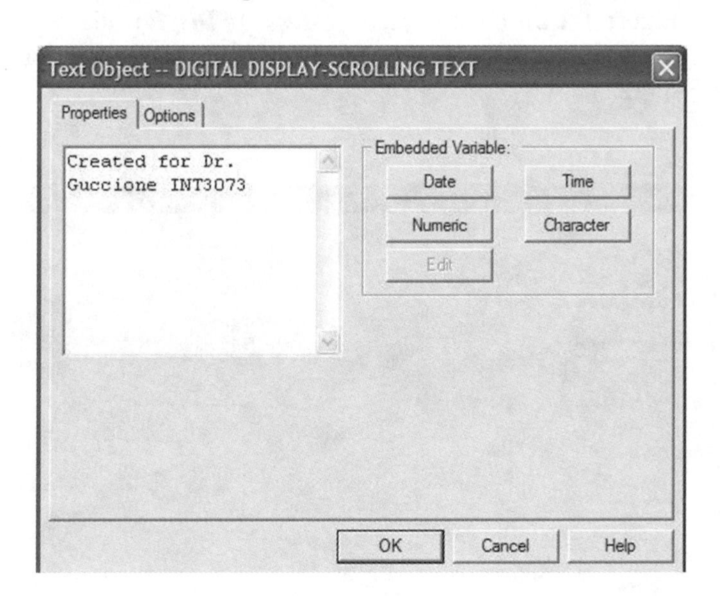

Figure 4-2-19: Creating Digital Display Scrolling Text message (Courtesy of Rockwell Automation, Inc.)

29. Be sure to save the HMI project by clicking ***File*** > ***Save As*** to the desired folder. Download the HMI to the PanelView. Check that it is connected to the computer and PLC. Click ***File*** > ***Download*** and click through the screens to start the download.

30. Verify that the HMI diagram works and the scrolling message is visible.

4.2.6 Creating a Message Display

31. Figure 4-2-20 shows another type of text display called a ***Message Display*** (the box with The Motor is not running).

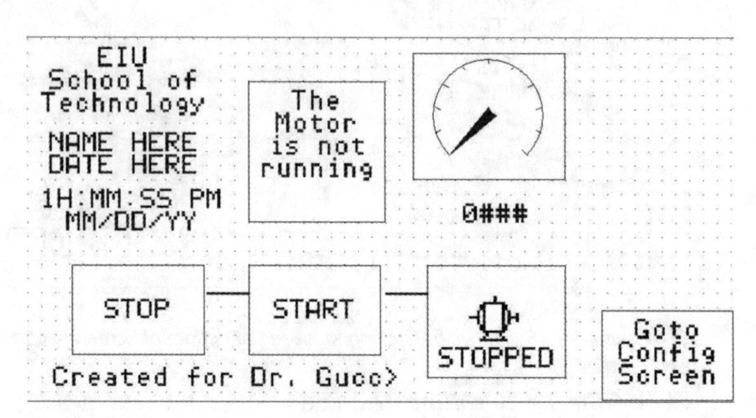

Figure 4-2-20:
Example of Message Display
(Courtesy of Rockwell Automation, Inc.)

32. To add the type of display shown in Figure 4-2-21, click ***Objects*** > ***Message Display***. Place the message on the screen by double clicking the box and enter values into the ***Properties*** and ***States*** screens, as shown in Figure 4-2-22.

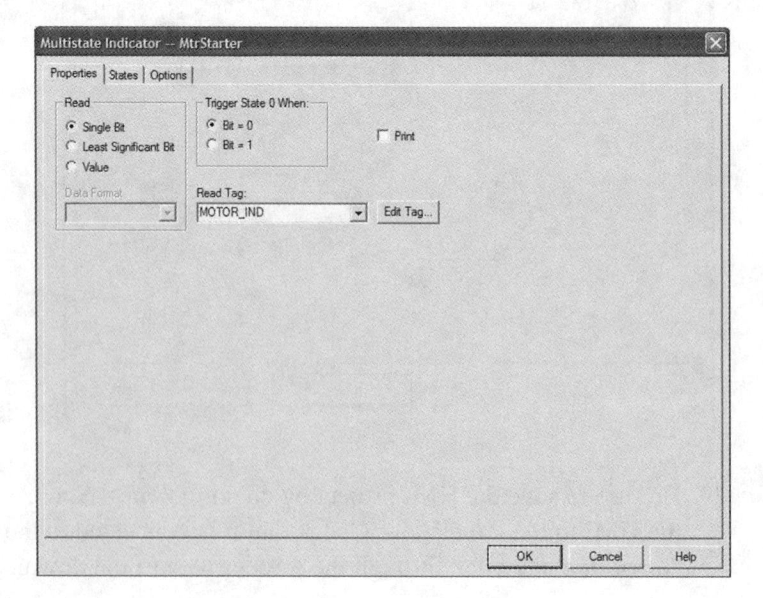

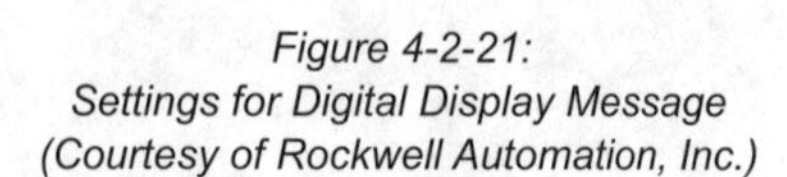
Figure 4-2-21:
Settings for Digital Display Message
(Courtesy of Rockwell Automation, Inc.)

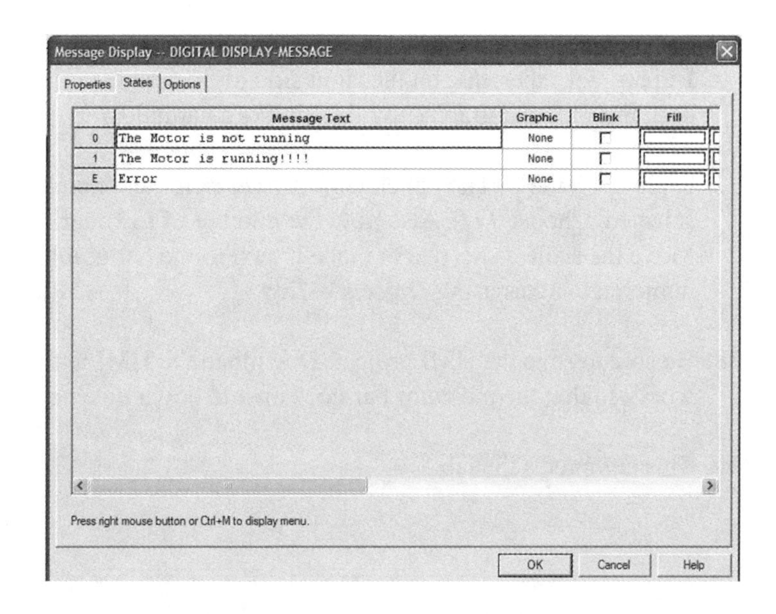

Figure 4-2-22:
Create Digital Display Message
(Courtesy of Rockwell Automation, Inc.)

33. Be sure to save the HMI project as needed. Then download the HMI to the PanelView. Verify that the HMI works and the message is displayed.

4.2.7 Creating a Bar Graph

34. Figure 4-2-23 shows a new display called a bar graph. Create a new screen by copying the existing screen, then deleting the message and analog gauge.

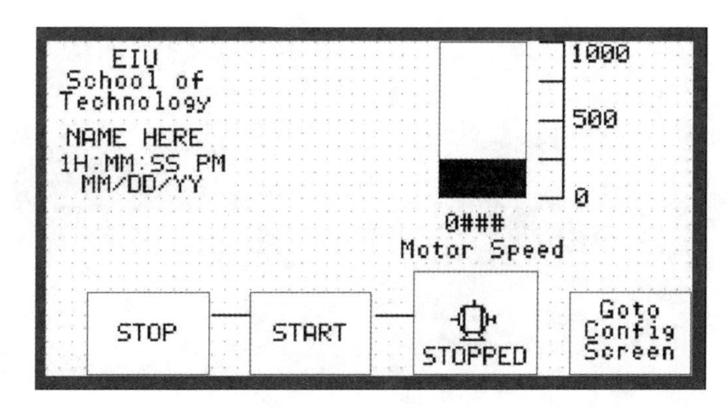

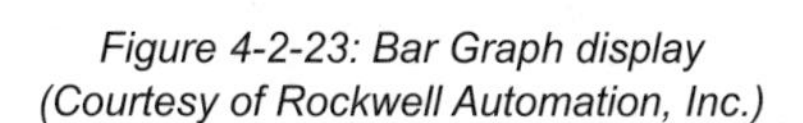
Figure 4-2-23: Bar Graph display
(Courtesy of Rockwell Automation, Inc.)

35. To create the bar graph, click ***Objects*** > ***Graphic Indicators*** > ***Bar Graph***. Size and place the bar graph as desired on the screen. Note that the black bar indicator is partially visible. This indicator will rise and fall with the data that is being measured.

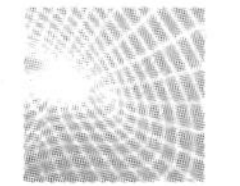

36. To create the scale beside the bar graph, click ***Objects*** > ***Graphic Indicators*** > ***Scales*** > **Linear**. Size the scale on the right side of the graph so that the scale aligns with the top and bottom of the bar graph. It will look like a capital E.

37. Right click the E. Then click ***Object Properties*** and note that four orientations can be selected. Choose ***Left***. Also note the number of ***tick*** marks that can be selected. Choose ***5***. Move the scale as needed to place it next to the bar graph, as shown in Figure 4-2-23. Add numerical values using ***Objects*** > ***Text***.

38. Be sure to save the HMI project. Download the HMI to the PanelView. Verify that the HMI works so that the indicator bar goes up and down in sync with the digital display.

39. This completes the lab.

PV Lab 4-3: Pump and Tank

4.3.1 Introduction

This lab illustrates some more complex graphic capabilities available in PanelBuilder32 and how multiple screens are used in Graphic Terminal Displays (especially small terminal sizes). It will build upon Labs 4-1 and 4-2 and their ladder logic diagrams.

4.3.2 Programming the MicroLogix PLC

Figure 4-3-1 shows the PLC program for this lab activity. Copy Lab 4-2's PLC program and modify as identified here. Note that internal documentation techniques were used to create a title, a program description with name and date, and symbol labels. This type of documentation is used in industrial ladder diagrams and should be included in all ladder diagrams for the practice.

This circuit uses a Timer on Delay HMI symbol to simulate a liquid being pumped into and out of the tank.

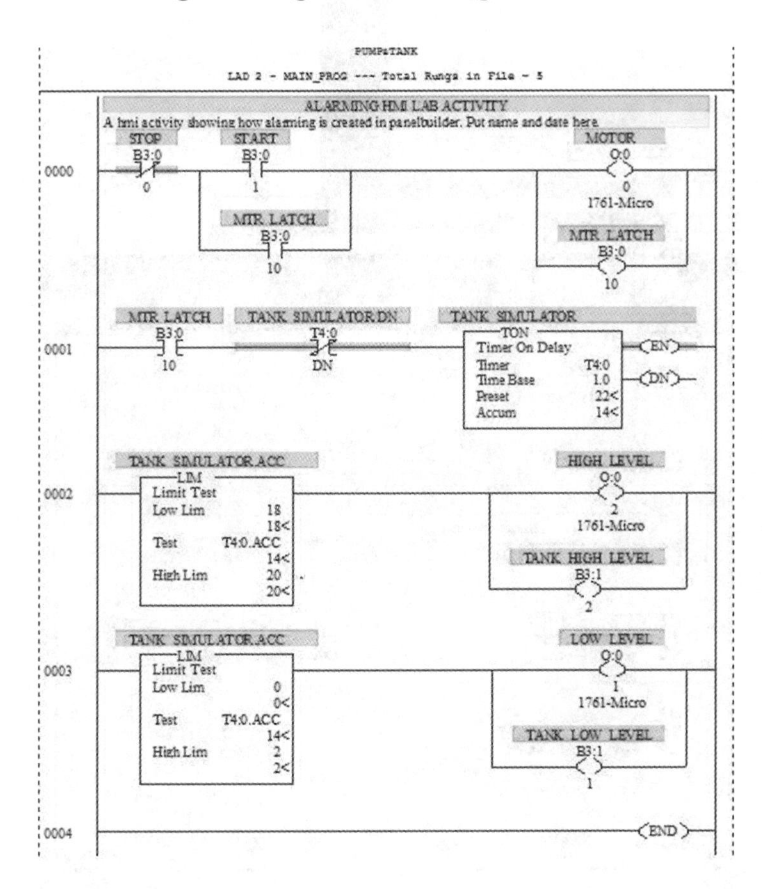

Figure 4-3-1:
Lab 4-2 PLC program for Lab 4-3
(Courtesy of Rockwell Automation, Inc.)

4.3.3 *Creating the HMI Pump and Tank Display*

1. Create and download the Lab 4-3-1 ladder diagram into the PLC and be sure to place it in ***RUN*** mode. Make sure that the PanelView is **plugged into** the computer with **Panel-Builder32** software and the PLC is **plugged into** the correct port on the **PanelView**.

2. ***Launch*** PanelBuilder32 and create a new HMI called ***Pump&Tank***. If needed, follow the steps you learned in **Labs 4-1 and 4-2**. The display shown in Figure 4-3-2 will appear on the PanelView, indicating it is ready to create the needed HMI screens for this lab.

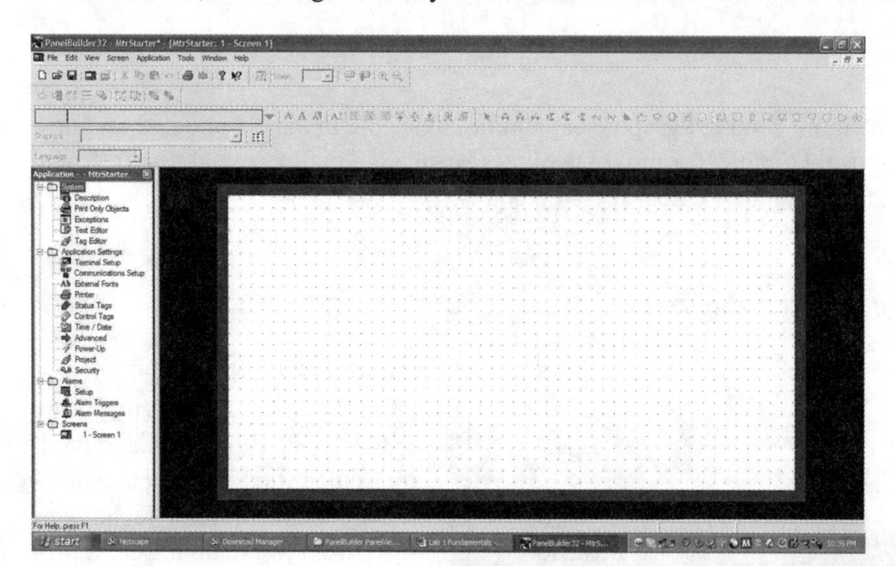

Figure 4-3-2: Blank screen for creating Pump&Tank HMI (Courtesy of Rockwell Automation, Inc.)

3. Don't forget to set the baud rate to ***19200*** by clicking ***Communications Setup*** (Figure 4-3-3). Also, remember to create the **Node Name, Node Address,** and **Node Type,** as used in **Lab 4-1**.

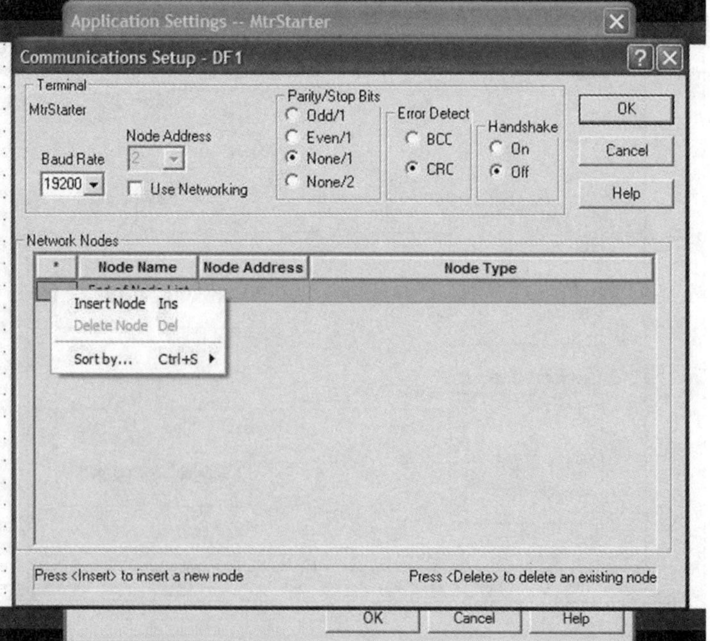

Figure 4-3-3: Communications Setup screen (Courtesy of Rockwell Automation, Inc.)

4. Figure 4-3-4 shows the ***first screen*** for this lab activity. Note that there are two screens, as shown in the circled highlight. Multiple screens are very common because graphic terminal screens are typically small and cannot display everything needed in one screen.

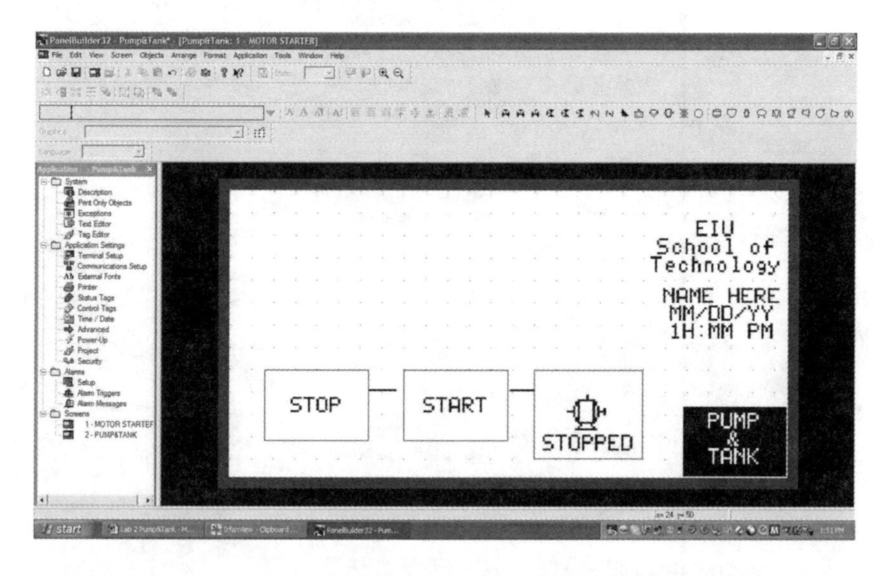

Figure 4-3-4: First HMI screen (Courtesy of Rockwell Automation, Inc.)

5. Using what was learned in **Lab 4-1**, construct the HMI screen seen in Figure 4-3-1, which is similar to the screen in Figure 4-1-14 in **Lab 4-1**. To change the screen name, highlight the screen name in the left window; then right click the highlighted name and select ***Properties.***

6. Don't forget that ***tags*** will need to be created in the ***Tag Editor,*** as done in Lab 4-1. Figure 4-3-5 shows what the tags should look like for this lab activity. Note the additional tags needed for this lab.

	Tag Name	Data Type	Array Size	Description	Node Name	Address	Initial Value
1	HIGH_LEVEL	Bit	1		ML1000	B3:1/2	0
2	LEVEL	Unsigned Inte	1		ML1000	T4:0.ACC	0
3	LOW_LEVEL	Bit	1		ML1000	B3:1/1	0
4	MOTOR	Bit	1		ML1000	B3:0/10	0
5	MOTOR_IND	Bit	1		ML1000	B3:0/10	0
6	PUMP	Bit	1		ML1000	B3:/10	0
7	START	Bit	1		ML1000	B3:0/1	0
8	STOP	Bit	1		ML1000	B3:0/0	0

Figure 4-3-5: Tag Editor (Courtesy of Rockwell Automation, Inc.)

7. Note also that the screen in Figure 4-3-5 looks very similar to the one in Figure 4-1-25 in **Lab 4-1**, except the ***Goto Config Screen*** has been replaced by a ***Pump&Tank screen*** selection button. Also note that the time is in **hours and minutes** only, NOT seconds.

8. To change the time setting, double click ***Terminal Setup*** under ***Lab Settings*** in the left window of Figure 4-3-4. Select the ***Time / Date tab*** as shown circled in Figure 4-3-6.

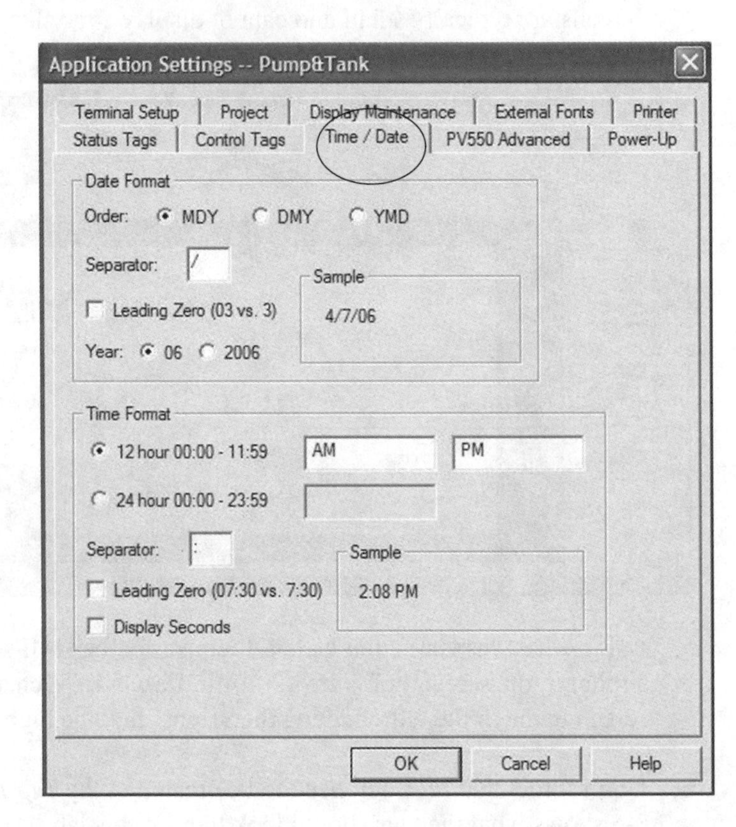

Figure 4-3-6: Changing Time and Date (Courtesy of Rockwell Automation, Inc.)

9. However, before adding the ***Pump& Tank*** screen selection to the first screen, the second screen must be created. Figure 4-3-7 shows how Screen 2 should appear.

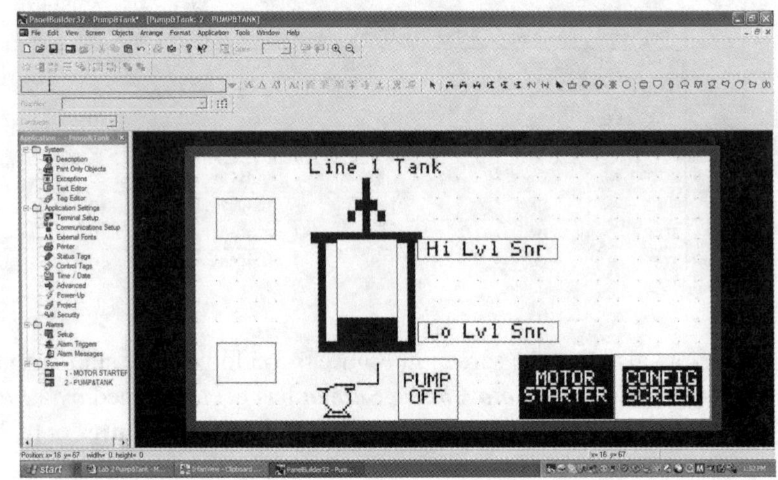

Figure 4-3-7: Pump&Tank HMI Screen 2 (Courtesy of Rockwell Automation, Inc.)

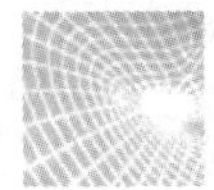

10. The ***first objects*** to put on the diagram are the ***tank and pump graphic*** as shown in step 2. To access the ***tank graphic***, choose ***Objects > Graphic > Graphic Image.*** Then, using the cross, place the box on the screen. Click the little down arrow under ***Graphics:*** on the left side of the screen and choose ***ISA Liquid*** from the list. Resize the graphic and place it in the proper location. Repeat this for the pump graphic using the ***ISA pump image***. Resize and place it in the proper location.

11. Create text for "***Line 1 Tank***", "***Hi Lvl Snr***", and "***Lo Lvl Snr***". Put rectangles around the last two text strings. The rectangles are under ***Objects > Graphics.***

12. Now create the indicator or display box by clicking ***Objects > Indicators > Multistate***. Size the box as needed; then **double click and enter** the values shown in Figure 4-3-8. One will be a **high level indicator,** as shown.

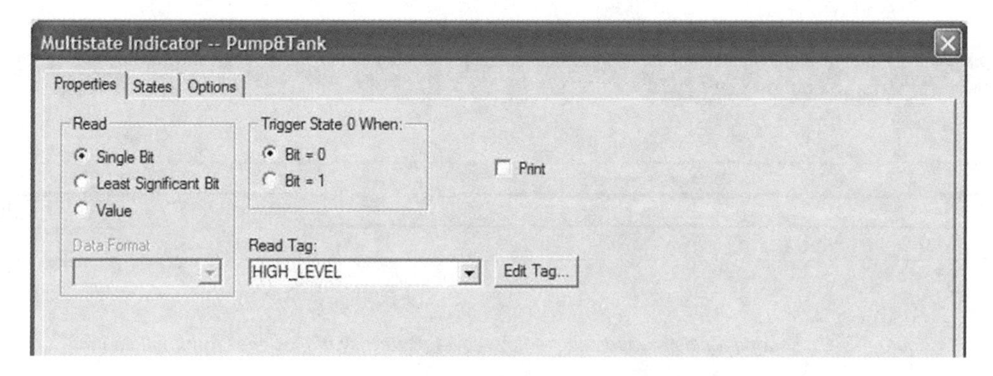

Figure 4-3-8: Create Multistate Indicator — High level (Courtesy of Rockwell Automation, Inc.)

13. In the ***States*** tab, ***delete states 2 and 3*** before entering the values for states ***0*** and ***1*** in the multistate indicator, as shown in Figure 4-3-9.

Multistate Indicator -- Pump&Tank

Properties | States | Options

	Message Text	Graphic	Blink	Fill
0		None	☐	
1	High	None	☑	
E	Error	None	☐	

Figure 4-3-9: Change States of Multistate Indicator — High level (Courtesy of Rockwell Automation, Inc.)

14. Create a ***second multistate indicator*** for the **low level indicator**, as shown in Figures 4-3-10 and 4-3-11.

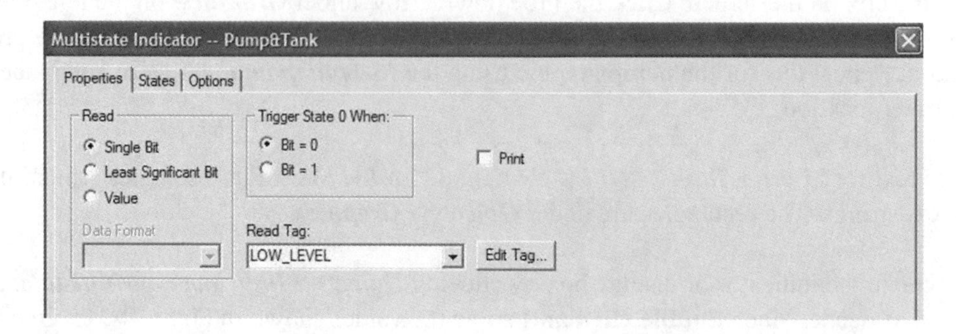

Figure 4-3-10: Second Multistate Indicator — Low level (Courtesy of Rockwell Automation, Inc.)

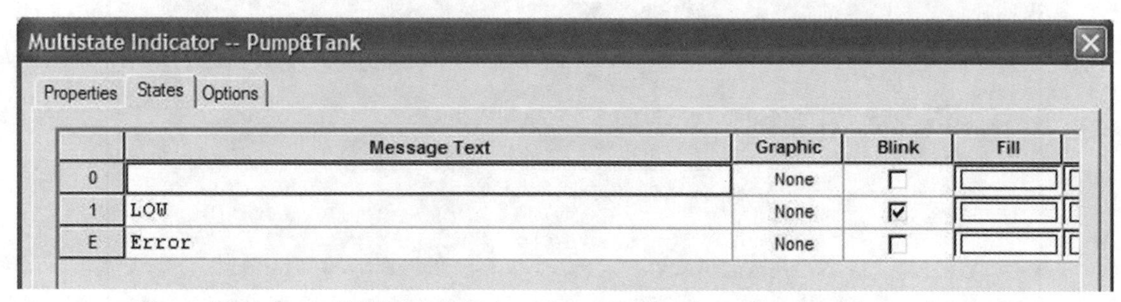

Figure 4-3-11: Second Multistate Indicator — Low level (Courtesy of Rockwell Automation, Inc.)

15. Finally, create a ***multistate indicator for the Pump Off indicator*** (Figures 4-3-12 and 4-3-13).

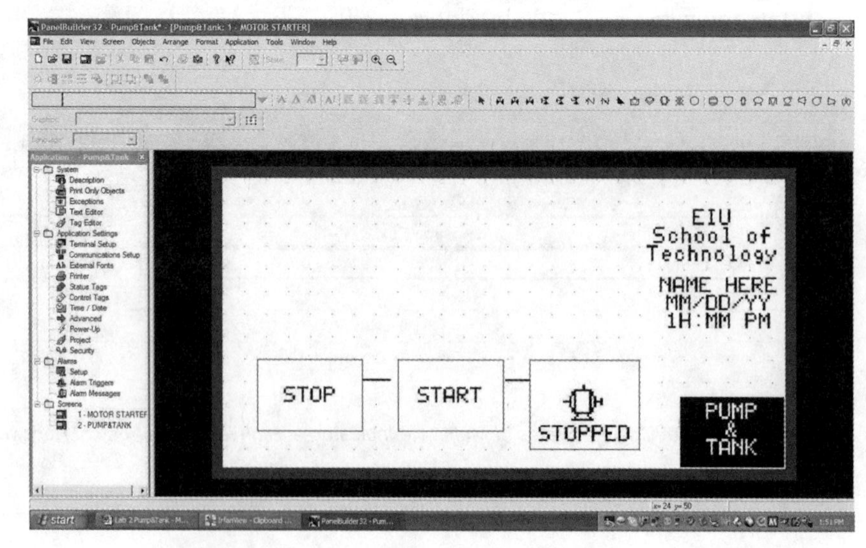

Figure 4-3-12: Pump OFF Multistate Indicator (Courtesy of Rockwell Automation, Inc.)

Multistate Indicator -- Pump&Tank

Properties | States | Options

	Message Text	Graphic	Blink	Fill
0	PUMP OFF	None		
1	ON	None		
E	Error	None		

Figure 4-3-13: Pump OFF Multistate Indicator (Courtesy of Rockwell Automation, Inc.)

16. Now a special graphic is needed to indicate the level of liquid in the tank. To simulate the liquid in the tank a Bar Graph will be created inside the tank in Figure 4-3-15. This Bar Graph is created using ***Objects > Graphic Indicators > Bar Graph***. Size the bar graph rectangle to fit inside the tank, as shown in Figure 4-3-15. Double click the rectangle and enter the values shown in Figure 4-3-14.

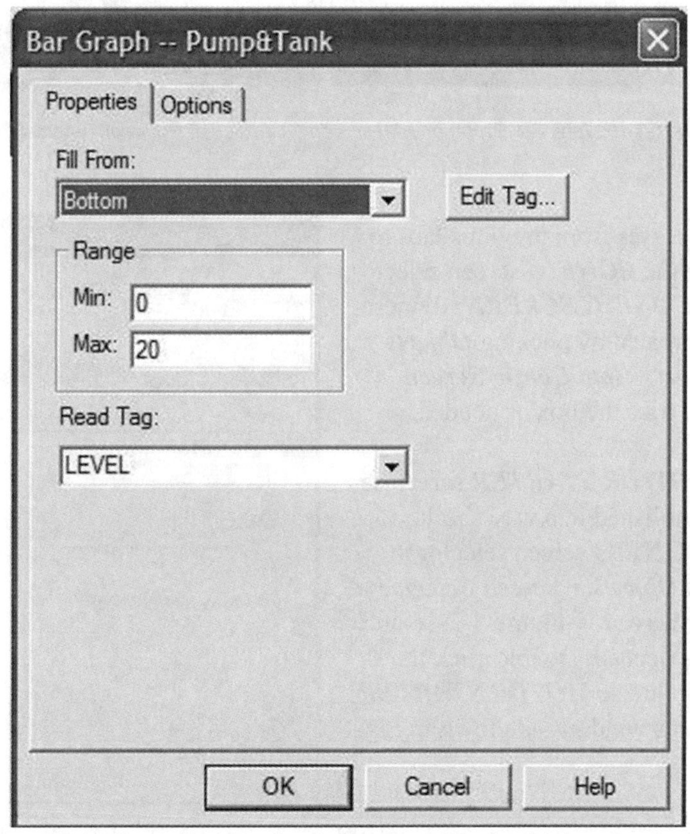

Figure 4-3-14: Create Tank Bar Graph (Courtesy of Rockwell Automation, Inc.)

4.3.4 Adding the Config and Motor Starter Screen Selectors to the HMI Display

17. The last two items that need to be added are the ***CONFIG*** and ***MOTOR STARTER*** screen selectors (Figure 4-3-15).

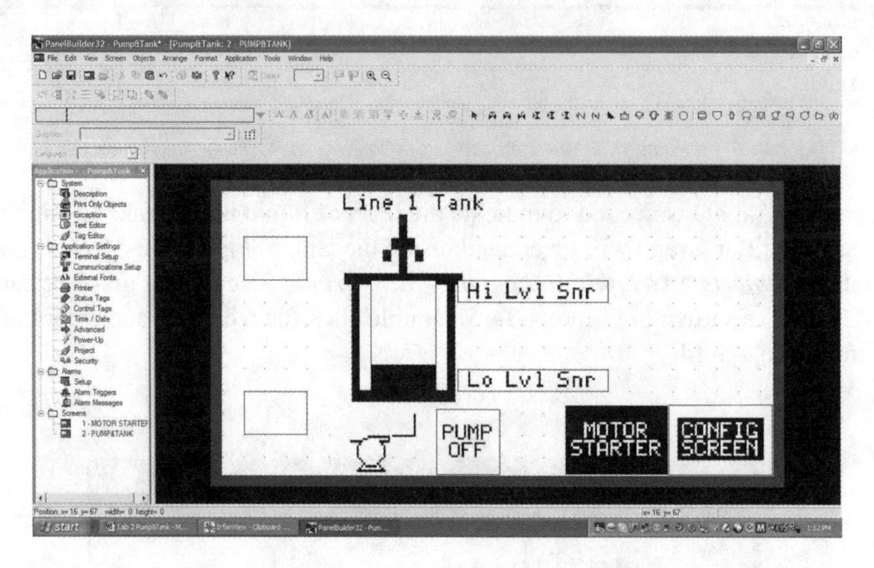

Figure 4-3-15: Tank Bar Graph on HMI screen (Courtesy of Rockwell Automation, Inc.)

18. Recall the processes from previous labs to begin creating the ***CONFIG*** screen selector, using ***GOTO CONFIG SCREEN***. Remember, it can be created by clicking ***Objects > Screen Selector > Goto Config Screen*** in Figure 4-3-16. Size the box as needed.

19. Creating the ***MOTOR STARTER*** screen selector is accomplished in a way similar to creating the **CONFIG** screen selector in Step 18. Click ***Objects > Screen Selector > Goto Specific Screen*** in Figure 4-3-16 and size the box as needed. Double click the **Goto Box** and choose ***MOTOR STARTER*** from the Screens window, as shown in Figure 4-3-16.

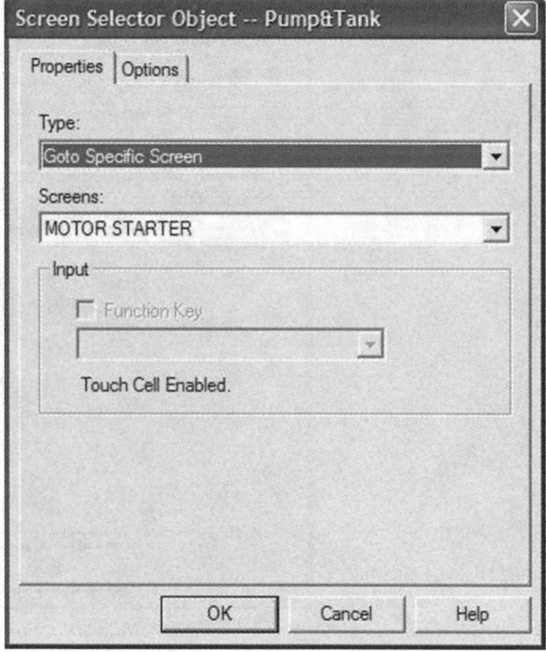

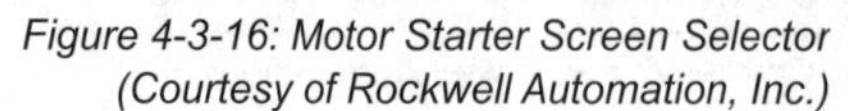

Figure 4-3-16: Motor Starter Screen Selector (Courtesy of Rockwell Automation, Inc.)

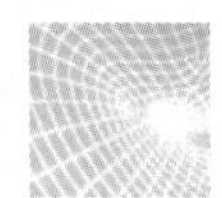

20. Be sure to save the HMI project by clicking ***File > Save As*** to the folder where this project is stored.

21. Download the HMI to the PanelView terminal. Check that the terminal is connected to the computer and PLC. Click ***File > Download*** and observe the screen shown in Figure 4-3-17.

Figure 4-3-17: HMI File Transfer setting (Courtesy of Rockwell Automation, Inc.)

22. Click **OK** to see the transfer screen shown in Figure 4-3-18.

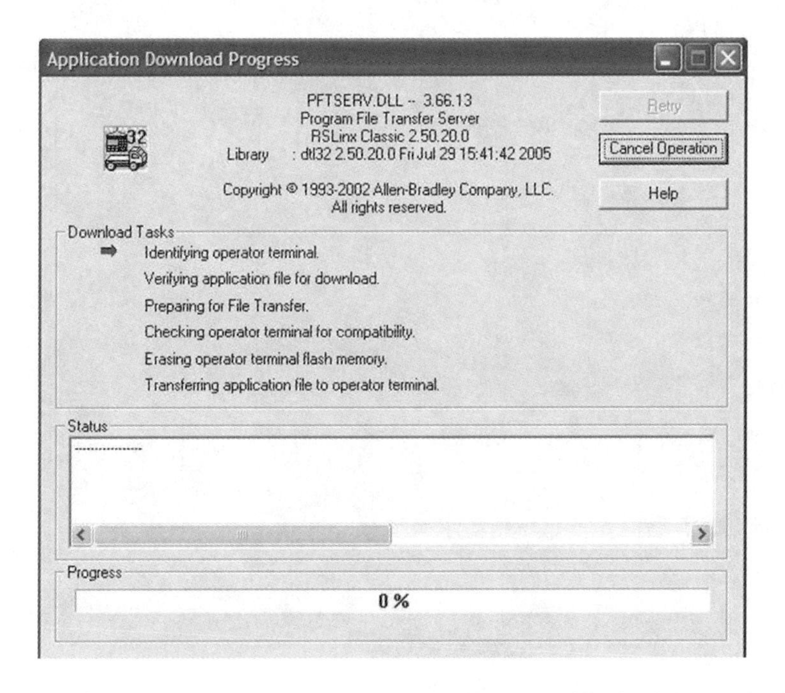

Figure 4-3-18: Application Download Progress screen (Courtesy of Rockwell Automation, Inc.)

23. If the communications are set up correctly, the ***six tasks*** listed on the transfer screen will get checked off and the ***Progress bar*** will begin moving. When the transfer is 100% complete, the **PV will restart itself**, which can take several minutes. The HMI screen created should appear on the PV.

24. Once again, if the **PLC is NOT connected,** an **offline error** will occur**.** Connect the PLC and verify that the screen works.

25. Touch the buttons to see that they work. Also, touch the Goto screens and see that they work as well.

26. This completes the lab.

PV Lab 4-4: Alarms

4.4.1 Introduction

A common use for Operator Terminals is to display alarms. An alarm is a screen that displays that a critical event has happened, such as exceeding a set value of temperature for a process. This lab will show how to create an alarm screen.

4.4.2 Programming the MicroLogix PLC

Figure 4-4-1 shows the PLC program for this lab activity. It uses a Timer on Delay to simulate a Heating Element rising and lowering in temperature.

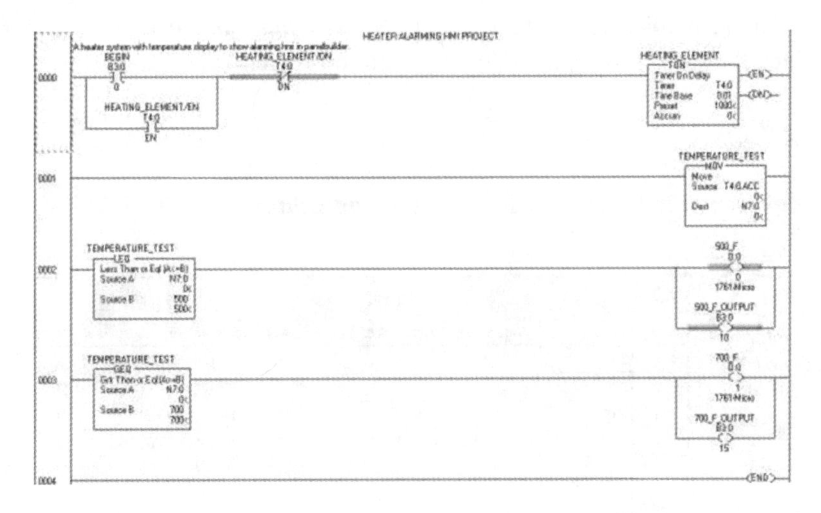

Figure 4-4-1: PLC program with TON (Courtesy of Rockwell Automation, Inc.)

1. Create and download the ladder shown in Figure 4-4-1 into the PLC and be sure to place it in ***RUN*** mode. Make sure that the PanelView is plugged into the computer with PanelBuilder32 software and the PLC is plugged into the correct port on the PanelView.

4.4.3 Begin Creating the HMI Diagram with PanelBuilder32

2. Launch PanelBuilder32 and create a new HMI called ***Alarm Display***. Follow what has been shown in Lab 4-3. Figure 4-4-2 shows the first screen for this multiple-screen HMI.

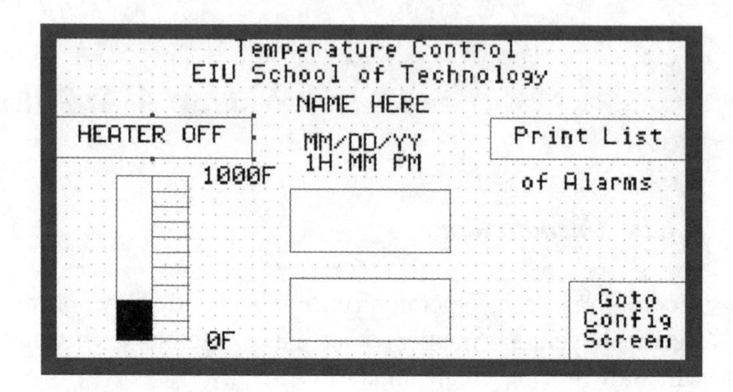

Figure 4-4-2: Alarm Display (Courtesy of Rockwell Automation, Inc.)

4.4.4 Creating a Second Screen for the Alarm

3. A second screen for the ***Alarm Display*** will need to be created. This screen is created in a special way, as described below.

4. Create the second screen as you did the first screen using Figure 4-4-2 as a guide to assembling this screen. Remember to create the ***Node Name, Node Address,*** and ***Node Type***. Also create and place the ***Goto Config Screen*** object on the screen.

4.4.5 Creating Tags

5. Create the ***Tags*** shown in Figure 4-4-3, using the ***Tag Editor***.

	Tag Name	Data Type	Array Size	Description	Node Name	Address	Initial Value
1	500F	Bit	1		ML1000	B3:0/10	0
2	700F	Bit	1		ML1000	B3:1/5	0
3	S1	Bit	1		ML1000	B3:0/0	0
4	TEMP	Unsigned Inte	0		ML1000	N7:0	0

Figure 4-4-3: Tag Editor (Courtesy of Rockwell Automation, Inc.)

6. Now make the ***HEATER OFF*** button, which is a momentary pushbutton. In Figure 4-4-4, note that the ***Properties*** tab is where the momentary pushbutton is configured.

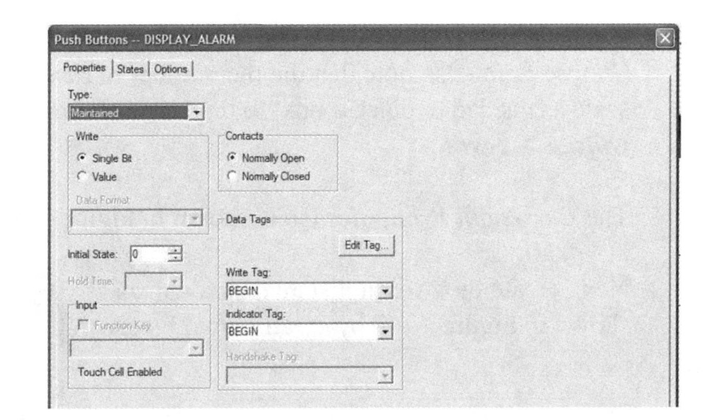

Figure 4-4-4:
Configure Properties of HEATER OFF button
(Courtesy of Rockwell Automation, Inc.)

7. The screen in Figure 4-4-5 shows the ***States*** tab for the ***HEATER OFF button***. Skip the ***Options*** tab.

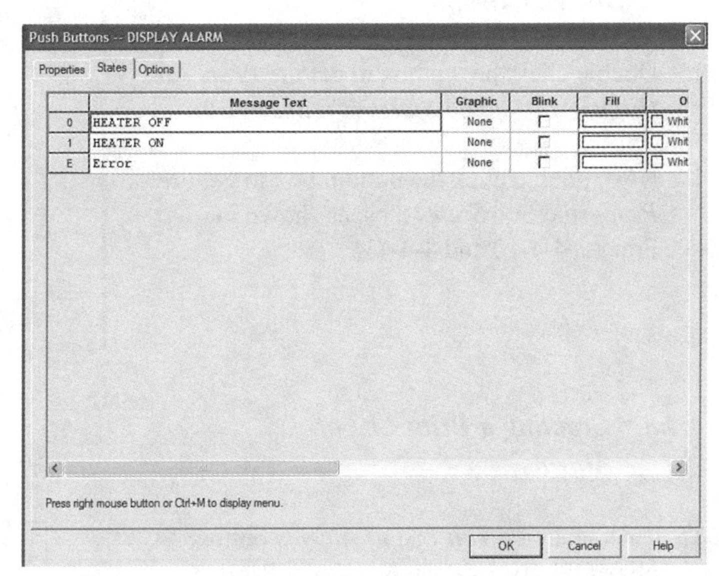

Figure 4-4-5:
Configure States of HEATER OFF button
(Courtesy of Rockwell Automation, Inc.)

4.4.6 Adding the Bar Graph

8. Now create the BarGraph; click ***Objects*** > ***Graphic Indicators*** > ***Bar Graph***. Size and place the bar graph on the screen, as shown in Figure 4-4-2. Note that the black bar indicator is partially visible. This indicator will rise and fall with the data that is being measured.

9. To create the scale beside to the bar graph, click ***Objects*** > ***Graphic Indicators*** > ***Scales*** > **Linear**. Size the scale on the right side of the graph in Figure 4-4-2 so that the scale aligns with the top and bottom of the bar graph. It will look like a capital E.

10. Right click the E; then click ***Object Properties*** and note that four orientations can be selected. Choose ***Left***. Also note that the the number of ***tick*** marks can be selected. Choose ***5***. Move the scale as needed to place it next to the bar graph, as shown. Add numerical values using ***Objects*** > ***Text***.

11. The ***BarGraph Properties*** tab is shown in Figure 4-4-6.

12. Next, create the two multistate display areas shown in Figure 4-4-7 by creating two boxes.

Bar Graph -- DISPLAY ALARM
Properties | Options
Fill From: Bottom
Edit Tag...
Range
Min: 0
Max: 1000
Read Tag: TEMP
OK | Cancel | Help

Figure 4-4-6: Bar Graph Properties (Courtesy of Rockwell Automation, Inc.)

4.4.7 *Setting the Properties and States of the Bar Graph*

13. Double click the top box to get the ***Properties (Figure 4-4-8)*** and ***States (Figure 4-4-9)***.

14. Next, double click the bottom box to get the ***Properties*** and ***States*** tabs, as shown in Figures 4-4-10 and 4-4-11.

4.4.8 *Creating a Print List of Alarms*

15. To create the ***Print List of Alarms*** button, click ***Objects > Alarm Buttons > Print Alarm List***. Then size and place on the screen as shown in Figure 4-4-12.

16. Add a ***text box*** under the ***Print List button*** to add the words "***of Alarms***", as shown in Figure 4-4-12.

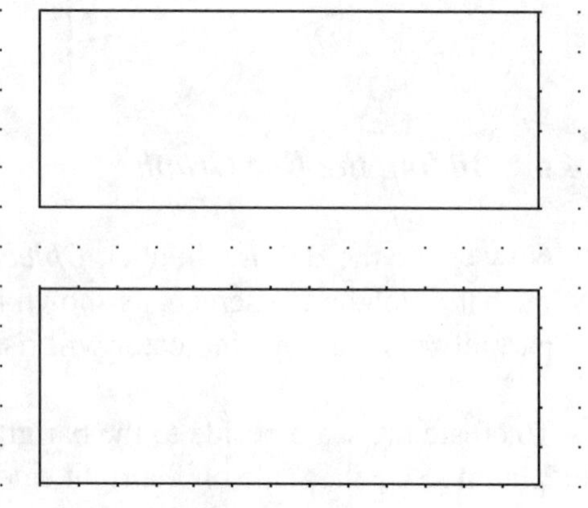

Figure 4-4-7: Boxes for Multistate display (Courtesy of Rockwell Automation, Inc.)

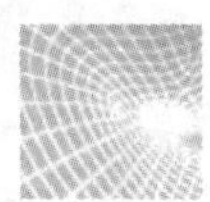

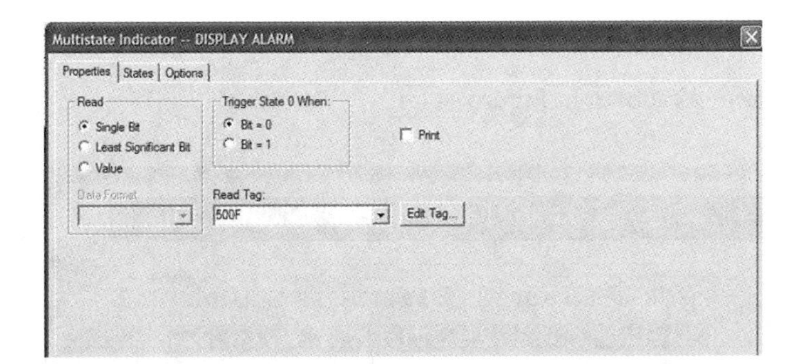

Figure 4-4-8: Properties of top box (Courtesy of Rockwell Automation, Inc.)

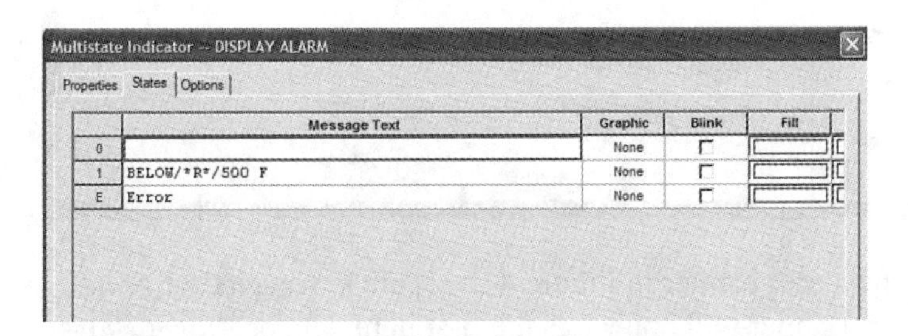

Figure 4-4-9: States of top box (Courtesy of Rockwell Automation, Inc.)

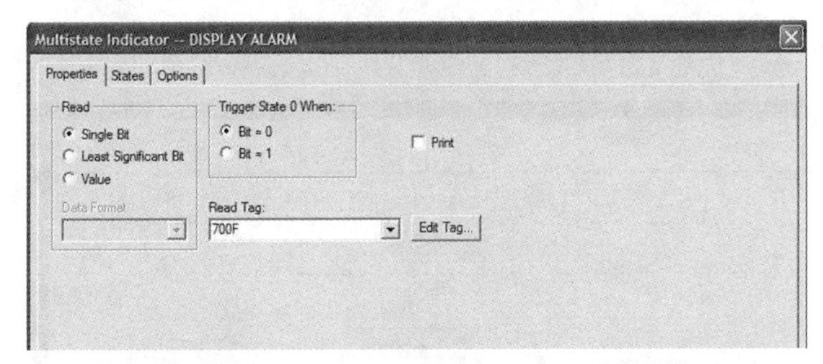

Figure 4-4-10: Properties of bottom box (Courtesy of Rockwell Automation, Inc.)

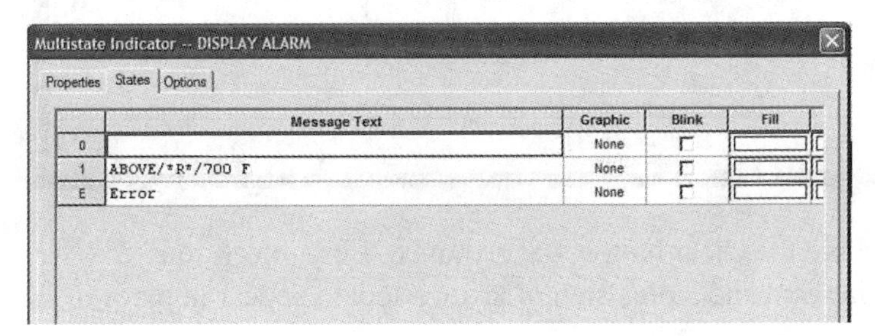

Figure 4-4-11: States of bottom box (Courtesy of Rockwell Automation, Inc.)

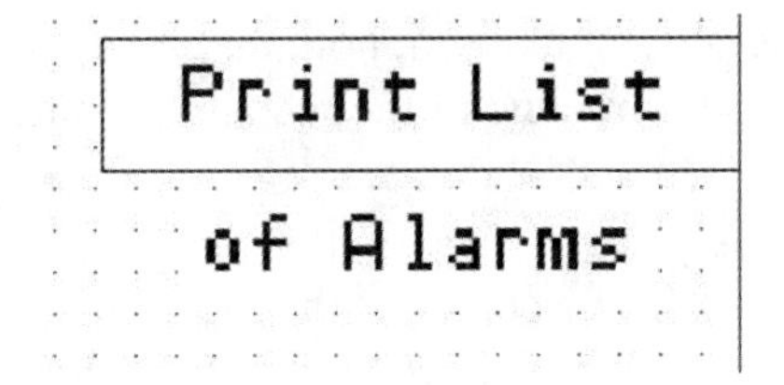

Figure 4-4-12: Print List of Alarms button (Courtesy of Rockwell Automation, Inc.)

4.4.9 Alarm Banner Screen

17. Next, create the ***Alarm Banner screen***, as shown in Figure 4-4-13.

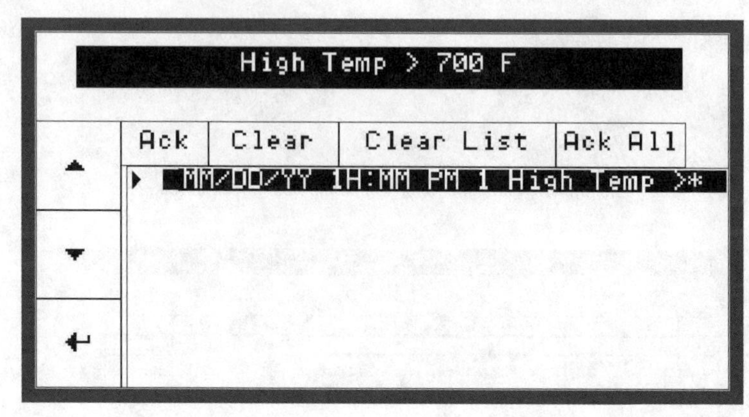

Figure 4-4-13: Alarm Banner screen (Courtesy of Rockwell Automation, Inc.)

18. To begin the process of creating the alarm banner in Figure 4-4-13, click ***Screens*** > ***Create Alarm Banner***; note that the Alarm Banner will appear as the first item. Under ***Alarm Banner Screens***, the screen in Figure 4-4-14 will appear.

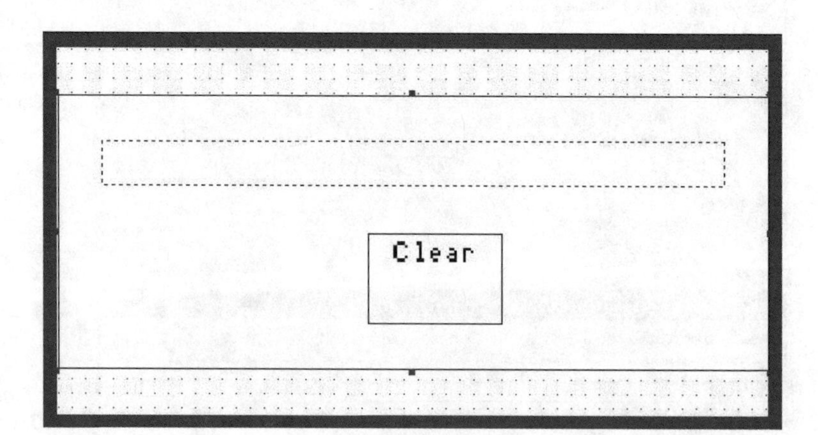

Figure 4-4-14: Clear Screen

19. Delete the ***Clear*** button. Click where the Clear button was removed. Little black squares should appear around a rectangular area and a plus sign of arrows should appear in place of the mouse pointer.

20. Holding the left mouse button down, move the plus sign down; note that the rectangle with the little black boxes moves with it. Move the bottom edge of the rectangle to the bottom of the screen, as shown in Figure 4-4-14.

21. Now move the plus sign up to the little black box in the center of the top edge of the rectangle. Carefully position the plus sign at the little black box until only a little up/down arrow

appears. While making sure the little up/down arrow is still displayed, hold down the left mouse button and move the mouse to the top of the screen area; then release the mouse button. The white screen (no background grid) below should appear, as seen in Figure 4-4-15.

Figure 4-4-15: White screen (Courtesy of Rockwell Automation, Inc.)

22. Create the alarm buttons ***Ack, Clear, Clear List,*** and ***Ack All*** as shown, by clicking ***Objects > Alarm Buttons***. Resize and place them, as shown in Step 17, Figure 4-4-13.

23. Move the dotted box to the top of the screen. Then create the three list keys on the left side of the screen by clicking ***Objects > List Keys***. The keys are the ***Move Up, Move Down,*** and ***Enter***. Stack them on top of each other, similar to Step 24 below.

24. In the left window, double click ***Alarms > Alarm Triggers***. Then add the values as shown in Figure 4-4-16.

Alarms -- DISPLAY ALARM

Setup | Alarm Triggers | Alarm Messages

	Trigger Tag	Trigger Typ
1	500F	Bit
2	700F	Bit

Figure 4-4-16: Display Alarm Triggers (Courtesy of Rockwell Automation, Inc.)

25. Click the ***Alarm Messages tab*** and add the values shown in Figure 4-4-17 and the Alarm Listing settings in Figure 4-4-18

Alarms -- DISPLAY ALARM

Setup | Alarm Triggers | Alarm Messages

	Message Text	Value/Bit	Trigger	Ack	Print
1	Temp < 500 F	0	500F	☑	☐
2	High Temp > 700 F	0	700F	☑	☐

Figure 4-4-17: Display Alarm Messages (Courtesy of Rockwell Automation, Inc.)

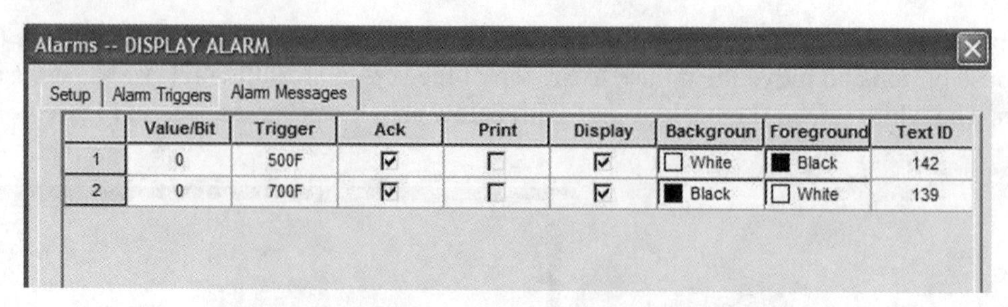

	Value/Bit	Trigger	Ack	Print	Display	Backgroun	Foreground	Text ID
1	0	500F	☑	☐	☑	White	Black	142
2	0	700F	☑	☐	☑	Black	White	139

Figure 4-4-18: Alarm List settings (Courtesy of Rockwell Automation, Inc.)

26. Now add the alarm listing by clicking ***Objects > Alarm List***. Then position the list, as shown in Figure 4-4-19. **Resize** the list to fill the space to the bottom and right side of the screen.

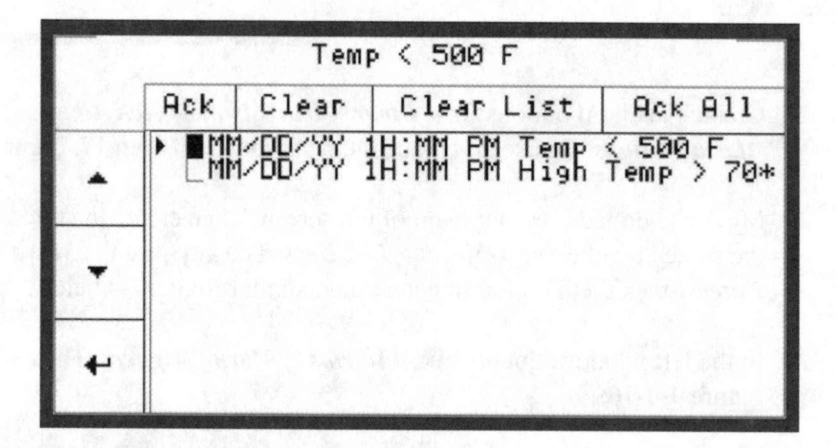

Figure 4-4-19: Alarm List display (Courtesy of Rockwell Automation, Inc.)

27. Now ***Download*** the HMI screens into the PanelView. Once the HMI is in the PanelView, connect the PLC and verify that the HMI works. ***Touch*** the ***Heater On*** and ***Off*** at least **four times** to establish a number of alarms. Do ***Ack*** or ***Ack All***, but do not **Clear** the alarms. Make sure the ***Config Screen*** button works as well.

4.4.10 Confirming that the Print List Button Works Correctly

28. To check that the ***Print List button*** works, follow these steps:

    a. To make sure the Print List button is correctly working requires special software called a Terminal Emulation Program, or TEP. (TEP emulates the data input to a real computer; it will decode and display the Print List data output.)

b. Many free TEPs are available on the Internet. A popular one is the DEC VT100. Download a TEP and install it on a computer. Connect the PLCs communication output to the computer's communication input where the TEP is located. Once installation is complete, continue following all of the remaining steps below.

c. Launch the TEP; on the right side of the computer's bottom task bar, find the icon that says ***RSLinx Classic Communciations Service***.

d. Right click this symbol and click ***Shutdown RSLinx Classic***. The icon should disappear.

e. Now go to the ***Config Screen*** of the PanelView. Select ***Printer Setup*** and ***Enable Printing*** in the ***Port Mode***. ***Exit***; then ***Run*** the HMI.

f. Launch the terminal emulation program.

g. Assign the name ***PV*** to the connection description in the program.

h. Choose ***COM1*** as the communication port.

i. Change the bits per second to ***19200*** and make sure the remaining entries are ***8 data bits, None parity, and 1 stop bit***.

j. Find the ***ASCII Setup*** in the terminal program. Then look for a setting that says something like ***append line feeds to incoming line ends***.

k. Now press the ***Print list***. The Alarms list should appear somewhat like the list in Figure 4-4-20.

29. This completes the lab.

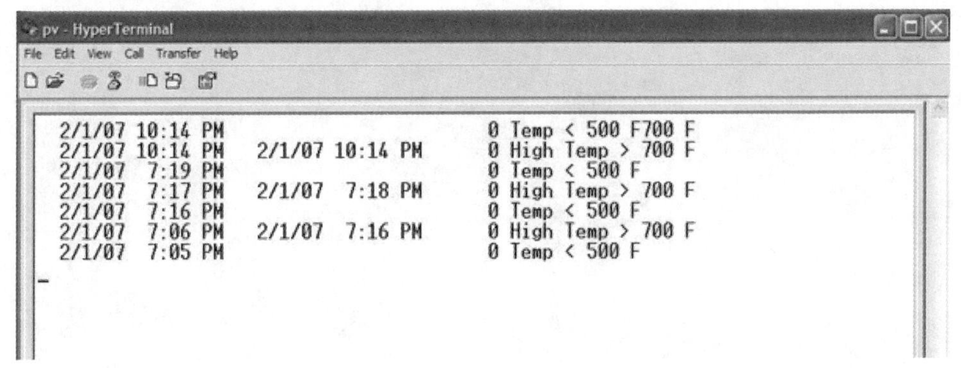

Figure 4-4-20: Example of Printed List of Alarm List (Courtesy of Rockwell Automation, Inc.)

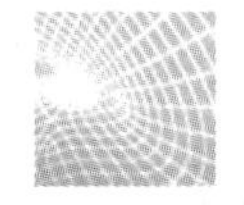

PV Lab 4-5: Analog Input and Math Operations

4.5.1 Introduction

Some PLCs have special inputs and outputs that are analog instead of digital. Analog inputs are used with sensors such as thermistors or thermocouples, which deliver a varying voltage versus temperature. This lab will show how these devices are used. Quite often, mathematical operations are needed to "scale" or format the voltage value of the sensor to display the correct physical quantity values.

For this lab, a MicroLogix 1100 PLC, which has two analog inputs, will be needed. In addition, a digital thermometer will need to be constructed and connected to one of the analog inputs. It delivers 10 mV of signal for each Celsius degree change in temperature.

The lab consists of the electronic thermometer sending an input temperature in tens of millivolts per deg Celsius. The temperature will be digitally displayed numerically, in both Fahrenheit and Celsius and through bar graphs.

4.5.2 Programming of the MicroLogix 1100 PLC and Subprogramming

Examine Figure 4-5-1 and note the information displayed. Specifically, the Program Files shows the list of programs that make up the Ladder Diagram. More than one diagram can be created. The Program Files from LAD3 and higher are called subprograms. Subprograms are often necessary because a large automated control circuit would be hard to enter as one diagram. Breaking the automated control circuit into several Program Files makes it easier to enter the program; it also makes the circuit easier to understand. In addition, PLCs are designed to use the subprogram technique.

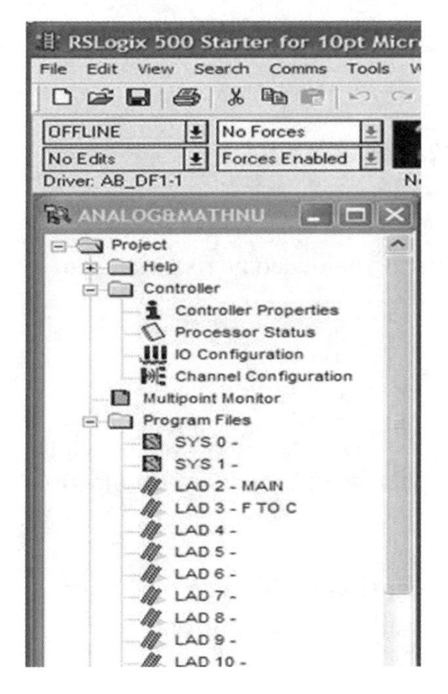

Figure 4-5-1: Program Files for PLC (Courtesy of Rockwell Automation, Inc.)

Program File LAD2 is always called the MAIN program. Note in this Ladder Diagram, sub-program LAD3 is called F TO C.

1. The Ladder Diagram for this lab is shown in Figures 4-5-2 and 4-5-3. Note it is in two separate diagrams.

Figure 4-5-2: LAD 2 main program (Courtesy of Rockwell Automation, Inc.)

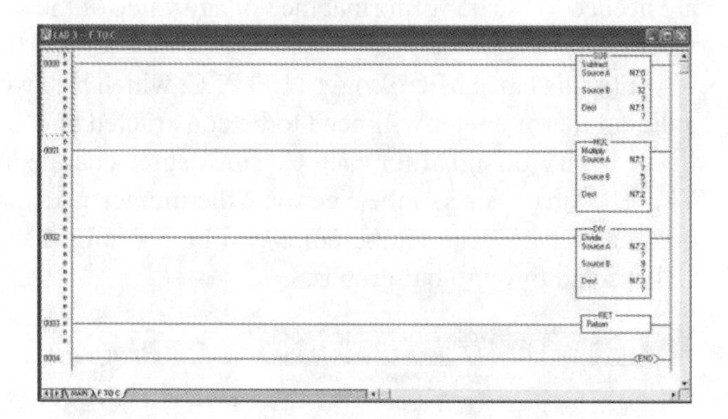

Figure 4-5-3: LAD 3 F to C subprogram (Courtesy of Rockwell Automation, Inc.)

4.5.3 Programming Math Operations, How PLC Math Works, and Subprogramming

2. The LAD3 subprogram ladder diagram shows how the equation below converts Celsius to Fahrenheit in a PLC.

$$°C = (°F - 32) \times 5 / 9$$

3. Be aware that the mathematical process in the MicroLogix PLC is integer-based, not decimal-based; therefore, the answer is rounded to give a whole number integer value for a temperature.

4. Create and download this ladder into the PLC and place the PLC in ***RUN*** mode. Make sure that the PanelView is plugged into the computer with PanelBuilder32 software and the PLC is plugged into the correct port on the PanelView.

4.5.4 Creating the HMI Display with a Temperature Bar Graph

5. Launch PanelBuilder32 and create a new HMI called ***Math&Analog***.

6. Create a ***Node Name*** of ***ML1100*** and enter a ***Node Address*** and ***Node Type***, as done in previous labs.

7. Create the screen shown in Figure 4-5-4 as ***Screen 1.*** Rename this screen to ***Main Temperature*** by right clicking the screen window and selecting ***Properties***.

Figure 4-5-4: Screen 1: Main Temperature (Courtesy of Rockwell Automation, Inc.)

4.5.5 Creating the Celsius HMI Display

8. Now create a second screen for temperature, as shown in Figure 4-5-5. Rename this screen as ***C***.

9. Next create Figure 4-5-6 as the digital display screen for Celsius temperature.

10. Finally create Figure 4-5-7 as the bar graph display screen for Celsius temperature.

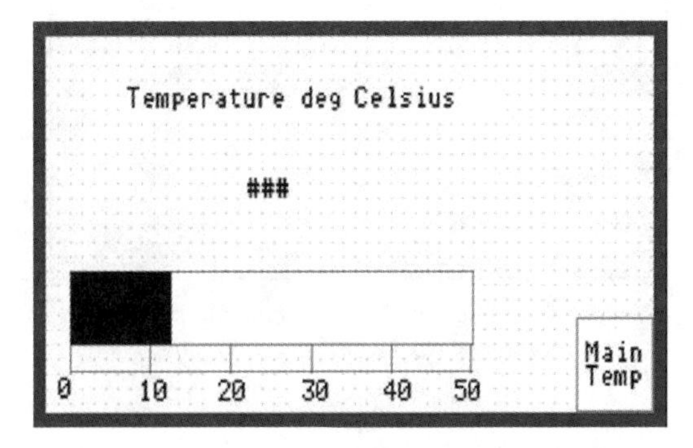

Figure 4-5-5: Screen 2: Celsius Temperature (Courtesy of Rockwell Automation, Inc.)

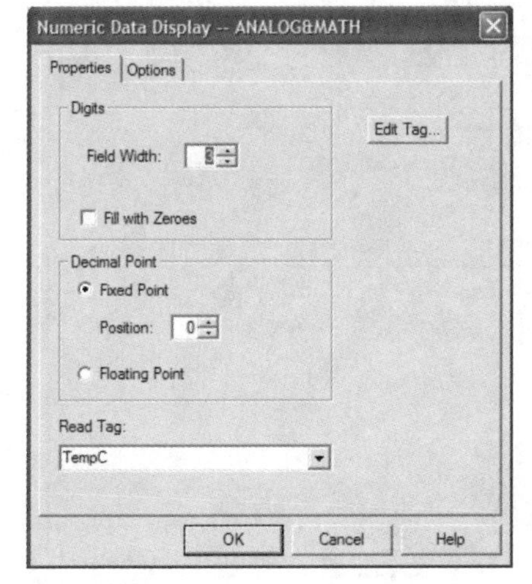

Figure 4-5-6: Create Numeric Data Display screen (Courtesy of Rockwell Automation, Inc.)

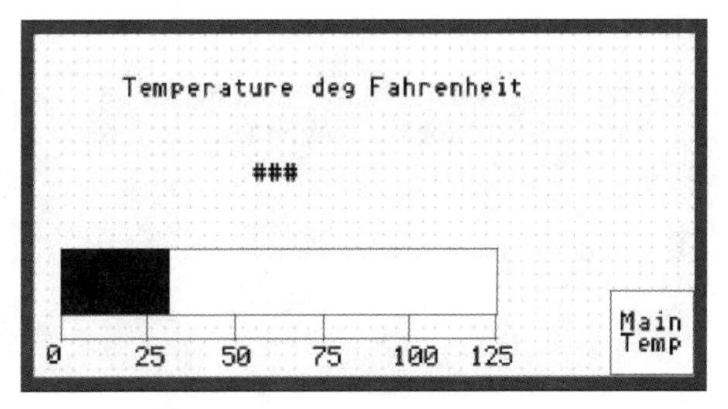

Figure 4-5-7: Bar Graph settings (Courtesy of Rockwell Automation, Inc.)

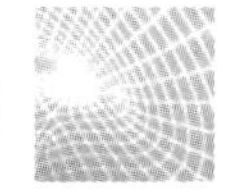

4.5.6 Creating the Fahrenheit HMI Display

11. Follow the steps in sections 4-6-5 and 4-6-6 as a guide to creating Fahrenheit display screens.

12. Begin by creating the temperature screen, as shown in Figure 4-5-8, and rename as ***F***.

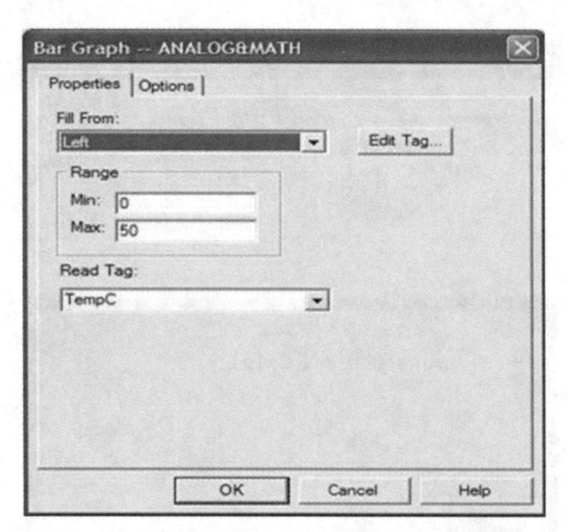

Figure 4-5-8: Screen 3: Fahrenheit Temperature (Courtesy of Rockwell Automation, Inc.)

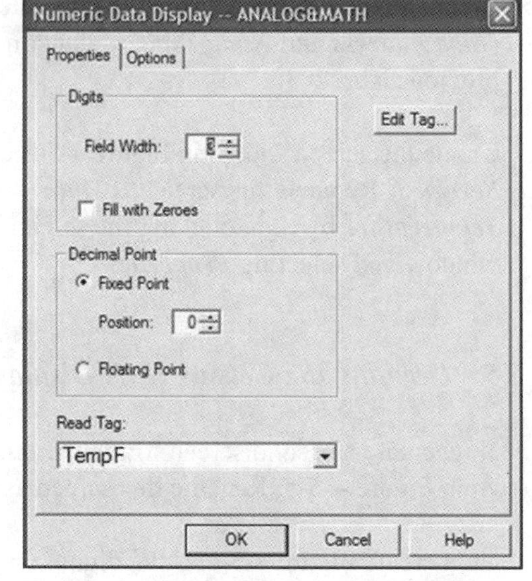

Figure 4-5-9: Create Numeric Data Display screen (Courtesy of Rockwell Automation, Inc.)

13. Next, similarly to the Celsius steps, create a digital data screen as shown in Figure 4-5-9.

14. Also, create a bar graph display screen in Figure 4-5-10 for Fahrenheit temperature.

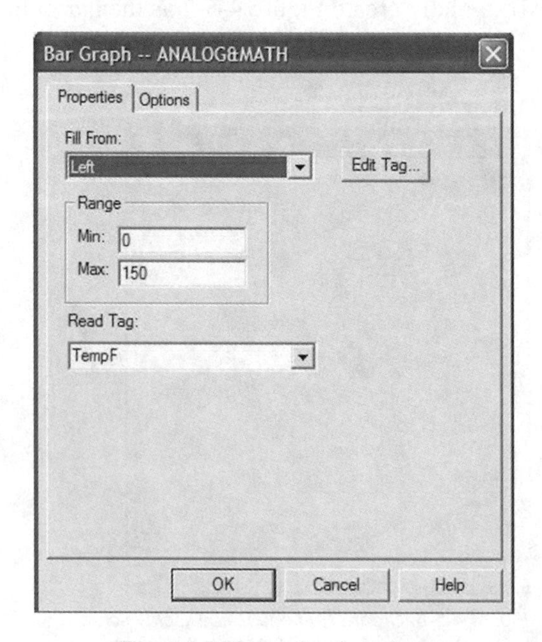

Figure 4-5-10: Bar Graph settings (Courtesy of Rockwell Automation, Inc.)

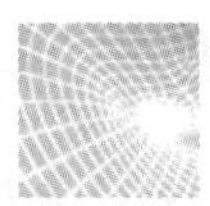

15. Using the *Tag Editor*, create the ***Tags*** shown in Figure 4-5-11.

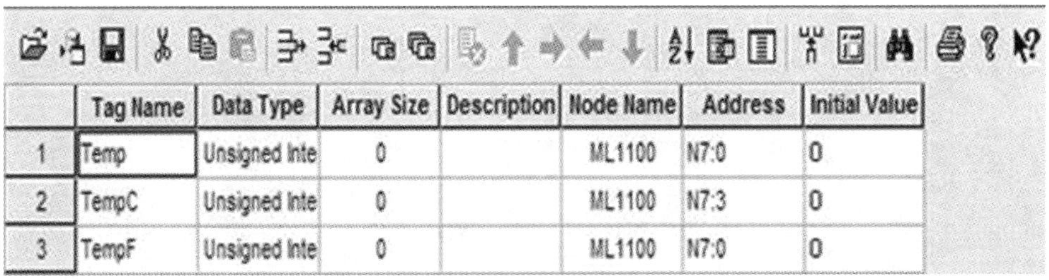

	Tag Name	Data Type	Array Size	Description	Node Name	Address	Initial Value
1	Temp	Unsigned Inte	0		ML1100	N7:0	0
2	TempC	Unsigned Inte	0		ML1100	N7:3	0
3	TempF	Unsigned Inte	0		ML1100	N7:0	0

Figure 4-5-11: Tag Editor (Courtesy of Rockwell Automation, Inc.)

16. Go back to each icon and associate the tag by clicking the value in the Address column.

17. Now create an ***ALARM BANNER*** to store alarms when the ***OVER TEMPERATURE*** value is reached. Follow the information in Lab 4-4 Alarm to create the alarm banner according to the specifications.

18. Download the HMI to the PanelView; then connect the PLC and verify that the screen works. ***Run*** the program enough times to ***establish a number of alarms***. Do not clear the alarms.

19. Make sure the ***Config Screen*** in the PV terminal works as well.

20. This completes this Lab.

PV Lab 4-6: Security

4.6.1 Introduction

Operator Terminals control the PLC which controls a machine within a process. Someone pressing a critical button on the Operator Terminal, whether accidentally or on purpose, could cause damage to the machine, worse damage to other machines, or even death to another human. Therefore, terminals are often secured through password-protect access. This lab will show how to set up security on the terminal.

4.6.2 Using Lab #2 to Add a Security Screen

1. Re-load the Digital Display PLC diagram from Lab 4-2 for this lab. Download it into the PLC and be sure to place it in RUN mode. Also, make sure that the PanelView is plugged into the computer with PanelBuilder32 software and the PLC is plugged into the correct port on the PanelView.

2. Launch PanelBuilder32. Three screens will need to be created for this lab as shown below. Figure 4-6-1 shows Screen 1, titled LOGIN.

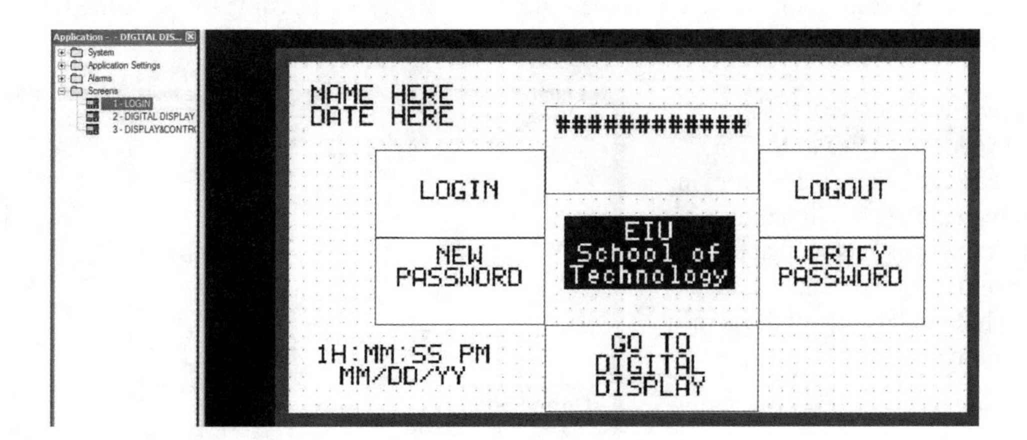

Figure 4-6-1: Screen 1: LOGIN (Courtesy of Rockwell Automation, Inc.)

3. Begin by creating the ***LOGIN*** screen as shown in Figure 4-6-1. ***Blink*** the EIU center display. Look under ***Format*** for blinking. (NOTE: You may not be able to complete some of the buttons and displays until all screens are available.)

4. Create the ***Login, Logout, New Password,*** and ***Verify Password*** buttons by clicking ***Objects > Security Keys*** , followed by the desired name, such as ***Login***.

5. Resize and place the buttons on the screen. Next click ***Objects > Security Keys > Select Operator*** button; then resize and position. There are no values to enter for these keys.

6. You may not be able to create some buttons such as the ***Goto Digital Display*** until that screen is completed.

7. The ***LOGIN*** screen will be revisited after all other screens are completed.

8. Next create the ***DIGITAL DISPLAY*** screen shown in Figure 4-6-2.

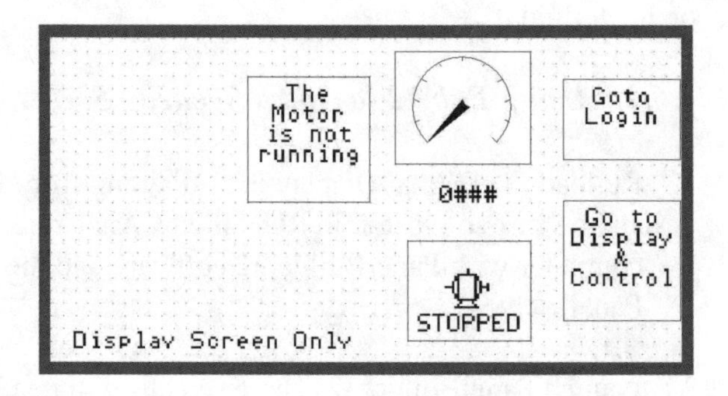

Figure 4-6-2: Screen 2: Digital Display (Courtesy of Rockwell Automation, Inc.)

9. Try to create the screen without referring to the guide or asking for help. Use Lab 4-2 as a guide if necessary for creating this screen. Again it may not be possible to create the ***Goto buttons*** yet.

10. Now create Screen 3: ***DISPLAY&CONTROL***. This screen can be created by copying Screen 2, then modifying it to make Screen 3. Note that a scrolling display will be added at the bottom left side of the screen (Figure 4-6-3).

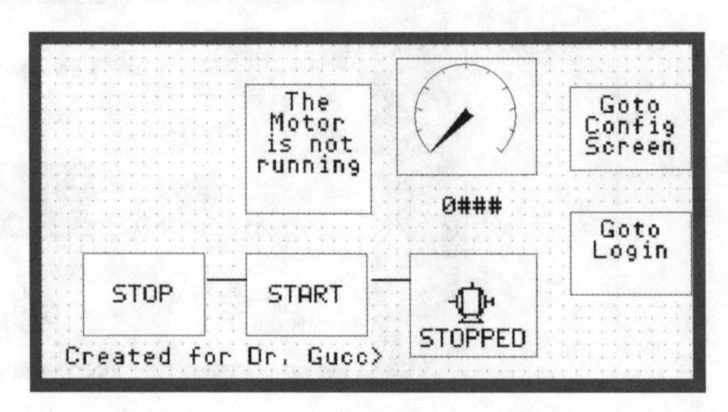

Figure 4-6-3: Screen 3: Display & Control (Courtesy of Rockwell Automation, Inc.)

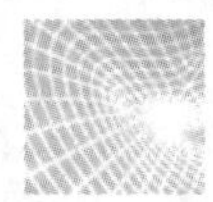

11. Create the ***Tags*** shown in Figure 4-6-4 using the ***Tag Editor***. Note the ***Operator*** tag. There is no address for it.

	Tag Name	Data Type	Array Size	Description	Node Name	Address	Initial Value
1	HIGH_LEVEL	Bit	0		ML1000	B3:1/2	0
2	LEVEL	Unsigned Inte	0		ML1000	T4:0.ACC	0
3	LOW_LEVEL	Bit	0		ML1000	B3:1/1	0
4	MOTOR	Bit	0		ML1000	B3:0/10	0
5	MOTOR_IND	Bit	0		ML1000	B3:0/10	0
6	Operator	Bit	1		ML1000		0
7	PUMP	Bit	0		ML1000	B3:/10	0
8	START	Bit	0		ML1000	B3:0/1	0
9	STOP	Bit	0		ML1000	B3:0/0	0

Figure 4-6-4: Tag Editor (Courtesy of Rockwell Automation, Inc.)

12. To define the users who are allowed to access this HMI and their properties, click ***Application > Security*** and the screen in Figure 4-6-5 will appear.

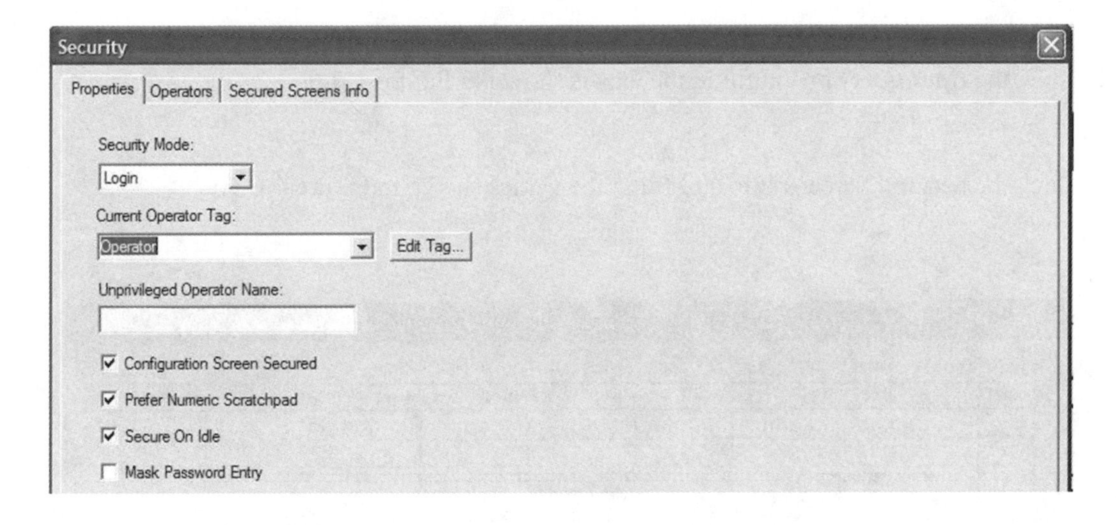

Figure 4-6-5: Security applications screen (Courtesy of Rockwell Automation, Inc.)

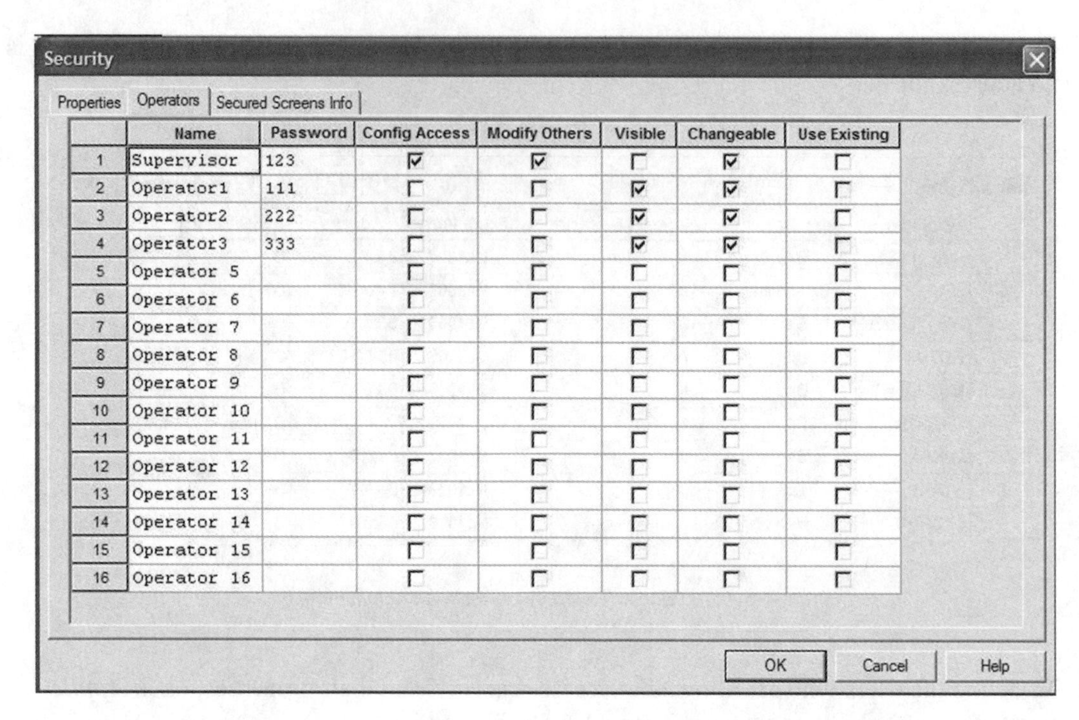

	Name	Password	Config Access	Modify Others	Visible	Changeable	Use Existing
1	Supervisor	123	☑	☑	☐	☑	☐
2	Operator1	111	☐	☐	☑	☑	☐
3	Operator2	222	☐	☐	☑	☑	☐
4	Operator3	333	☐	☐	☑	☑	☐
5	Operator 5		☐	☐	☐	☐	☐
6	Operator 6		☐	☐	☐	☐	☐
7	Operator 7		☐	☐	☐	☐	☐
8	Operator 8		☐	☐	☐	☐	☐
9	Operator 9		☐	☐	☐	☐	☐
10	Operator 10		☐	☐	☐	☐	☐
11	Operator 11		☐	☐	☐	☐	☐
12	Operator 12		☐	☐	☐	☐	☐
13	Operator 13		☐	☐	☐	☐	☐
14	Operator 14		☐	☐	☐	☐	☐
15	Operator 15		☐	☐	☐	☐	☐
16	Operator 16		☐	☐	☐	☐	☐

Figure 4-6-6: Operators tags in Security application screen (Courtesy of Rockwell Automation, Inc.)

13. Click the ***Operators tab*** and fill in the values shown in Figure 4-6-6.

14. Click the Secured Screens Info and fill in the values shown in Figure 4-6-7.

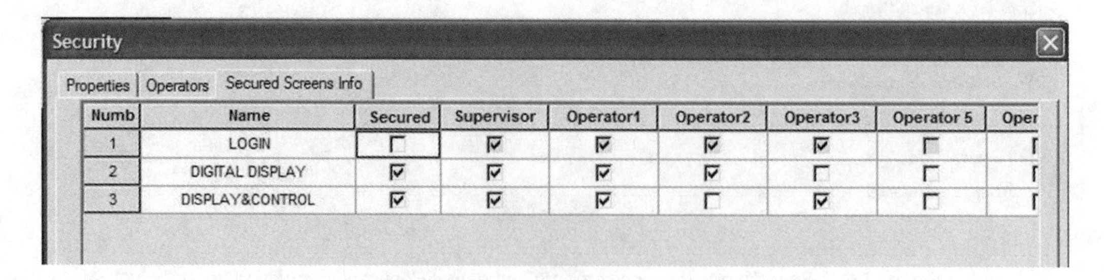

Numb	Name	Secured	Supervisor	Operator1	Operator2	Operator3	Operator 5	Oper
1	LOGIN	☐	☑	☑	☑	☑	☐	☐
2	DIGITAL DISPLAY	☑	☑	☑	☑	☐	☐	☐
3	DISPLAY&CONTROL	☑	☑	☑	☐	☑	☐	☐

Figure 4-6-7: Secured Screens Info screen (Courtesy of Rockwell Automation, Inc.)

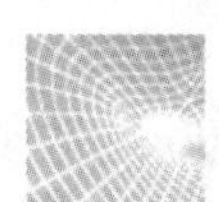

15. Note the screens that each operator and supervisor can access.

16. Download the HMI to the PanelView, connect the PLC, and verify that the security system is working correctly.

17. Check all operators and their accessibility to the screens and buttons. Which ones have access to the ***Goto CONFIG*** screen?

18. This completes this lab.

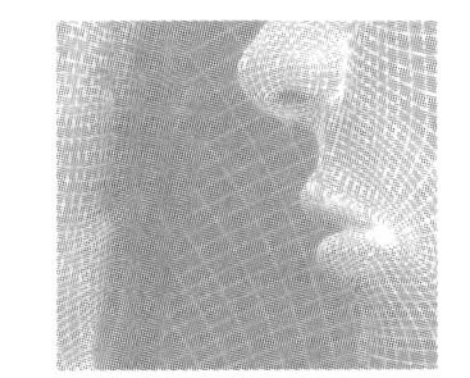

C H A P T E R   F I V E

RSView32 Labs

Introduction

This chapter contains step-by-step procedures for the eight laboratories listed below. These labs demonstrate how the computer-based HMI display is programmed. It also demonstrates that some types of HMI are created to show an animation of machine processes that can be controlled by machine operators, automation engineers, and managers.

- ***Lab 5-1: Basic Motor Starter***
- ***Lab 5-2: Analog and Digital Displays***
- ***Lab 5-3: Pump and Tank***
- ***Lab 5-4: Alarms***
- ***Lab 5-5: Math Operations, Subprograms, and Alarms***
- ***Lab 5-6: Security***
- ***Lab 5-7: TrendX and Data Logging***
- ***Lab 5-8: Moving Animation***

5.1 RSView Lab 5-1: Basic Motor Starter

This laboratory demonstrates how RSView32 software is used to create animated graphic displays of a process on a computer screen. The programming and actuation of a simple latching motor circuit will serve as the basis to create an initial animated HMI.

This lab utilizes RSView32 programming software along with an Allen-Bradley Micrologix programmable logic controller (PLC) to create a computer display to interact with, actuate, and control a motor-latching circuit with lights for outputs. The RSLogix 500 software is used to program the PLC.

RS232 serial communications allows RSView32, RS Logix500, and the PLC to "speak" to each other via RSLinx. But to establish the correct communications among these devices, the PLC must be connected to the computer and must be powered on! This is most critical.

Follow the steps below to complete this laboratory.

5.1.1 Creating the PLC Program in RSLogix and Transferring to the PLC

1. If it has not already been done, create a folder on the *Local Disk (C:)* drive entitled *RSLinx_RSLogix500 Programs Storage Folder*, as shown in Figure 5-1-1. Store all of the files for this laboratory in this folder. Also, create another folder for specific Lab use and save it with a name like *Lab_1_XXX* (XXX = the initials).

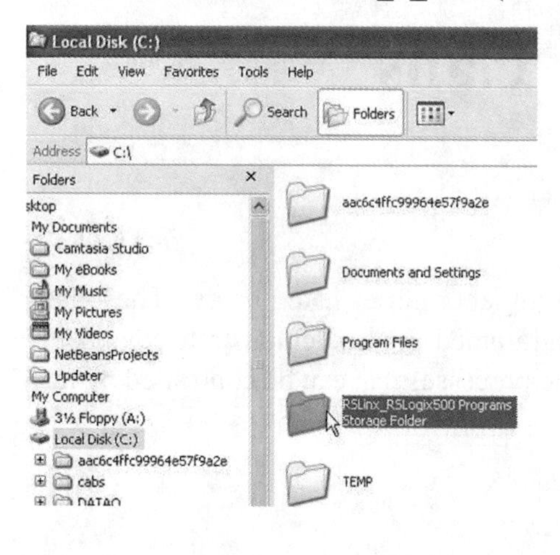

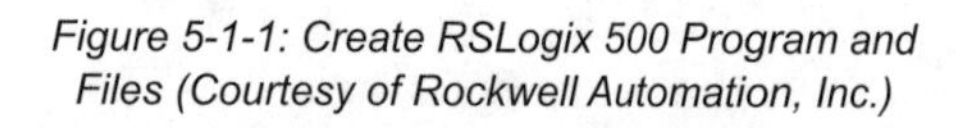

Figure 5-1-1: Create RSLogix 500 Program and Files (Courtesy of Rockwell Automation, Inc.)

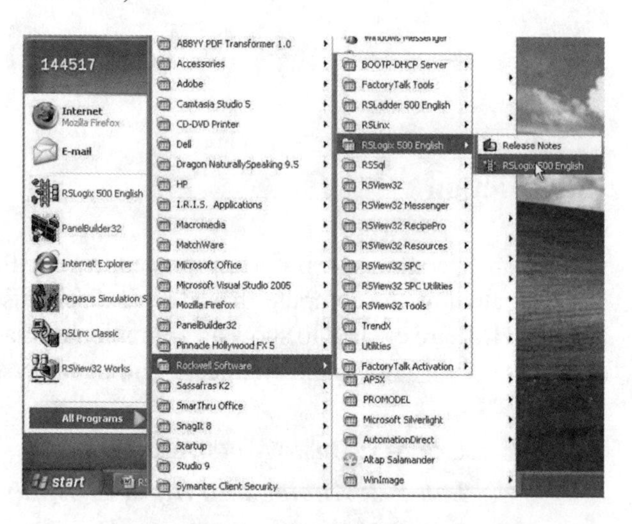

Figure 5-1-2: Open RSLogix 500 (Courtesy of Rockwell Automation, Inc.)

2. Ensure the **PLC is connected** to the computer through a serial port connection and powered on. Then open the ***RSLogix 500*** (English) PLC software, as shown in Figure 5-1-2.

3. After clicking ***File***, then ***New***, in ***RSLogix 500*** (as shown in Figure 5-1-3), select the appropriate ***Processor Type*** to communicate with ***RSLinx*** through to the computer.

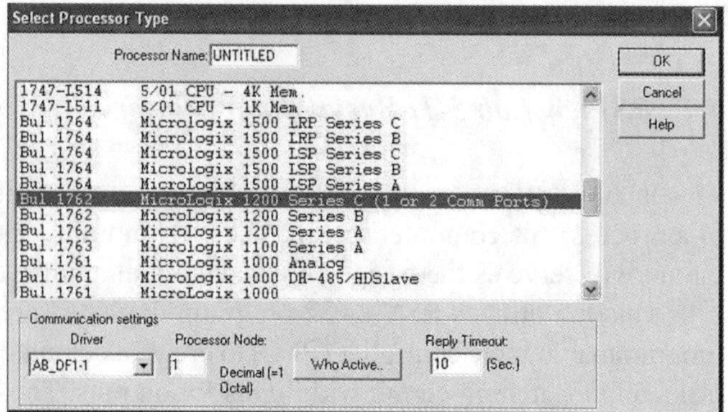

Figure 5-1-3: Select PLC Type (Courtesy of Rockwell Automation, Inc.)

4. Input the ***PLC program*** into ***RSLogix***, as shown in Figure 5-1-4.

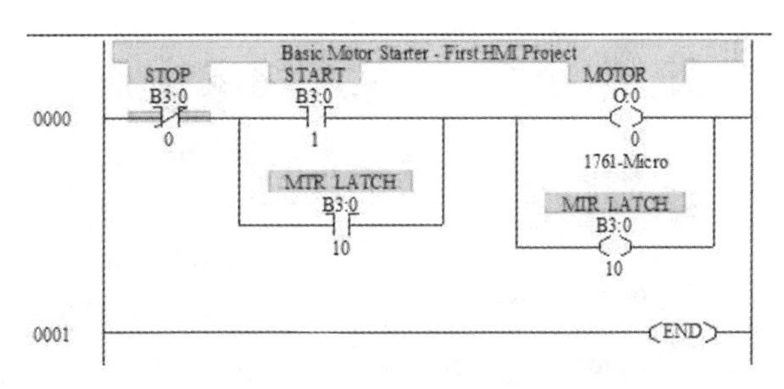

Figure 5-1-4: PLC program for Lab 5-1 (Courtesy of Rockwell Automation, Inc.)

5. For field applications, it is normal to use traditional physical input addresses (ex. I2: 0/0) to identify applications inputs. However, for animated processes — in other words, processes where animated representations or icons are used to control outputs — it is necessary to program with binary addresses (ex. B3: 0/0). Examine the screen in Figure 5-1-5.

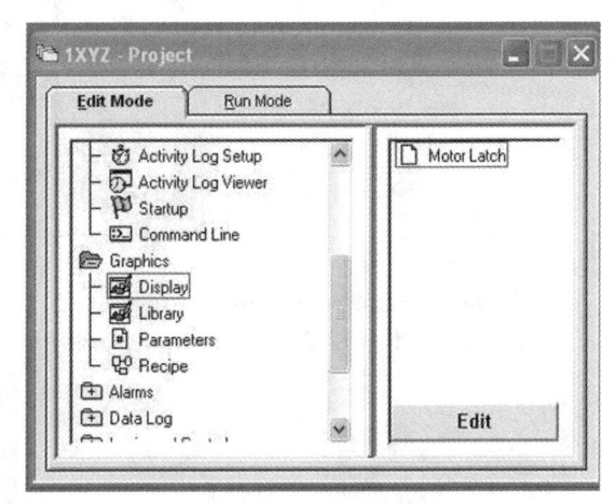

Figure 5-1-5: PLC program in RSLogix 500 (Courtesy of Rockwell Automation, Inc.)

6. From the **Data Files** folder in Figure 5-1-5, click the B3-BINARY icon once. Then click and drag a single binary address (Figure 5-1-6) onto the box highlighted in Figure 5-1-7.

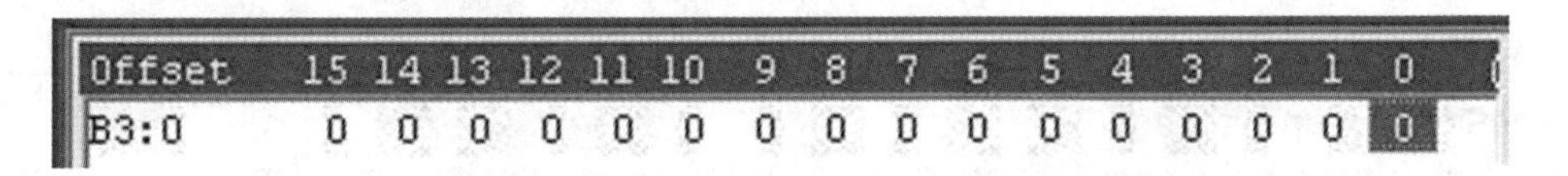

Offset	15	14	13	12	11	10	9	8	7	6	5	4	3	2	1	0
B3:0	0	0	0	0	0	0	0	0	0	0	0	0	0	0	0	0

Figure 5-1-6: Binary Values (Courtesy of Rockwell Automation, Inc.)

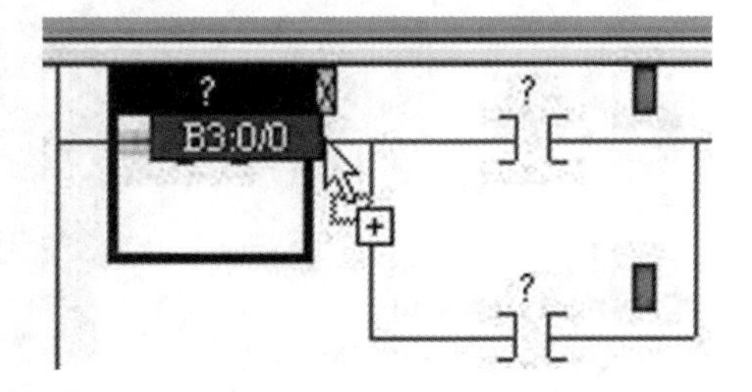

Figure 5-1-7: Enter Binary Value on Contact (Courtesy of Rockwell Automation, Inc.)

7. Use the procedures and figures in Steps 5 and 6 as a guide to label outputs.

8. Save the PLC program with the file name ***Motor Latching Circuit.rss***; saving it to local (C:) drive is best. Then minimize ***RSLogix 500.*** Do not download the PLC program into the PLC.

9. Next, open the Rockwell communication software ***RSLinx***, as shown in Figure 5-1-8.

Figure 5-1-8: Open RSLinx Communications software (Courtesy of Rockwell Automation, Inc.)

a. A pop-up screen, as shown in Figure 5-1-9, will appear. Configure RS232 communications as follows: Click ***Communications*** and select ***Configure Drivers***.

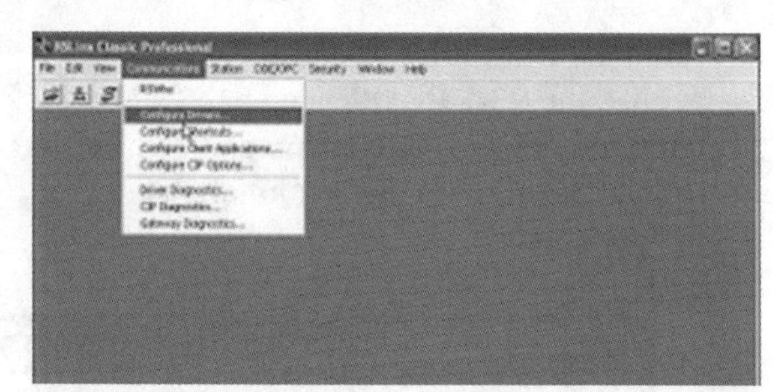

Figure 5-1-9: Configure Drivers (Courtesy of Rockwell Automation, Inc.)

b. Another pop-up screen will appear, similar to the one shown in Figure 5-1-10.

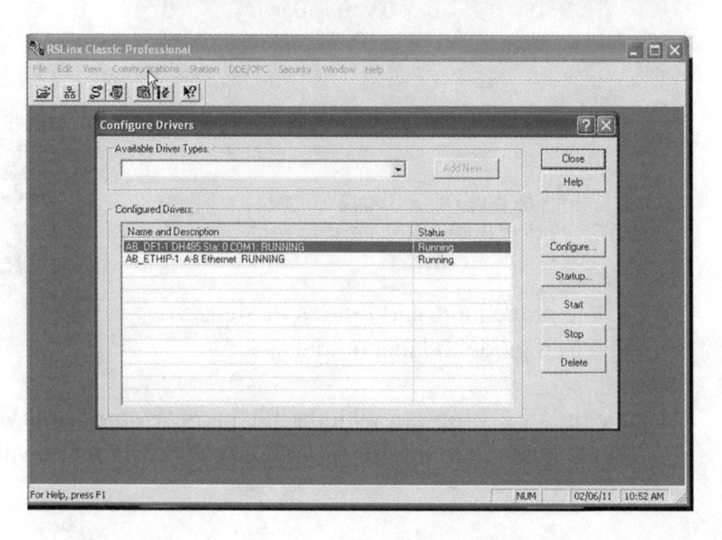

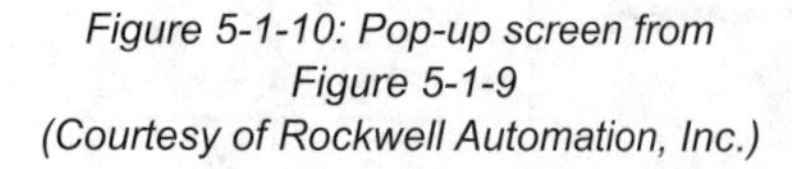

Figure 5-1-10: Pop-up screen from Figure 5-1-9 (Courtesy of Rockwell Automation, Inc.)

c. Under ***Configured Drivers: Name and Description***, click the ***AB_DF1-1 DH-485*** selection, as illustrated in Figure 5-1-10. Then click the ***Configure…*** button to configure this RSLinx driver. Make sure that a direct connection exists through the serial port cable to the computer before making the selection. Click ***OK*** and ***Close***. Skip to **Step 10**. If there is no ***AB_DF1-1 DH-485*** selection to choose, proceed to Step 9d below.

d. If under ***Configured Drivers: Name and Description*** another driver selection besides ***AB_DF1-1 DH-485*** appears, click the pull down arrow next to ***Available Driver Types:*** as shown in Figure 5-1-11.

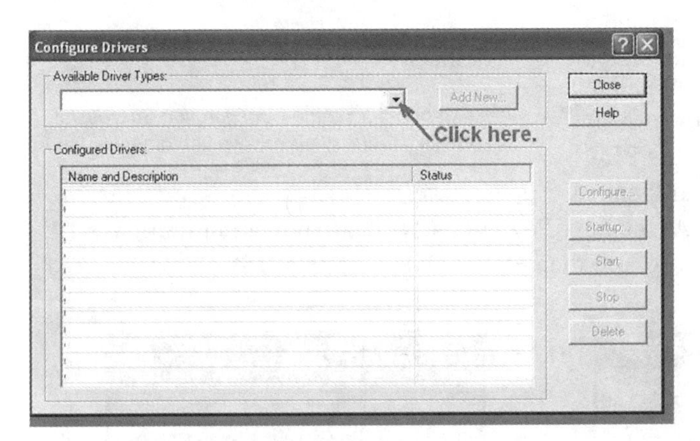

Figure 5-1-11: Continue Configure Drivers (Courtesy of Rockwell Automation, Inc.)

e. Select the ***RS232 DF1*** driver. Then click the ***Add New…*** button immediately to the right side of this selection.

f. Next, click ***Configure…***If ***AB_DF1-1*** appears, and ***RS232 DF-1 devices*** appear within the ***Available Driver Types:*** window, click ***OK*** and ***Close.*** Proceed to **Step 10**. If no devices appear, repeat Steps 9a to 9f.

10. Now ***RS-232*** communications network needs to be configured. Doing so will allow ***RSLinx*** to very rapidly create the communication link to the **PLC via RSLogix and RSView**. As Figure 5-1-12 indicates, click the ***Auto-Configure*** button.

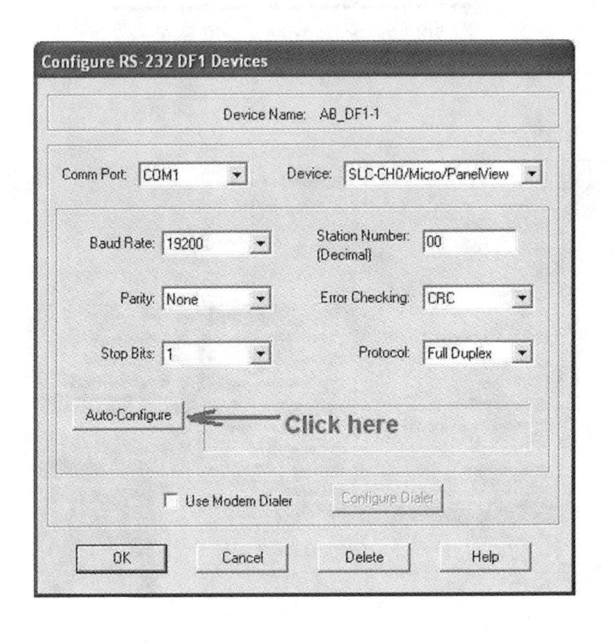

Figure 5-1-12: Use Auto Configure button (Courtesy of Rockwell Automation, Inc.)

11. A message appears, as shown in Figure 5-1-13, indicating that auto configuration is successful.

12. Click ***OK*** and then click ***Close*** at the ***Configure Drivers*** pop-up screen.

13. If a rapid response toward successful auto-configuration does not occur, either the PLC was **not plugged into the computer** or the PLC was **not turned on**. Remedy the situation by going back through **Steps 10–13**; if errors occur, then reinstall the ***RS232 DF-1*** driver.

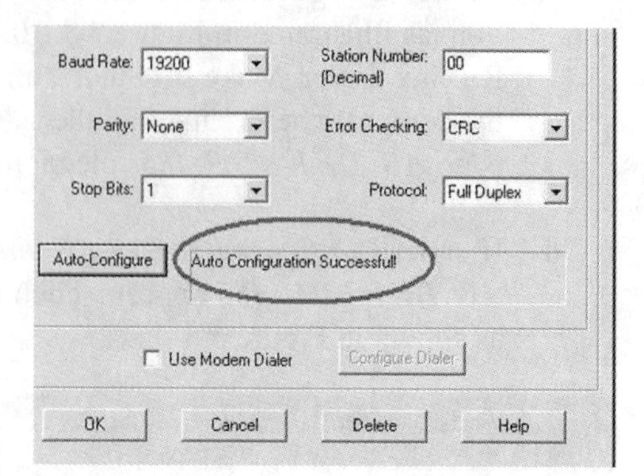

Figure 5-1-13: Auto Configuration Successful screen (Courtesy of Rockwell Automation, Inc.)

14. Re-open RSLogix, if not already open, and maximize the motor-latching circuit that was programmed. Download the program to the PLC by clicking ***Comms***, then ***Download***, and then ***OK*** at the ***Revision Note*** pop-up screen. At the next ***RSLogix*** screen, as shown in Figure 5-1-14, click ***Yes*** to finish downloading the program; then go online (Figure 5-1-15).

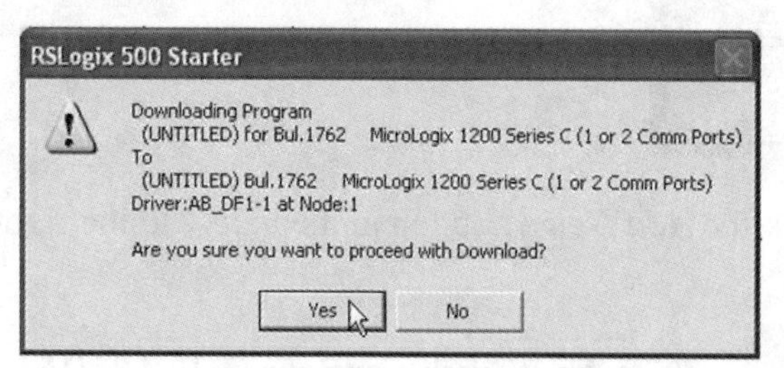

Figure 5-1-14: Download Program (Courtesy of Rockwell Automation, Inc.)

Figure 5-1-15: Connecting Online (Courtesy of Rockwell Automation, Inc.)

15. Upon going online, be sure the PLC program is in the ***RUN*** mode.

5.1.2 *Launching and Configuring RSView to Begin the HMI Creation Process*

16. Launch **RSView32**, as shown in Figure 5-1-16.

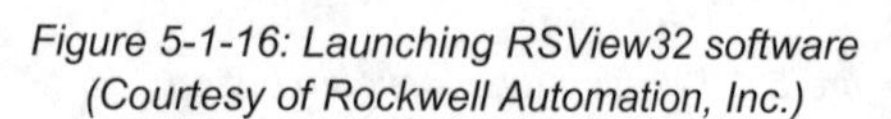

Figure 5-1-16: Launching RSView32 software (Courtesy of Rockwell Automation, Inc.)

17. The pop up screen in Figure 5-1-17 will appear. Maximize this pop-up to fill the entire display.

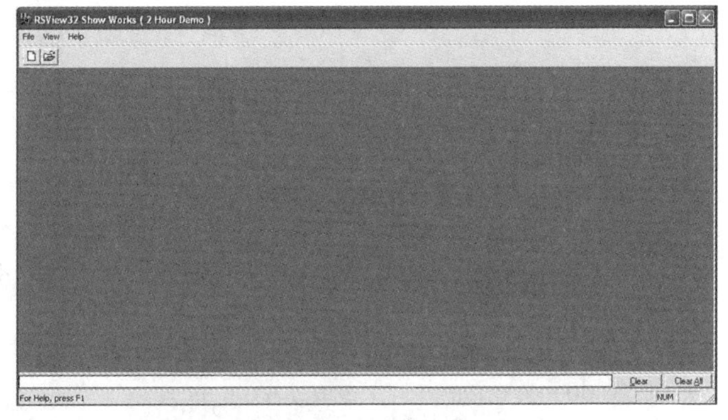

Figure 5-1-17: Pop-up for Figure 5-1-16 (Courtesy of Rockwell Automation, Inc.)

18. Click ***File***, then ***New***. Be sure ***Local Disk (C:)*** is selected in the ***Look in:*** dialog box when the Create New Project screen pops up (Figure 5-1-18). Select the ***RSLinx_RSLogix500 Programs Storage Folder*** before creating the HMI.

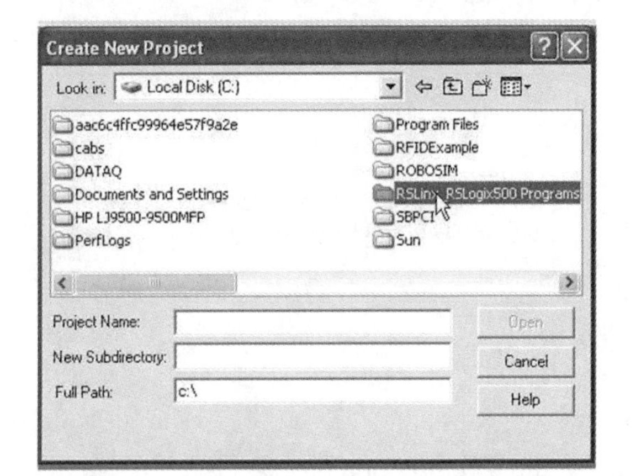

Figure 5-1-18: Select RSLinx_RSLogix500 Programs (Courtesy of Rockwell Automation, Inc.)

19. Type ***Project_1_XXX*** in the ***Project Name:*** portion of this screen. Then click ***Open***.

20. After a short period of time, a pop-up screen appears, as shown in Figure 5-1-19. In completing this step, remember that a ***database*** folder was created for associated files needed by the animated HMI program.

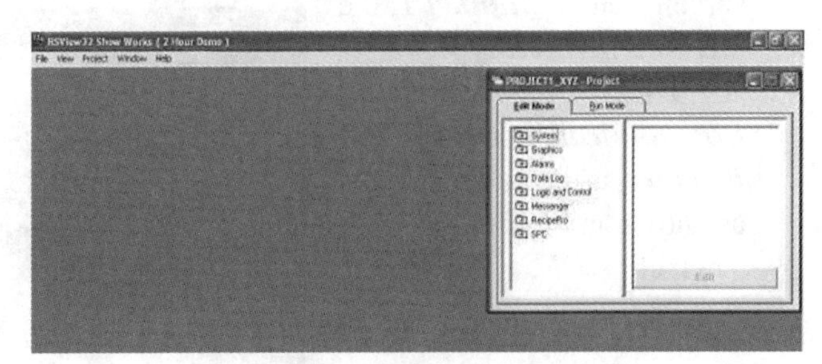

Figure 5-1-19: Pop-up for Figure 5-1-18 (Courtesy of Rockwell Automation, Inc.)

21. Now configure ***RSView*** for creating and running an animated PLC program. Make sure the small project screen with the ***Edit Mode*** and ***Run Mode*** **tabs** is situated at the upper right side of the screen.

22. The screens created will appear on the left side, so the ***Edit Mode*** and ***Run Mode*** pop-up screens will be on the right side.

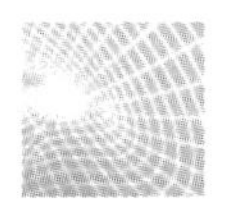

23. Double click the ***System*** folder, as shown in Figure 5-1-20.

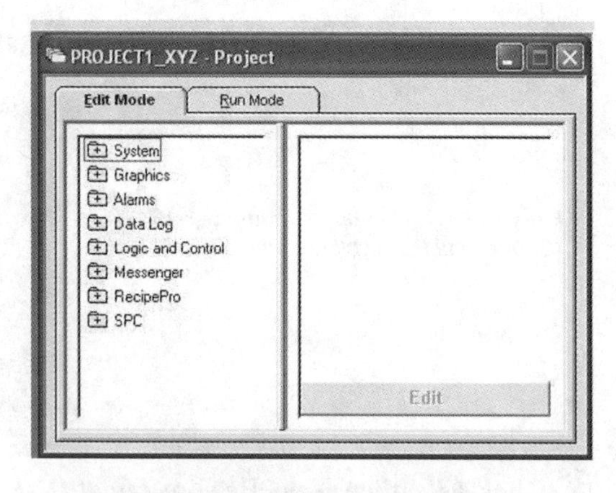

Figure 5-1-20: Select System Folder (Courtesy of Rockwell Automation, Inc.)

24. Next, double click the ***Channel*** icon, as shown in Figure 5-1-21.

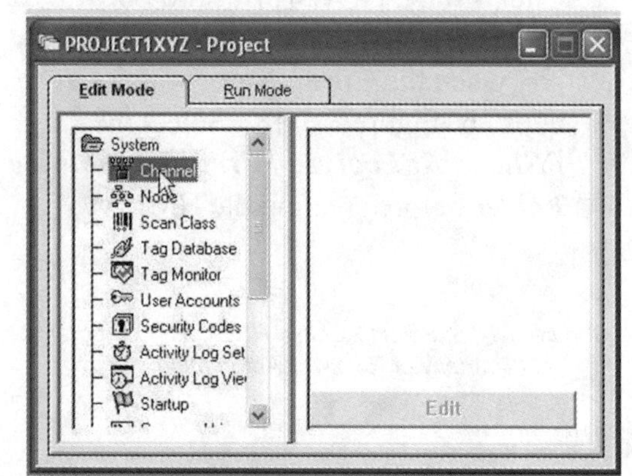

Figure 5-1-21: Select Channel pop-up screen (Courtesy of Rockwell Automation, Inc.)

25. At the ***Channel*** pop up screen, select ***Channel 1***, the ***DH-485*** network type, ***AB_DF1-1*** from the ***Primary Communication Driver***, and ***Primary*** as the active driver, as shown in Figure 5-1-22. Then click ***OK***.

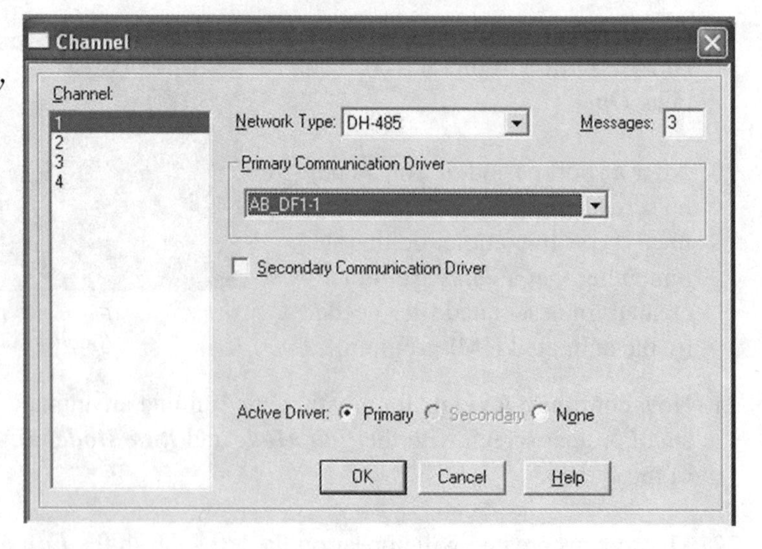

Figure 5-1-22: Select Communication Driver (Courtesy of Rockwell Automation, Inc.)

26. Double click ***Node*** as shown in Figure 5-1-23.

27. At the ***Node*** pop-up screen, enter a name like ***RSVLAB1_Intro*** in the ***Name:*** dialog box, as shown in Figure 5-1-24. Next, double click the button to the right of the ***Station*** dialog box.

28. Make sure the other dialog boxes appear as shown in Figure 5-1-23 and Figure 5-1-24.

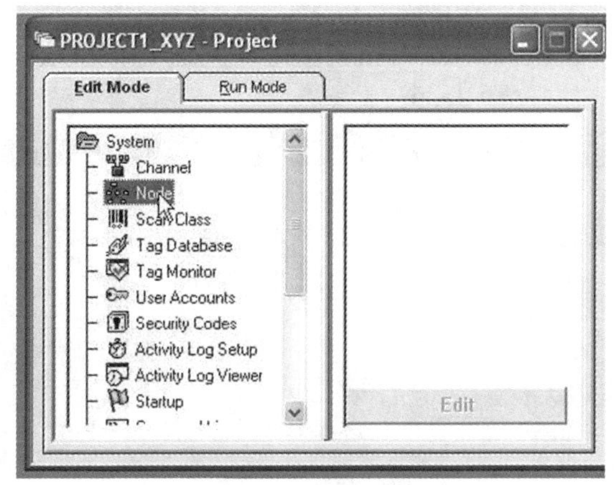

Figure 5-1-23: Node (Courtesy of Rockwell Automation, Inc.)

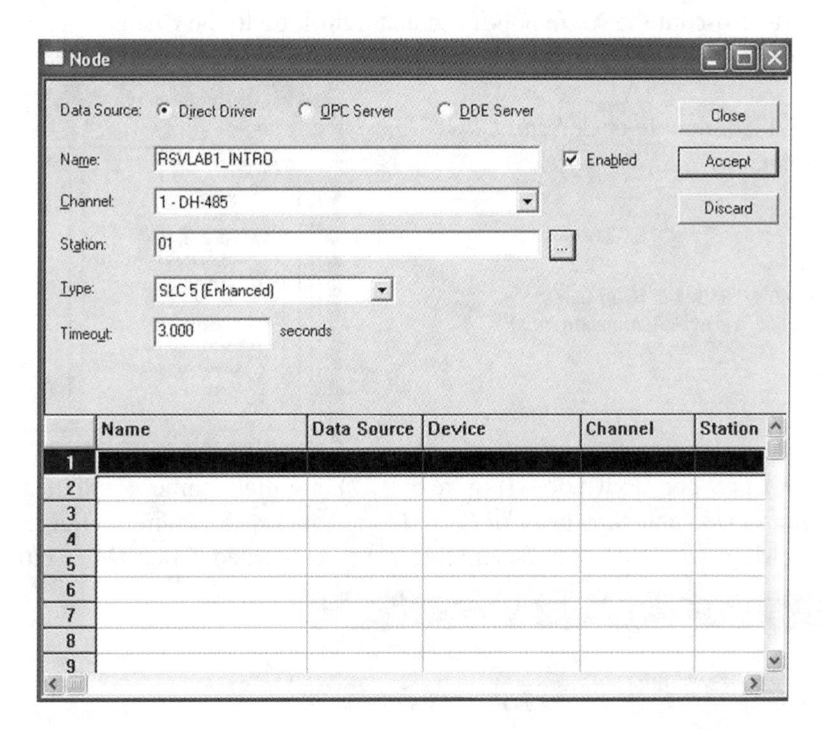

Figure 5-1-24: Pop-up for Figure 5-1-23 (Courtesy of Rockwell Automation, Inc.)

29. As the ***RSWho – Stations*** pop-up screen launches, be sure to select the appropriate PLC and node, as shown in Figure 5-1-25. Click ***OK*** to close out this screen and return to the screen in Figure 5-1-24.

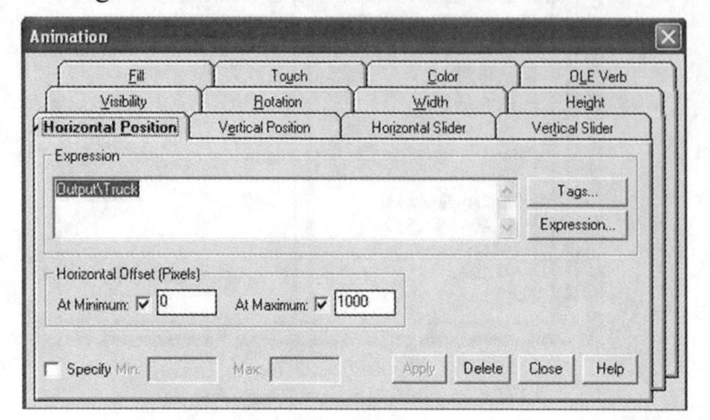

Figure 5-1-25: RSWho Pop-up
(Courtesy of Rockwell Automation, Inc.)

30. Click ***Accept*** and notice that ***Line 1*** is highlighted with the information just entered. Next, click ***Close*** to closeout the ***Node*** pop-up screen, returning to the Channel screen already seen in Figure 5-1-21.

31. In Figure 5-1-26, double click ***Scan Class*** under ***System***.

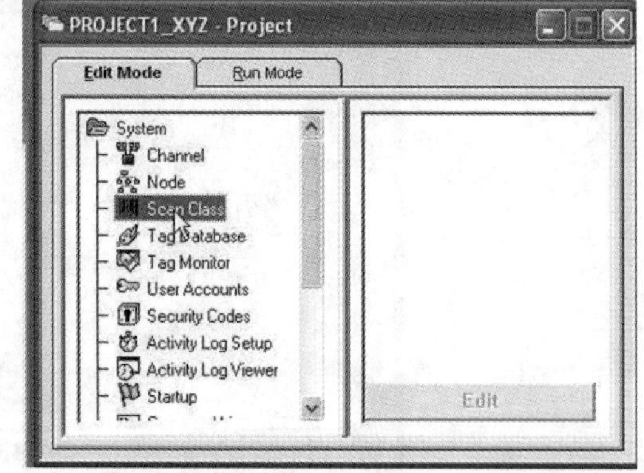

Figure 5-1-26: Select Scan Class
(Courtesy of Rockwell Automation, Inc.)

32. At the ***Scan Class*** pop-up window (Figure 5-1-27), highlight name ***A*** and then change the ***Foreground Period*** and ***Background Period*** to the values shown in the figure. Click ***B*** to record the changes and then ***A*** to ensure the changes recorded. Click ***OK*** to close this screen.

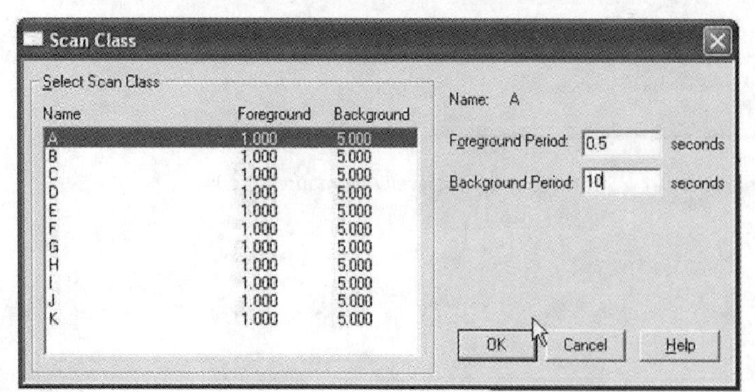

Figure 5-1-27: Scan Class Time Periods
(Courtesy of Rockwell Automation, Inc.)

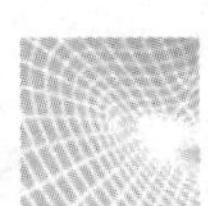

5.1.3 Creating Tags for Inputs and Outputs in the Tag Database

33. The next step is to set up the ***Tag Database***. Once the graphics are in place, the ***tags*** link the PLC ladder diagram to graphic display icons.

34. To begin, double click ***System*** and then ***Tag Database*** (Figure 5-1-28).

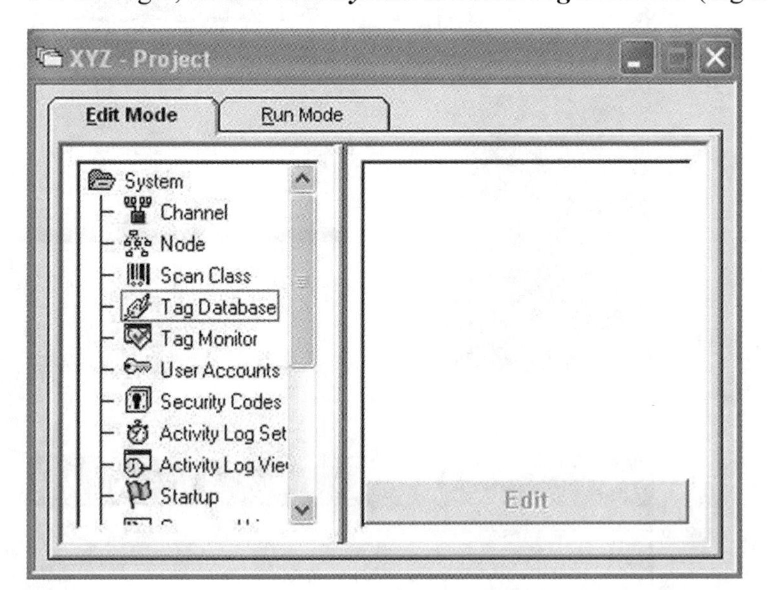

Figure 5-1-28: Activate Tag Database (Courtesy of Rockwell Automation, Inc.)

35. The screen in Figure 5-1-29 will appear.

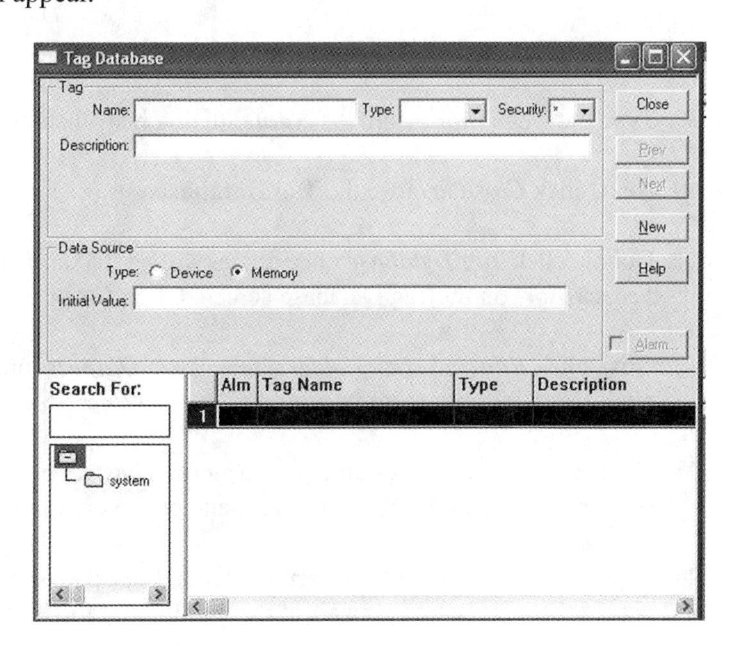

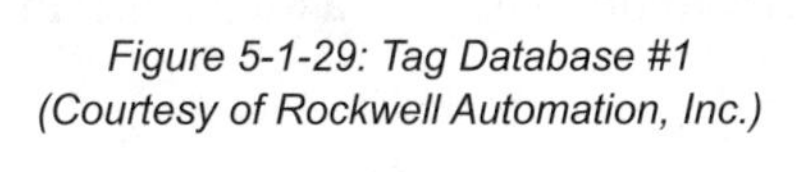

Figure 5-1-29: Tag Database #1 (Courtesy of Rockwell Automation, Inc.)

36. The ***Tag Database*** folder organizes the type of **tags**—such as **inputs, outputs,** and **alarms**—that will be used along with the PLC program.

37. Click ***Edit***, then ***New Folder***, as shown in Figure 5-1-30.

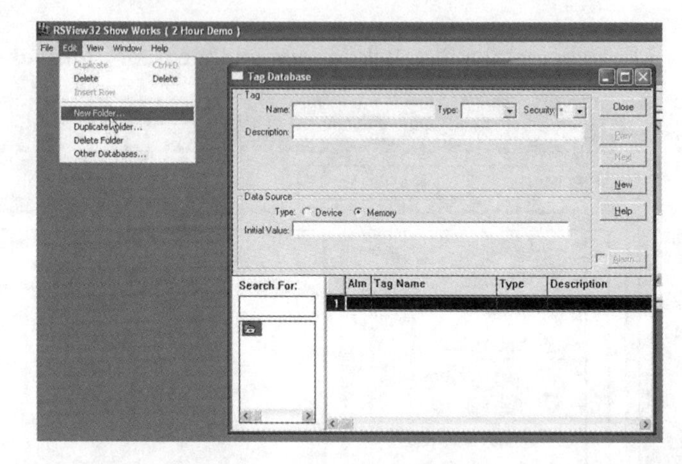

Figure 5-1-30: RSView32 (Courtesy of Rockwell Automation, Inc.)

38. The pop-up screen in Figure 5-1-31 will appear.

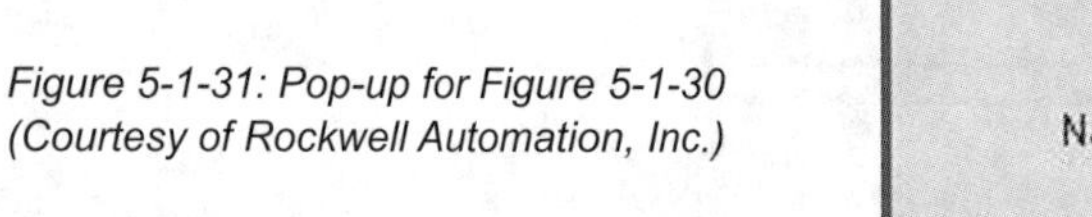

Figure 5-1-31: Pop-up for Figure 5-1-30 (Courtesy of Rockwell Automation, Inc.)

39. Type the word ***Inputs*** into the ***Name:*** dialog box; then click ***OK*** to close the ***Inputs*** window.

40. Next, click ***Close*** to close the **Tag Database**.

41. Double click ***Tag Database*** once again. Notice that an ***Inputs*** folder has been added beneath ***Search For:*** on the Tag Database screen. Click ***Close***.

42. Now, click ***Edit*** and ***New Folder*** again. Type ***Outputs*** into the ***Name:*** dialog box and then click ***OK***. Close the ***Outputs*** window.

43. Click ***System*** and then double click ***Tag Database*** once again. Then double click ***Inputs*** at the bottom left side in the ***Search For:*** menu box shown in Figure 5-1-32.

44. Type the information into the screen dialog boxes, as shown in Figure 5-1-32. Click ***Accept*** to ensure the information appears at the highlighted ***line 1***; then click ***Close.***

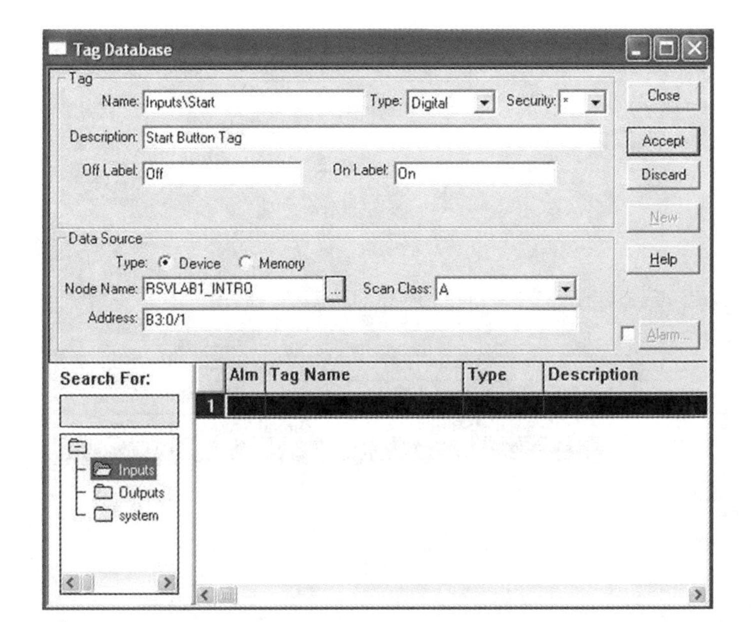

Figure 5-1-32: Tag Database #2 (Courtesy of Rockwell Automation, Inc.)

45. In the ***Search For:*** menu box, click in ***2*** the highlighted ***line 2***, as shown in Figure 5-1-33.

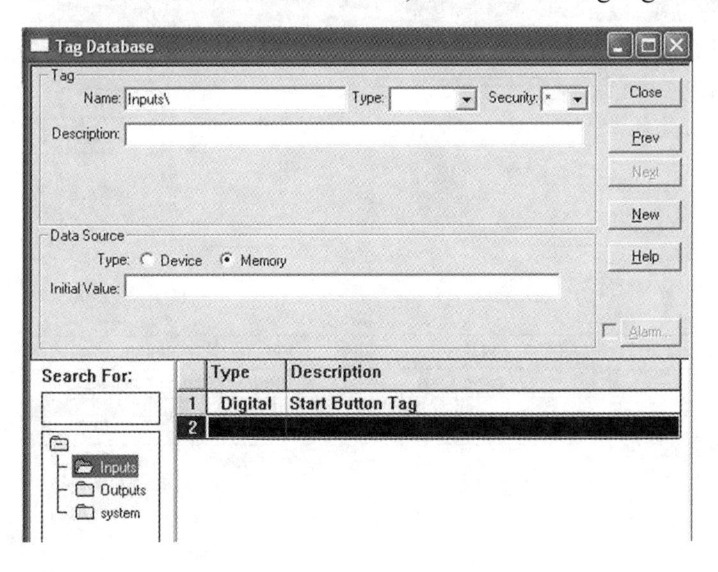

Figure 5-1-33: Tag Database #3 (Courtesy of Rockwell Automation, Inc.)

46. Enter the information onto the ***Tag Database*** screen, as shown in Figure 5-1-34.

47. After inputting the information, click ***Accept.*** Notice that the inputting information has been added to the second line (Figure 5-1-35); then click ***Close***.

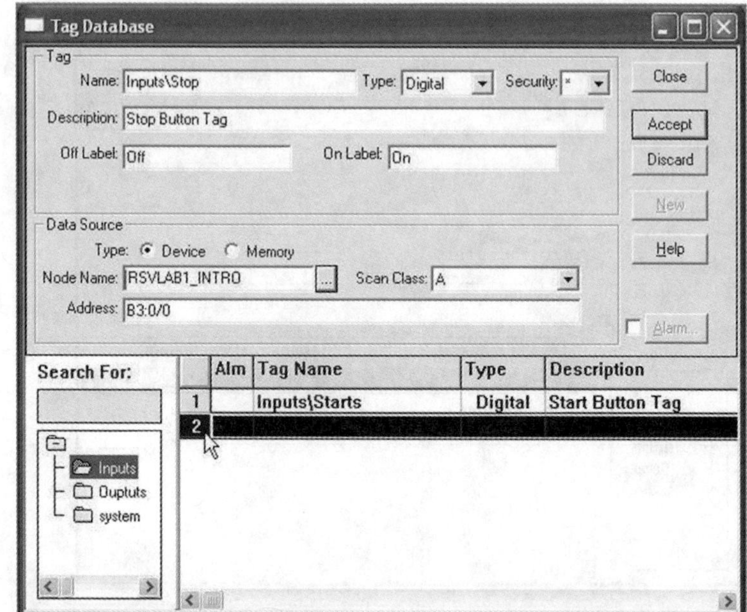

Figure 5-1-34: Tag Database #4
(Courtesy of Rockwell Automation, Inc.)

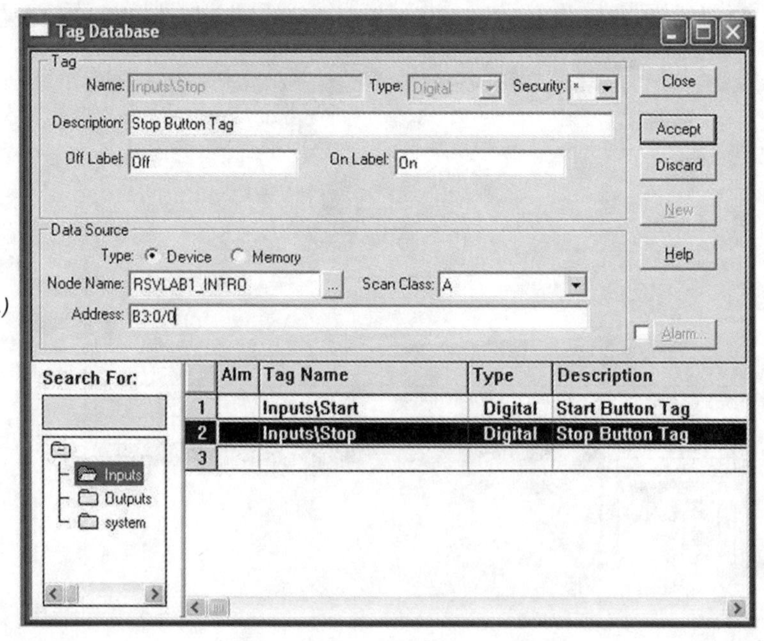

Figure 5-1-35: Tag Database #5
(Courtesy of Rockwell Automation, Inc.)

48. Re-launch ***Tag Database*** and double click ***Outputs*** in the ***Search For:*** menu box. Enter the information onto the Tag database screen, as shown in Figure 5-1-36.

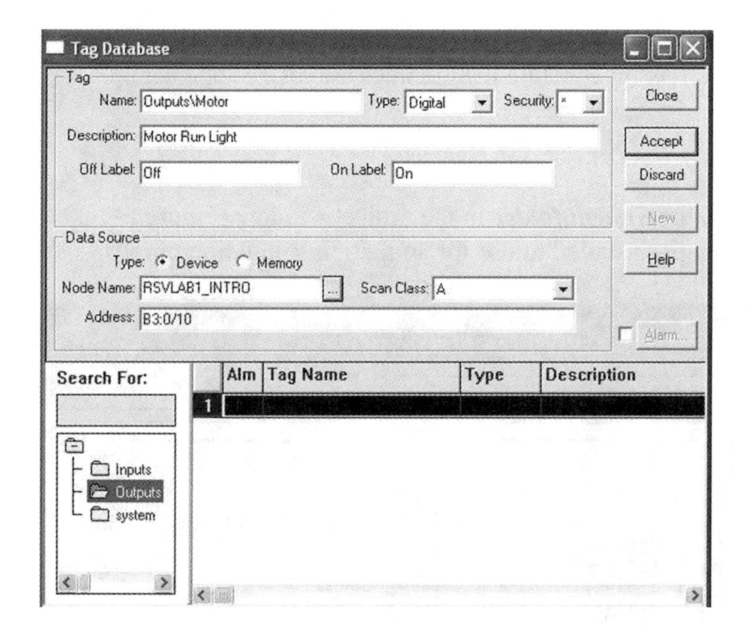

Figure 5-1-36: Tag Database #6 (Courtesy of Rockwell Automation, Inc.)

49. Click ***Accept***. Notice that the inputting information has been added to ***line 1***; then click ***Close***.

5.1.4 Creationing Graphic Images for the Control Process

50. The next step is to create the graphics icons that will control the inputs and the outputs. In the project window that appears in the upper right of the screen shown in Figure 5-1-37, select the ***Edit Mode*** Tab menu. Double click ***Graphics*** and then double click ***Display***.

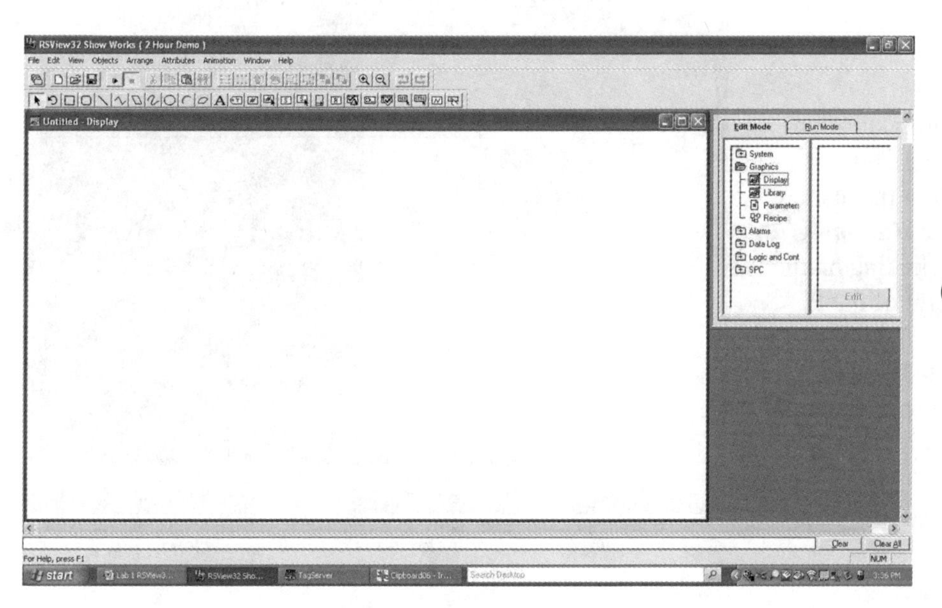

Figure 5-1-37: RSView32 (Courtesy of Rockwell Automation, Inc.)

51. Resize this screen to fill most of the display, but do not overlap the small project window. The project window may need to be resized a few times while the graphics are resized and set up.

5.1.5 Button and Motor Creation

52. Click ***Library*** once (**ONLY ONCE!!**) in the ***System folder*** in the project window (Figure 5-1-38). A list of graphic icon categories has been preinstalled to use for animation simulation purposes.

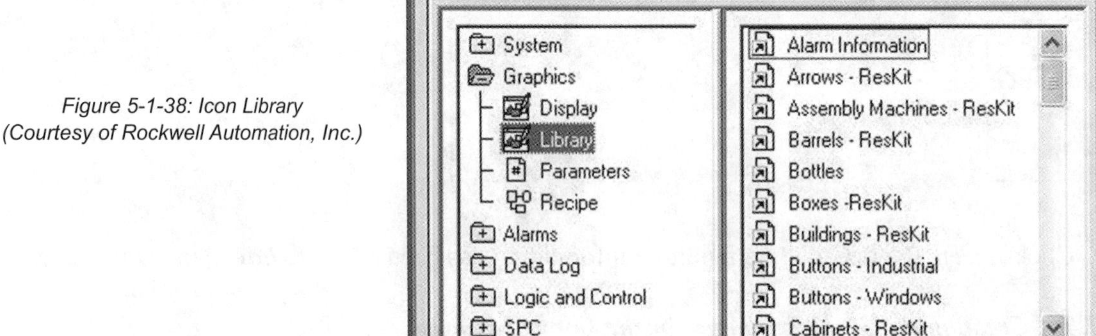

Figure 5-1-38: Icon Library
(Courtesy of Rockwell Automation, Inc.)

53. Double click ***Buttons Industrial***. The graphics menu in Figure 5-1-39 will appear.

54. Click, drag, and drop an Allen-Bradley ***Start*** button icon and ***Stop*** button icon into the display envelope. Close the ***Buttons – Library*** menu by clicking on the ***red X*** at the top right of the screen. If prompted, **DO NOT SAVE!!!** changes to the library. Saving may change permanent icons in this library.

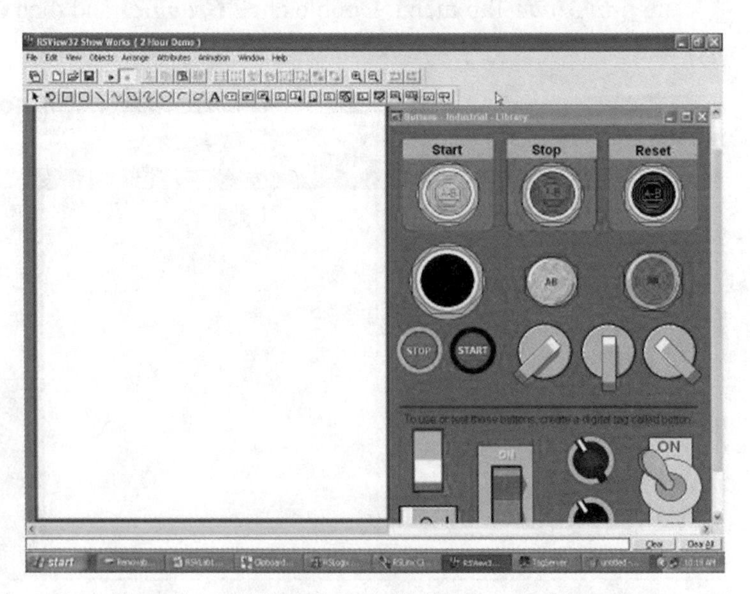

Figure 5-1-39: Library for Industrial Buttons, Switches, etc.
(Courtesy of Rockwell Automations, Inc.)

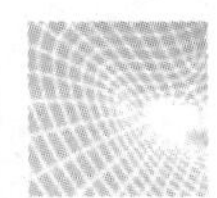

55. Open the ***Motors*** selection. Click, drag, and drop a motor icon into the display envelope. Close the ***Motors – Library*** menu. If prompted, **DO NOT SAVE** changes to the library. Saving may change permanent icons in this library.

56. The ***Untitled – Display*** screen should appear, similar to that shown in Figure 5-1-40.

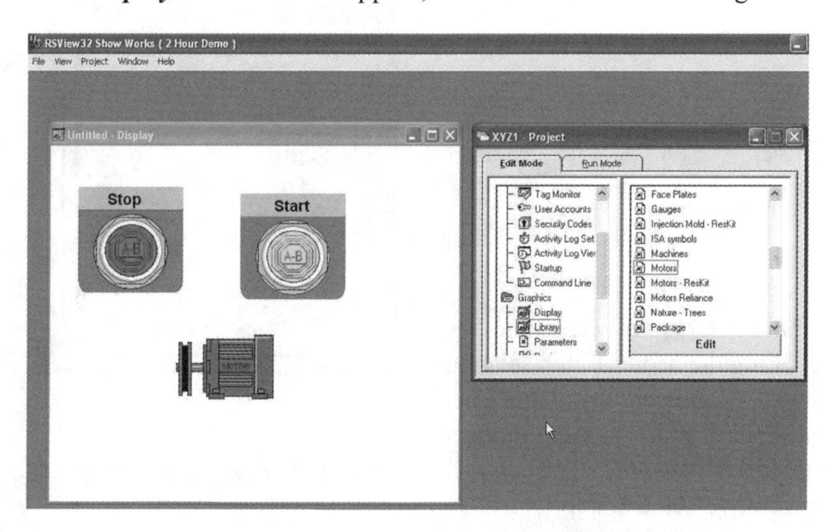

Figure 5-1-40: Create Motor Display (Courtesy of Rockwell Automation, Inc.)

57. As shown in Figure 5-1-41, click the ***red X*** at the top right of the ***Untitled – Display*** screen to close this screen.

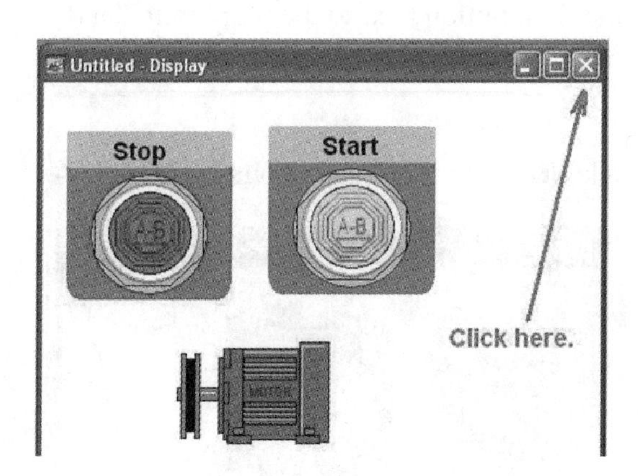

Figure 5-1-41: Close Untitled Display (Courtesy of Rockwell Automation, Inc.)

58. Click ***Yes.*** When prompted, and in the ***RSLinx_RSLogix500 Programs Storage Folder,*** save the display graphic with a file name — like ***Motor Latch.gfx*** — in the ***Lab_1_XXX*** folder.

59. Click ***Display*** in the ***Graphics*** folder, which is located in ***Edit Mode*** tab under the **Systems** folder.

60. A graphic representation file, **Motor Latch**, will appear, as shown in Figure 5-1-42.

Figure 5-1-42: Graphic for Motor Latch Display (Courtesy of Rockwell Automation, Inc.)

61. The ***Motor Latch – Display*** screen can be edited by clicking the Motor Latch selection on the right side of the pop-up screen. Then click the ***Edit*** button found at the bottom of the screen to edit the ***Display***. By clicking and then right clicking each graphic icon in the screen, icons can be deleted, duplicated, and so forth.

62. From the menu at the top of the RSView32 screen, other graphics can be added to the ***Motor Latch – Display*** screen.

63. Practice resizing and moving the buttons. Move the Stop button next to the Start button and try to resize them. Right click to delete any objects not needed; place the buttons and a motor as shown.

64. Remember to **NOT** save the library screen, so click ***NO*** when prompted. Saving will permanently modify the library, possibly deleting copy images from it.

65. As shown in Figure 5-1-43, click the ***red X*** at the top right of the ***Motor Latch – Display*** screen to close this screen.

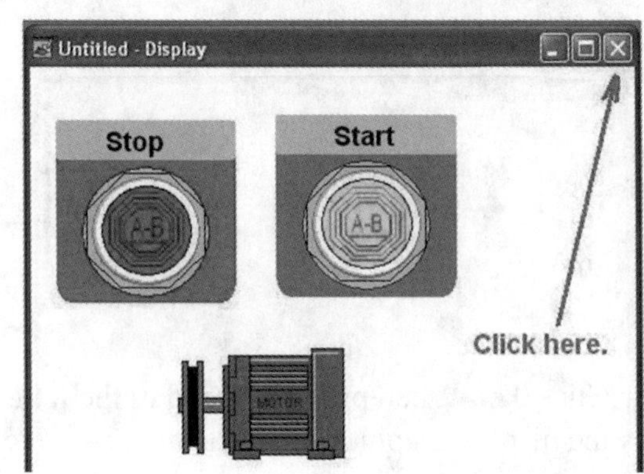

Figure 5-1-43: Close Display (Courtesy of Rockwell Automation, Inc.)

5.1.6 Adding Date and Time to Display

66. The next step is to add text and time to the ***Motor Latch – Display*** screen. Reopen the ***Motor Latch – Display*** screen. Click ***Objects*** and then ***Text***, as shown in Figure 5-1-44, from the ***RSView32*** *screen.*

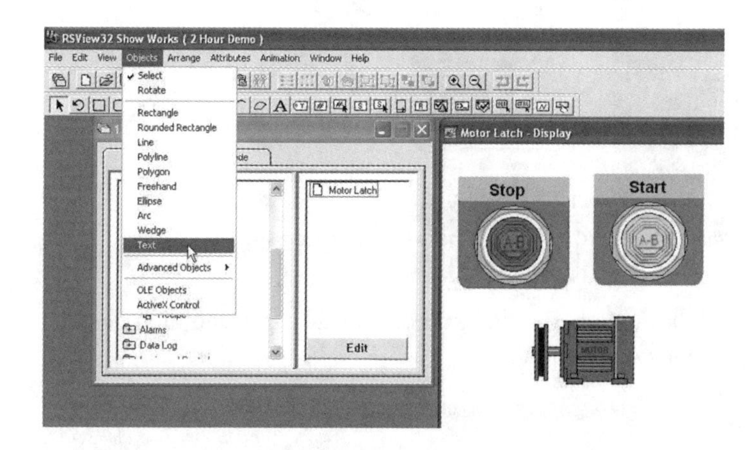

Figure 5-1-44: Adding Objects and Text (Courtesy of Rockwell Automation, Inc.)

67. Place the computer cursor somewhere in the display screen other than a graphic icon in the ***Motor Latch – Display*** screen place. As shown in Figure 5-1-45, the text "Motor Latching Circuit" was added to the display area.

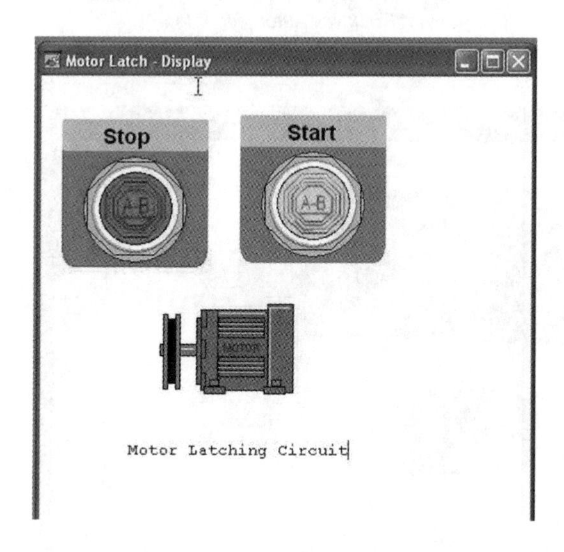

Figure 5-1-45: Text Added to Screen (Courtesy of Rockwell Automation, Inc.)

68. Next, place both the **date** and the **time** in the display area. As shown in Figure 5-1-46, double click ***Graphics***, then single click ***Library***, and then double click ***Clocks*** in the ***Project*** screen.

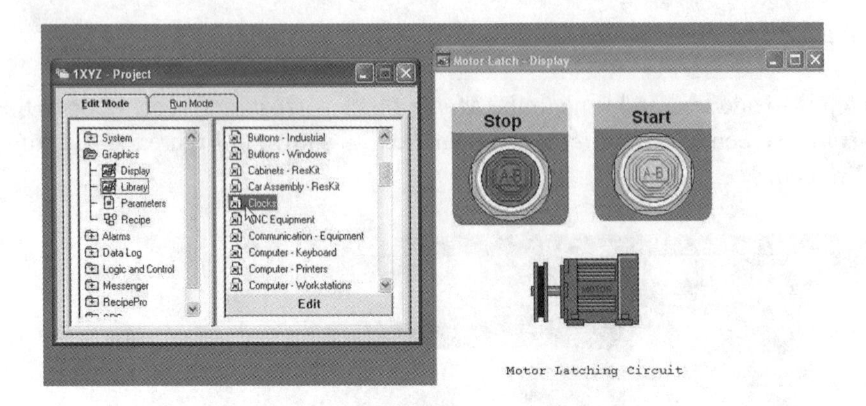

Figure 5-1-46: Adding Other Graphics (Courtesy of Rockwell Automation, Inc.)

69. As shown in Figure 5-1-47, right click the boxed string ***ssssssss***, click ***copy***, and then ***paste*** the string in the ***Start-Stop-Motor*** icons display area.

Figure 5-1-47: ssssssss String for Time
(Courtesy of Rockwell Automation, Inc.)

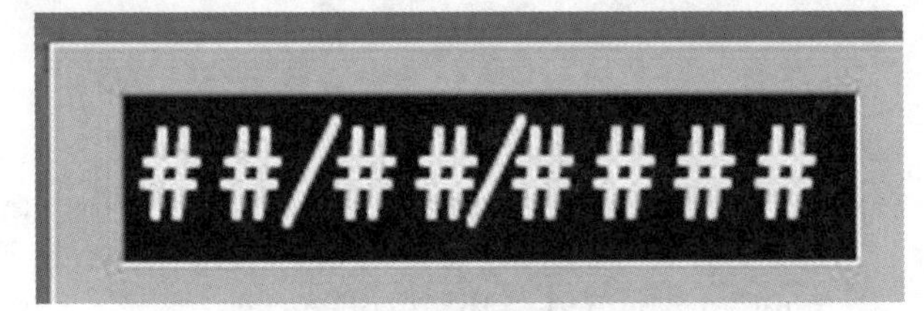

Figure 5-1-48: Date Display
(Courtesy of Rockwell Automation, Inc.)

70. From the ***Library***, as shown in Figure 5-1-48, right click the ***8-digit date box string***, click **copy**, and then **paste** the string in the ***Start-Stop-Motor*** icons display area.

71. The display should appear similar to the representation provided in Figure 5-1-49.

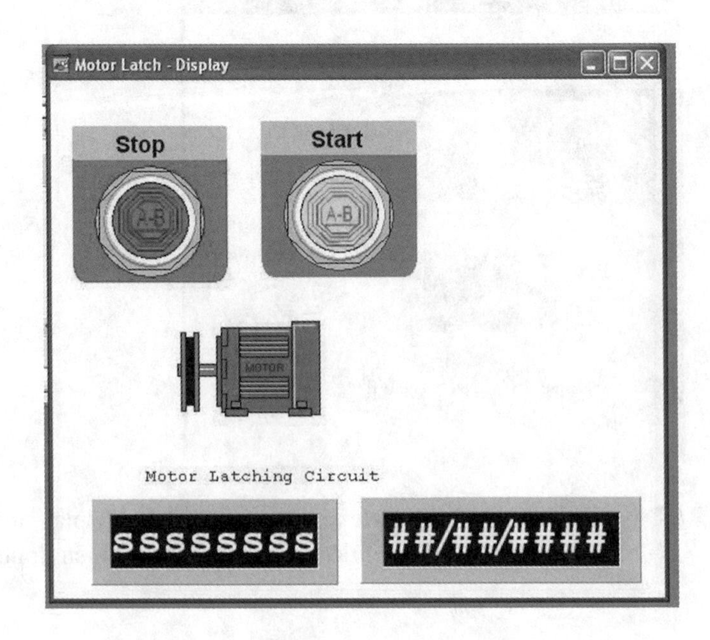

Figure 5-1-49: Complete Display
(Courtesy of Rockwell Automation, Inc.)

72. Click the ***red X*** at the top right of the library screen to close the screen. Remember to **NOT** save the library screen; therefore, click ***NO*** when prompted. Saving will permanently modify the library possibly deleting copy images from it.

5.1.7 Animating Inputs and Outputs

73. The next step is to animate display inputs and outputs. Open the Motor Latch Display screen with the procedure discussed in Step 35. Right click the **Start** button, highlight ***Animation***, and then click ***Touch***, as demonstrated in Figure 5-1-50.

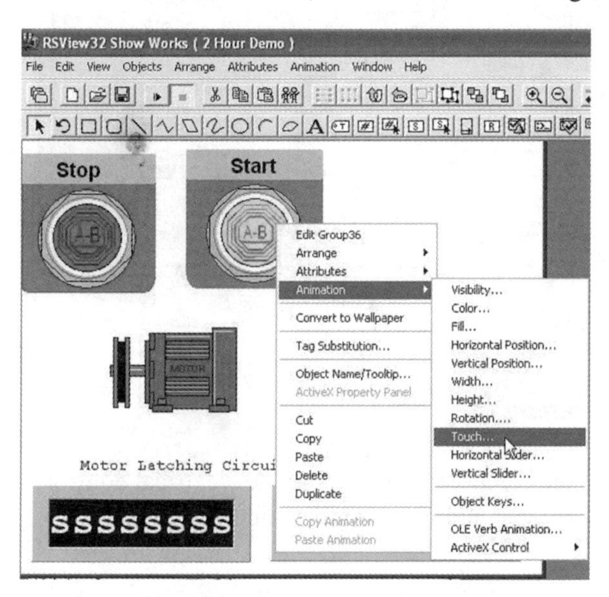

Figure 5-1-50: Animating Inputs and Outputs (Courtesy of Rockwell Automation, Inc.)

74. As an ***Animation*** pop-up screen appears, **input information** to the dialog boxes, as shown in Figure 5-1-51. Be careful to input the information exactly as it is shown. Incorrect spaces and characters will cause the animation to not function properly.

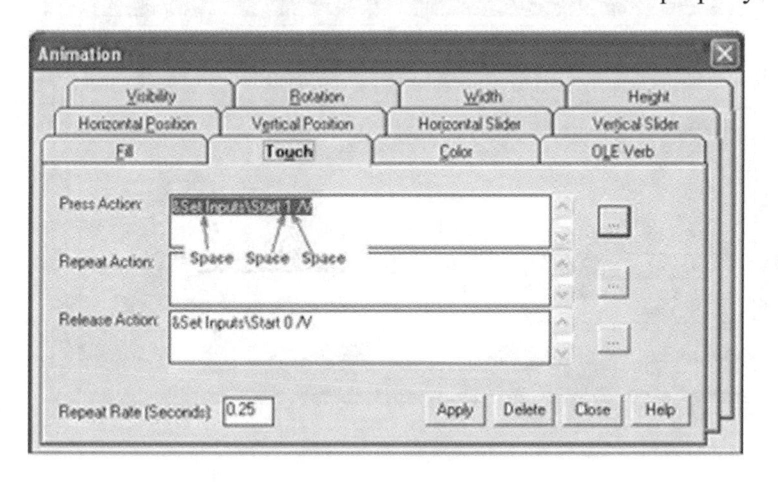

Figure 5-1-51: Adding Touch (Courtesy of Rockwell Automation, Inc.)

75. Click ***Apply*** and ***Close*** to close this screen.

76. Right click the ***Stop*** button, highlight ***Animate***, and then click ***Touch***, inputting values and characters exactly as shown in Figure 5-1-52.

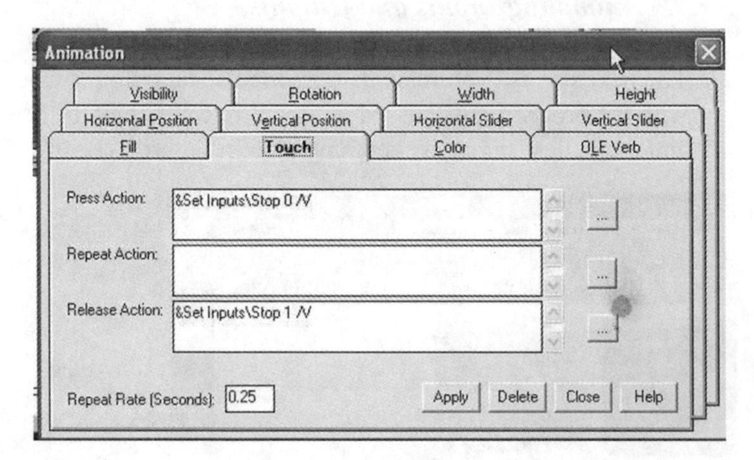

Figure 5-1-52: Completion of Touch (Courtesy of Rockwell Automation, Inc.)

77. Click ***Apply*** and ***Close*** to close this screen.

5.1.8 Programming Animation of Inputs and Outputs

78. The next step is to **program** inputs that will "see electrical values" and the electrical control values for outputs. Also, the use of **color control** can add much to the animation provided by the input and output electrical data values..

79. To begin this process, right click the output ***Motor*** icon and highlight ***Animation***. Then highlight and click **Color**.

80. When the ***Animation*** pop-up screen appears, enter the **expression** and change the **colors**. Enter the logic expression exactly as it is shown in Figure 5-1-53 in the ***Expression*** dialog box.

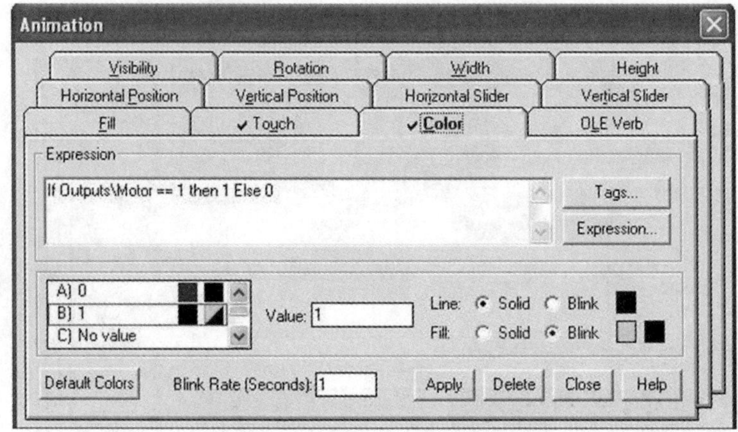

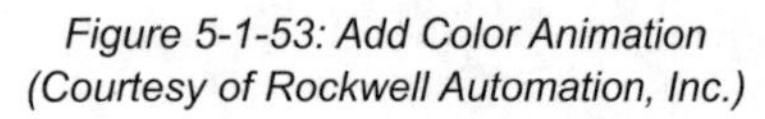

Figure 5-1-53: Add Color Animation (Courtesy of Rockwell Automation, Inc.)

81. Click the ***A) 0*** selection, as shown in Figure 5-1-53.

82. Next, left click the ***black box*** on the opposite side of the ***Animation*** screen to access the **hidden color palette** (Figure 5-1-54). Select a color other than black from the palette to determine how ***A) 0*** is affected.

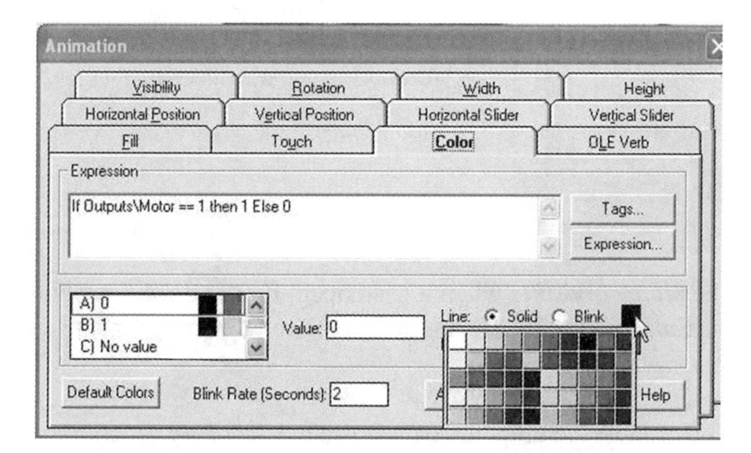

Figure 5-1-54: Animation Screen (Courtesy of Rockwell Automation, Inc.)

83. Note that if the ***Blink*** dialog, rather than the **Solid** dialog, were to be selected, the output would blink black icons.

84. Now click ***B) 1***. Use the color pallet on the opposite side of the screen to ***Blink*** colors and use ***Green*** and ***Black*** colors, as shown in Figure 5-1-55. This means the motor icon will **blink the colors of Green to Black** when actuated.

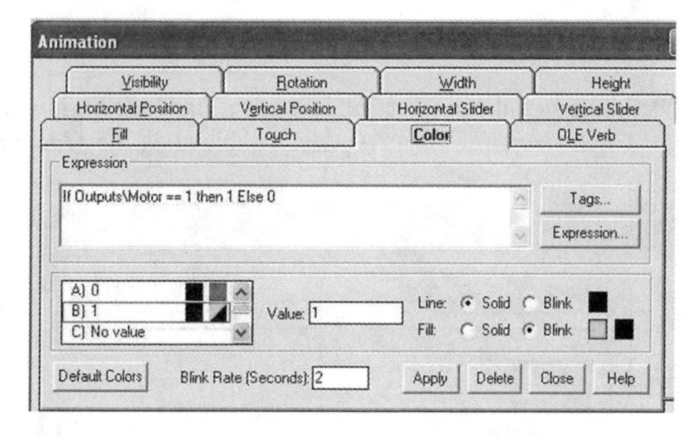

Figure 5-1-55: Change Colors (Courtesy of Rockwell Automation, Inc.)

85. Click ***Apply*** and ***Close*** to save this color animation.

5.1.9 Adding Invisibility to Animation

86. The next step is to add invisibility to the animation. Right click the ***Motor*** icon, highlight ***Animation***, and then highlight and click ***Visibility***. Input information as shown in Figure 5-1-56 in the ***Expressions*** dialog box.

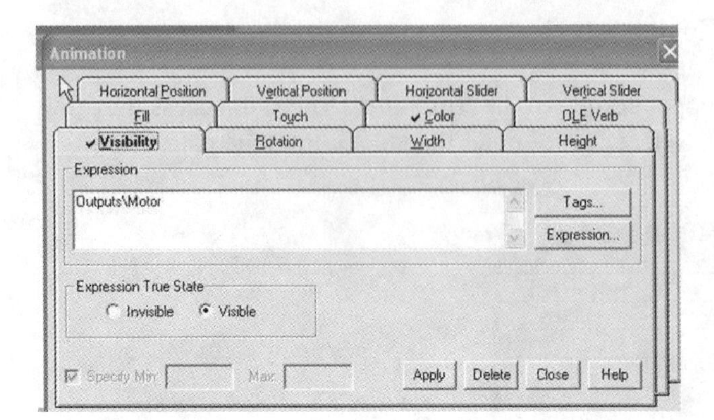

Figure 5-1-56: Visibility and Color
(Courtesy of Rockwell Automation, Inc.)

87. This animation choice will make the ***motor invisible*** when it is stopped. Pressing the ***Start*** button will make the ***motor visible*** and ***Color*** will blink the motor from Black to Green repeatedly. Click ***Apply*** and ***Close*** to close the ***Animation*** screen.

88. The final step is to test the animation. To test the graphic display, use the ***Test Run*** function in ***Edit Mode*** or run the graphic display in ***Run Mode***. ***Save*** and ***Close*** the project display screen.

89. Test both methods for this test case.

90. Make sure the project window has ***Edit Mode*** highlighted in bold black letters (Figure 5-1-57). Click ***Graphics***, then ***Display***. Double click the ***Motor Latch*** file in the right column.

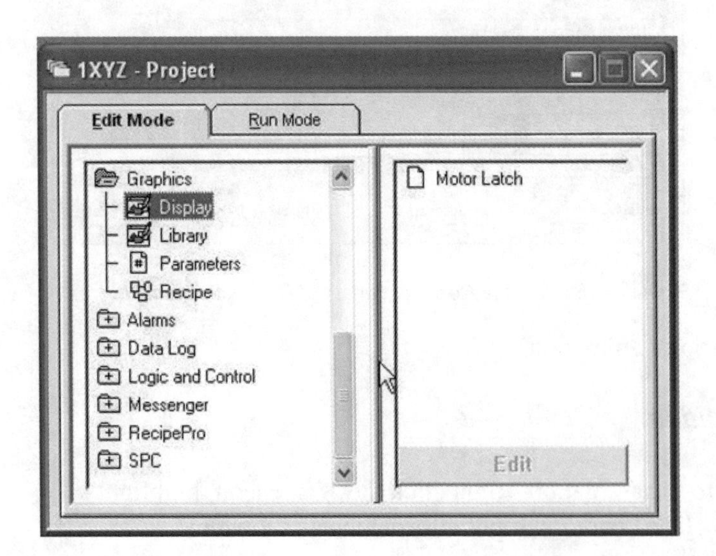

Figure 5-1-57: Edit and Run Mode screen
(Courtesy of Rockwell Automation, Inc.)

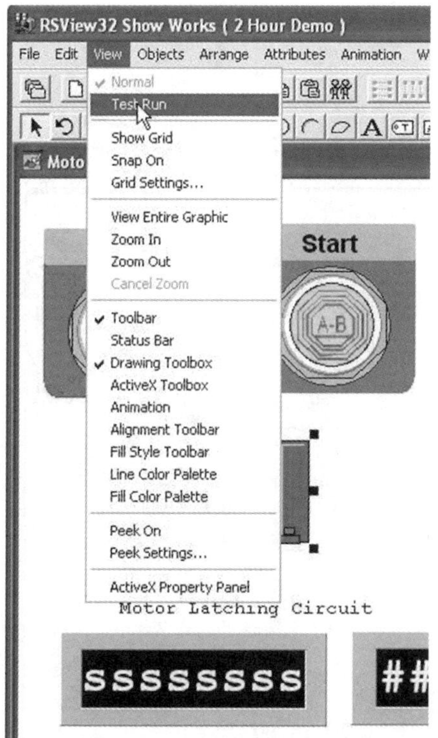

91. Click ***View*** at the top of the screen, then ***Test Run***, as shown in Figure 5-1-58.

Figure 5-1-58: Test Run (Courtesy of Rockwell Automation, Inc.)

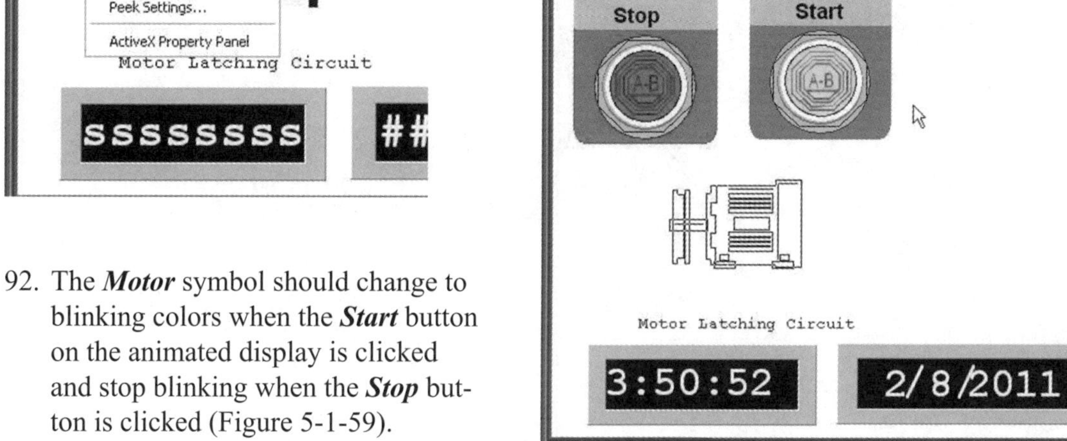

92. The ***Motor*** symbol should change to blinking colors when the ***Start*** button on the animated display is clicked and stop blinking when the ***Stop*** button is clicked (Figure 5-1-59).

Figure 5-1-59: Final Display (Courtesy of Rockwell Automation, Inc.)

93. Close out the display window if it is opened.

94. Click the ***Run Mode*** tab and watch the words change to bold.

95. Click ***Graphics***, then ***Display***. Double click the ***Motor Latch*** file on the right side.

96. Click the ***Start*** button. The Motor symbol should change to blinking colors when the Start button is clicked and stop blinking when the Stop button is clicked.

97. This completes this lab.

RSView Lab 5-2: Analog and Digital Displays

5.2.1 Introduction

This lab illustrates some of the more complex graphic capabilities available in computerized human machine interface displays. There are two ways to create Lab 5-2. One way is to copy the project from Lab 5-1, rename it as Lab 5-2, then modify as needed by using the steps below. The other way is to create the HMI from the start by following the steps below.

5.2.2 Creating and Downloading the PLC Diagram

1. Enter the PLC program from Figure 5-2-1 into ***RSLogix 500 (English)***. Notice that Binary Addressing, Limit Tests, and a Time On-Delay (TON) timer are used.

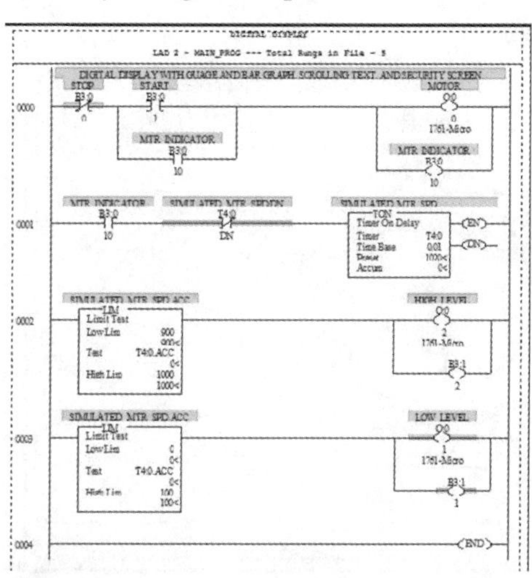

Figure 5-2-1: Lab 5-2 PLC program (Courtesy of Rockwell Automation, Inc.)

2. It is typical for ***RSLogix 500*** to provide only a limited number of binary addresses. In the ***Data File B3*** pop-up menu, as shown in Figure 5-2-2, only addresses from ***B3:0/0 to B3:0/15*** are offered.

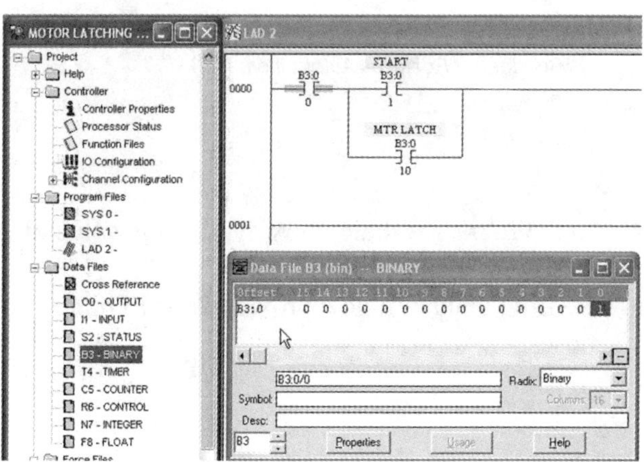

Figure 5-2-2: Data File B3 screen (Courtesy of Rockwell Automation, Inc.)

3. To increase the number of address selections, click the ***Properties*** button at the bottom of the ***Data File B3*** pop-up screen (Figure 5-2-3).

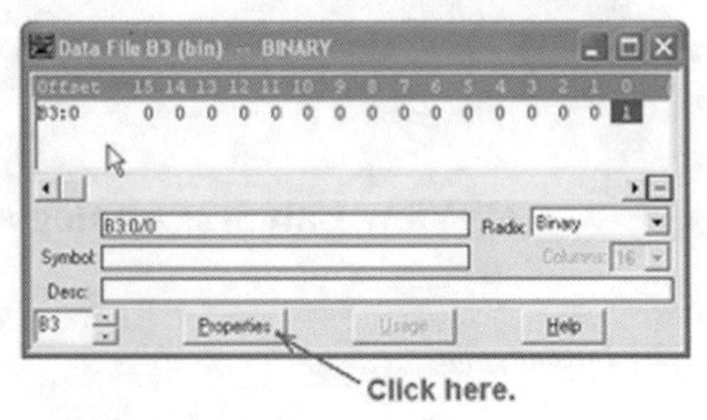

Figure 5-2-3: Data File B3 Address Selection (Courtesy of Rockwell Automation, Inc.)

a. The pop-up screen in Figure 5-2-4 will appear.

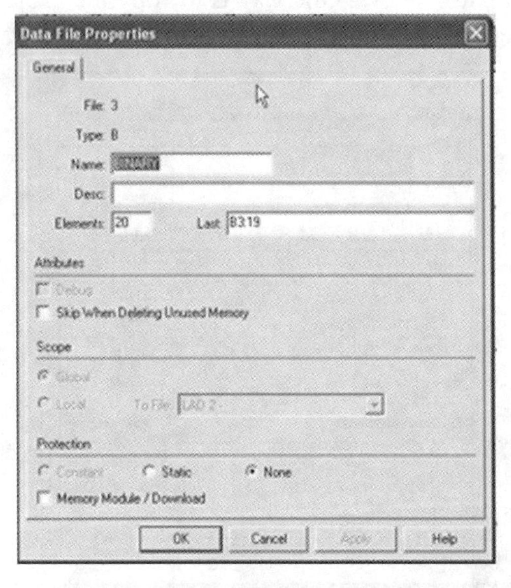

Figure 5-2-4: Data File B3 Properties (Courtesy of Rockwell Automation, Inc.)

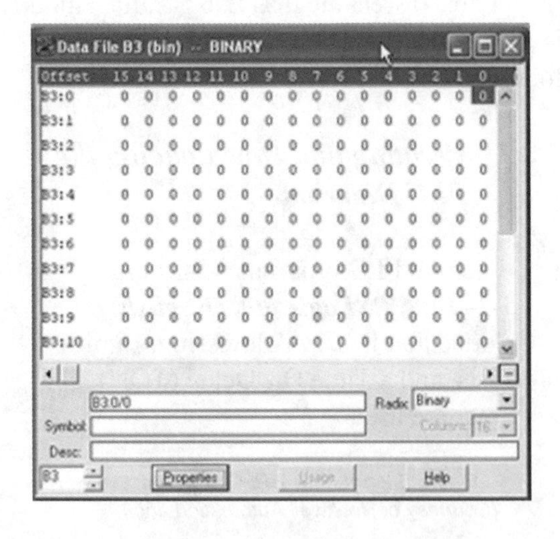

Figure 5-2-5: Data File B3 Elements (Courtesy of Rockwell Automation, Inc.)

4. In the ***Elements*** dialog box, type in the value of ***20*** (Figure 5-2-5). Click ***Apply***, then ***OK***.

5. **RSLogix 500** will now provide ***20 x 15*** binary addresses for input and output labeling purposes.

5.2.3 *Creating and Configuring the Timer PLC*

6. To label the ***Time On-Delay (TON)*** timer, double click the ***T4-Timer*** icon underneath the ***Data Files*** Folder in RSLogix 500. A ***Data File T4*** pop-up screen will appear (Figure 5-2-6).

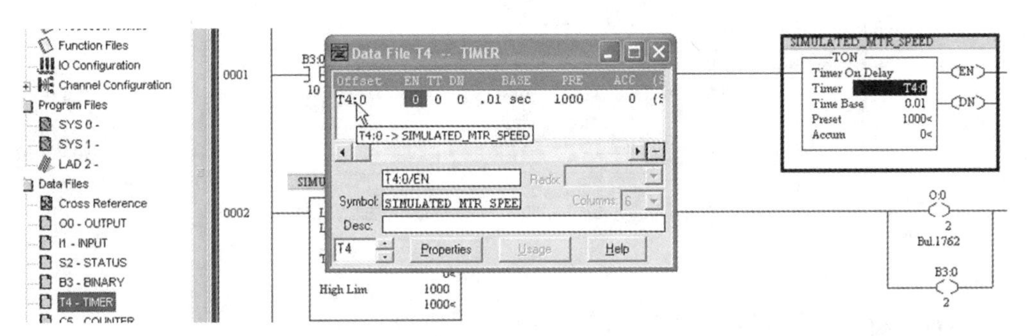

Figure 5-2-6: Time On-Delay (TON) Timer (Courtesy of Rockwell Automation, Inc.)

7. Click the ***T4:0*** in the pop-up menu and then drag it over to the ***Timer labeling*** area (that's shown darkened in Figure 5-2-6).

8. On the ***TON*** icon in the PLC program, click the current ***Time Base*** value. Change the value to ***0.01*** (seconds), as shown in Figure 5-2-7, and then click somewhere off of the icon to store that value.

9. Double click the ***Preset*** value on the ***TON*** icon to change the preset value to ***1000***.

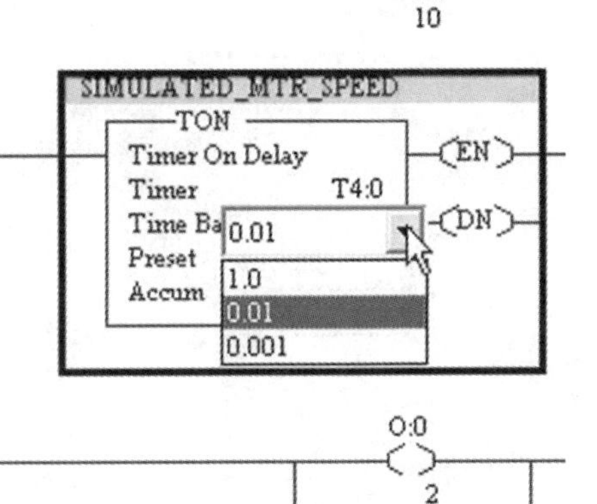

Figure 5-2-7: TON Time Base (Courtesy of Rockwell Automation, Inc.)

10. Next, label the normally closed input, as shown in rung ***0001*** of the PLC program seen in Figure 5-2-8, so that it is actuated/linked to the ***T4:0 timer***. To do so, click the ***T4 –Timer*** icon from underneath ***Data Files*** (Figure 5-2-9). Next, click the ***DN*** from the ***Data File T4*** pop-up window. Drag the ***DN*** symbol over and then place it on top of the normally closed input to label it.

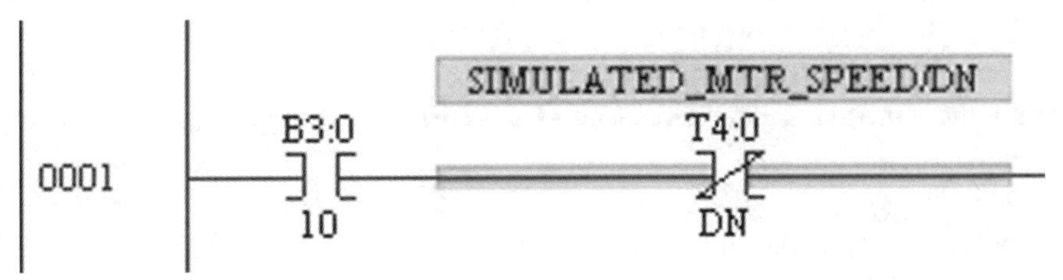

Figure 5-2-8: DONE bit Normally Closed Contact (Courtesy of Rockwell Automation, Inc.)

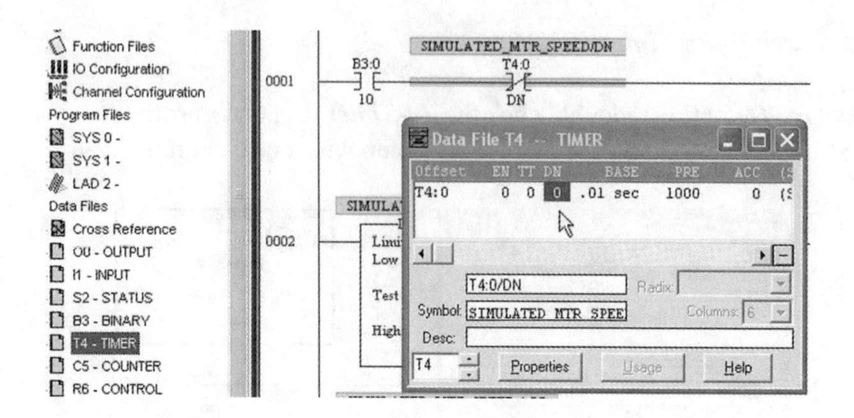

Figure 5-2-9: TON Timer Data File (Courtesy of Rockwell Automation, Inc.)

5.2.4 Creating the PLC Limit Test Symbols

11. To label a ***Limit Test***, click the ***question mark (?)*** located next to the ***Low Lim*** symbol on the icon. A dialog box appears, as shown in Figure 5-2-10.

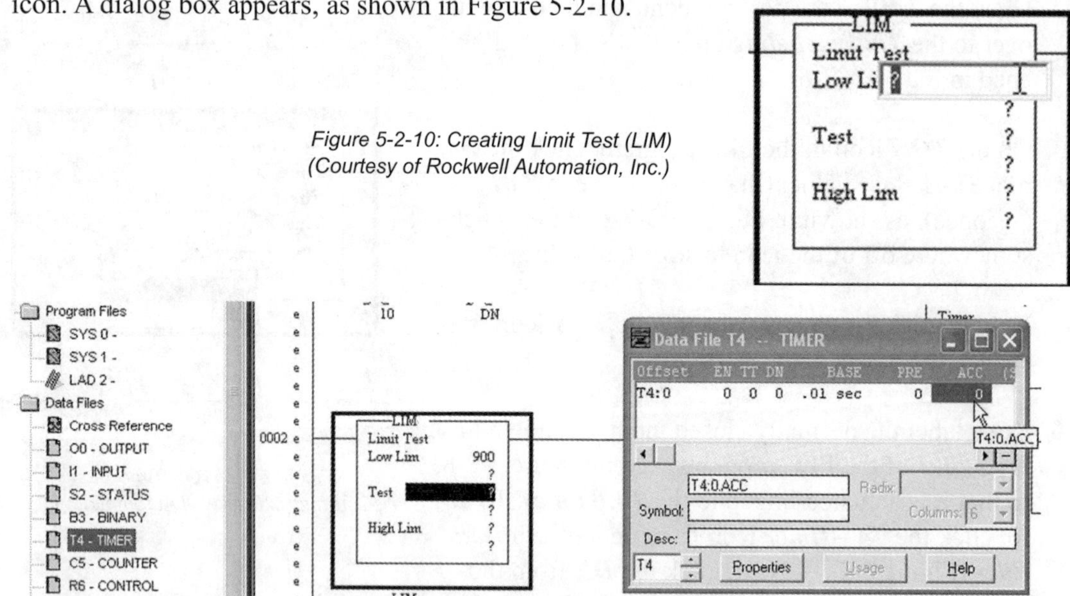

Figure 5-2-10: Creating Limit Test (LIM) (Courtesy of Rockwell Automation, Inc.)

Figure 5-2-11: LIM connected to TON (Courtesy of Rockwell Automation, Inc.)

a. Type in the ***count value*** needed for the **Low Lim** test. Repeat this for the ***Test*** value and for the ***High Lim*** value.

b. To label the ***Test*** portion of the ***Limit Test*** icon, note the graphic provided in Figure 5-2-11. Click the ***T4 –Timer*** icon from underneath ***Data Files***.

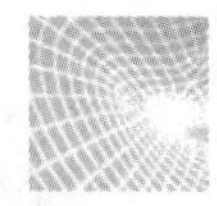

c. Next, click ***ACC*** from the ***Data File T4*** pop-up window, as shown in Figure 5-2-11. Drag the ***ACC*** symbol over and then place on top of the ***Test*** field inside the icon to label it.

d. To save the changes to the ***Limit Test*** icon, click the program somewhere other than a program icon.

12. ***Download*** the ladder program into the PLC. Be sure to place the program in the ***RUN*** mode to continue.

5.2.5 Starting RSView in Preparation to Build the HMI

13. Launch ***RSView32*** as demonstrated in Figure 5-2-12.

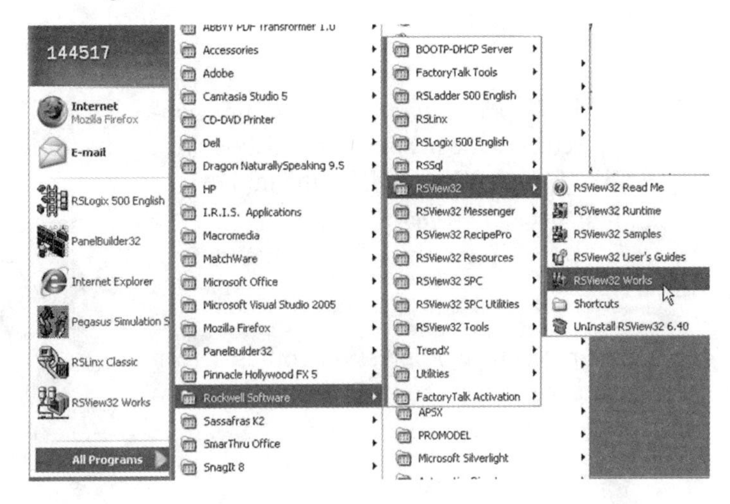

Figure 5-2-12: Launching RSView32 (Courtesy of Rockwell Automation, Inc.)

14. The pop-up screen in Figure 5-2-13 will appear. Maximize this pop-up screen to fill the entire display.

Figure 5-2-13: Pop-up screen (Courtesy of Rockwell Automation, Inc.)

15. Click ***File*** and then ***New***. In the ***Create New Project*** pop-up screen that appears (Figure 5-2-14), in the ***Look in:*** dialog box, be sure ***Local Disk (C:)*** is selected and the ***RSLinx_RSLogix 500 Programs Storage Folder*** is selected before creating an animated project using ***RSLinx***.

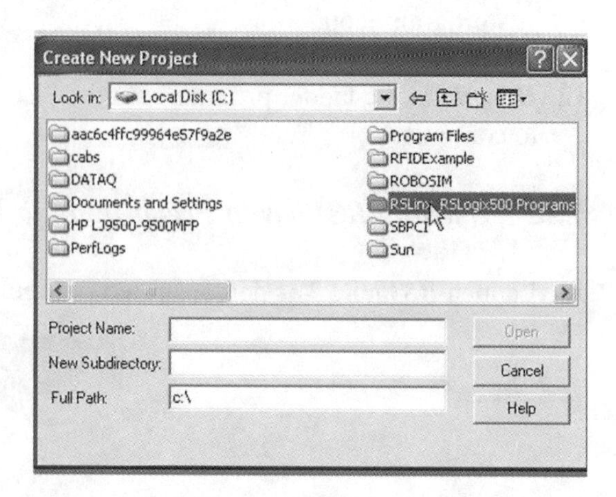

Figure 5-2-14: Create New Project
(Courtesy of Rockwell Automation, Inc.)

a. In the ***Project Name:*** portion of this pop-up screen, type in something distinctive, like ***Lab_2_XXX*** (XXX = the initials), and then click ***Open***.

16. After a percentage progression pop-up screen appears, the screen shown in Figure 5-2-15 appears.

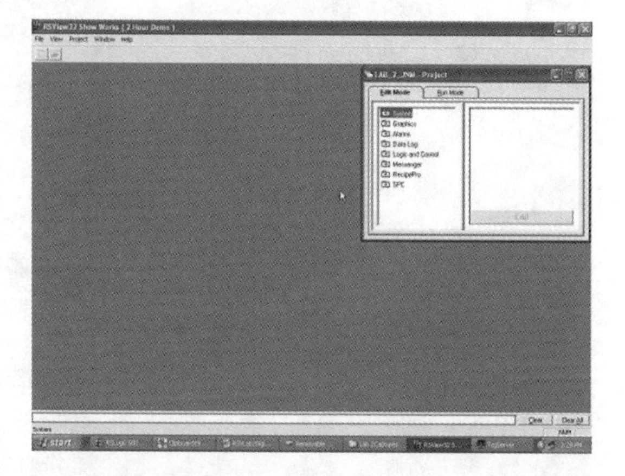

Figure 5-2-15: Beginning a Project
(Courtesy of Rockwell Automation, Inc.)

17. Double click the ***System*** folder, as shown in Figure 5-2-16.

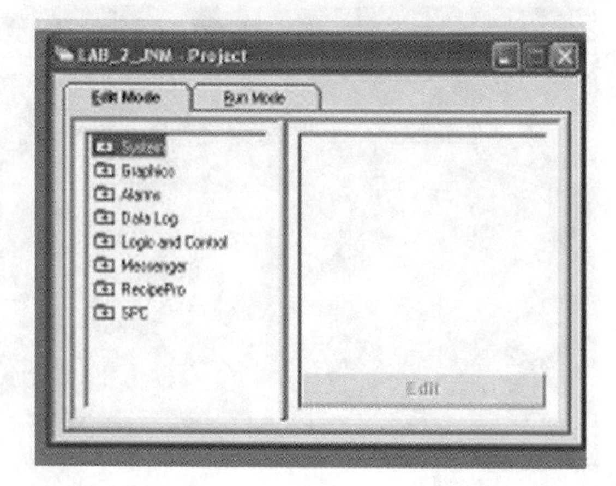

Figure 5-2-16: System Edit
(Courtesy of Rockwell Automation, Inc.)

18. Next, double click the ***Channel*** icon, as shown in Figure 5-2-17.

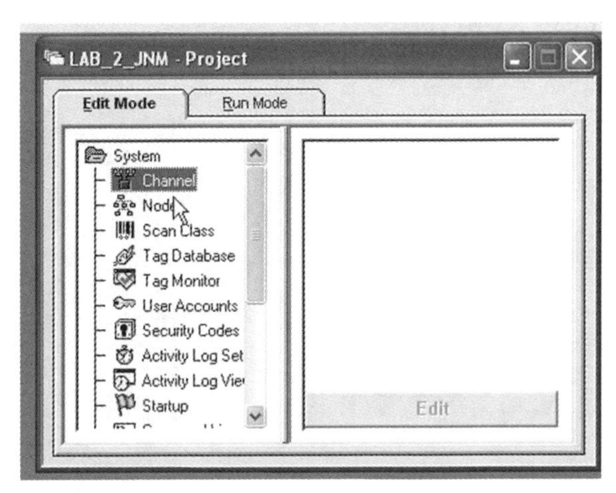

Figure 5-2-17: Configure Channel (Courtesy of Rockwell Automation, Inc.)

19. Enter the following values, as shown in Figure 5-2-18. At the ***Channel*** pop up screen, select the number **1** in the channel column. In the ***Network Type*** selections, choose ***DH-485.*** Then in the ***Primary Communication Driver***, select ***AB_DF1-1*** as the active driver. Finally, click ***Primary*** selection as the **Active Driver.** Don't forget to click ***OK***.

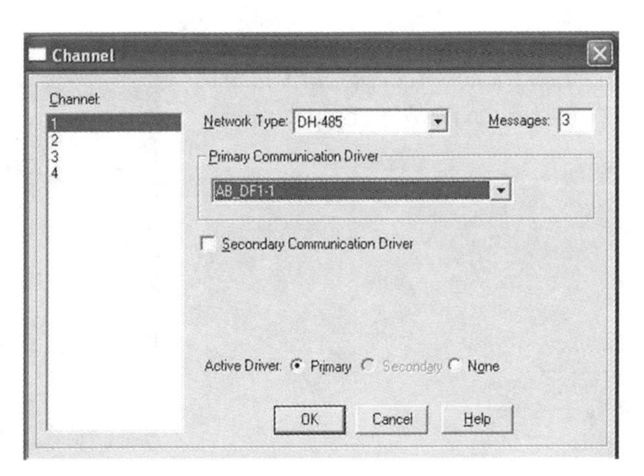

Figure 5-2-18: Values for Channel Settings (Courtesy of Rockwell Automation, Inc.)

20. Double click ***Node*** located in the ***System*** folder, which is in the ***Edit Mode***, as displayed in ***Step 14***.

    a. At the ***Node*** pop-up screen, enter a name like ***LAB_2_XXX***, as shown in Figure 5-2-19 in the ***Name:*** dialog box. Next, double click the ☐ button on the ***Station*** dialog box. Check that the other dialog boxes appear as shown in Figure 5-2-19.

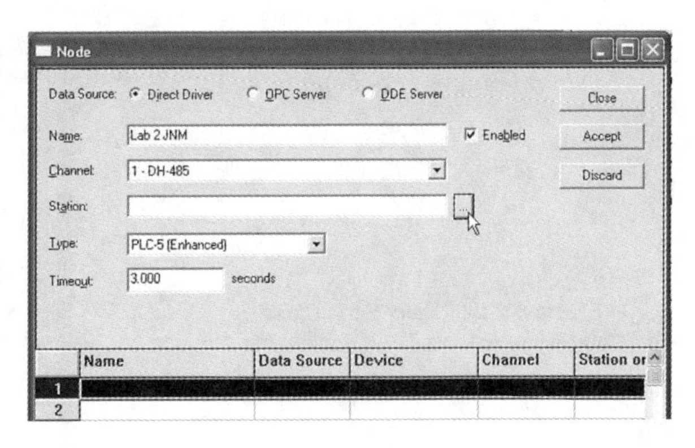

Figure 5-2-19: Node Screen (Courtesy of Rockwell Automation, Inc.)

21. As the ***RSWho – Stations*** pop-up screen launches, be sure to select the appropriate PLC and node, as shown in Figure 5-2-20. Click ***OK*** to close out this screen.

    a. Click ***Accept***. Notice line ***1*** highlighted at the bottom with the information just inputted (Figure 5-2-21).

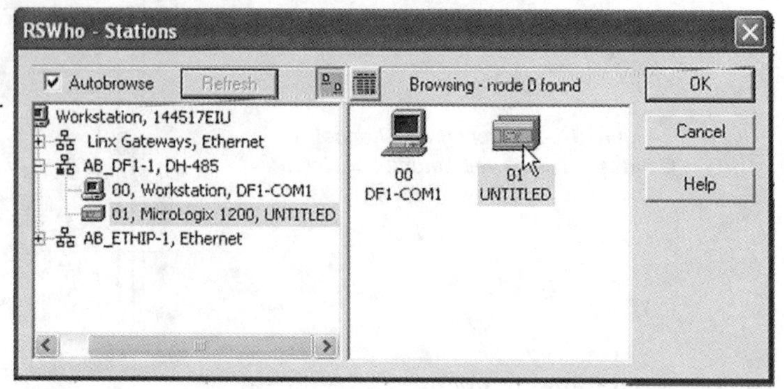

Figure 5-2-20: RSWho Screen (Courtesy of Rockwell Automation, Inc.)

22. Next, click ***Accept and Close*** to close the ***Node*** pop up screen.

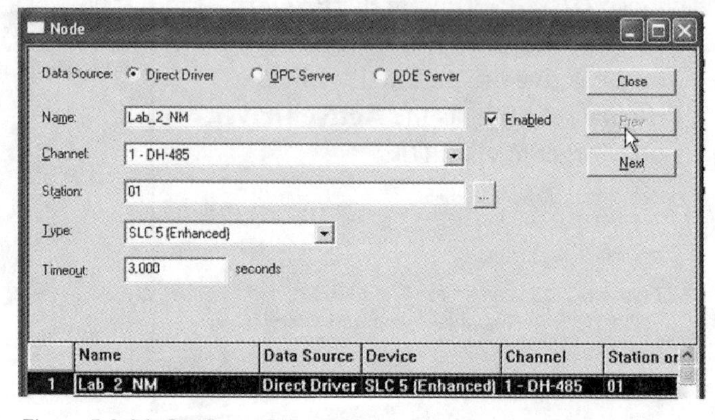

Figure 5-2-21: Configure Drivers (Courtesy of Rockwell Automation, Inc.)

23. Double click ***Scan Class*** under ***System*** (Figure 5-2-22).

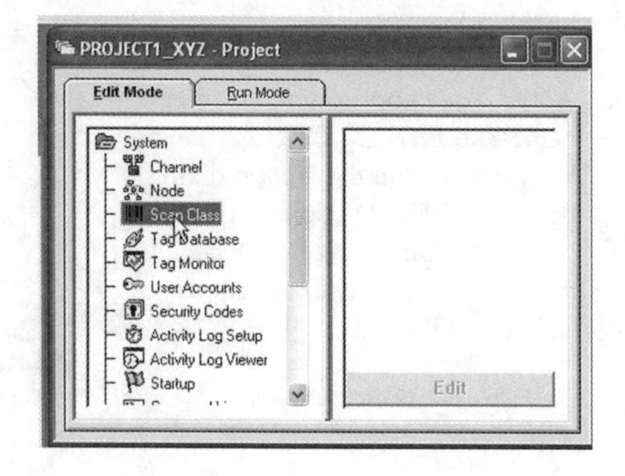

Figure 5-2-22: Select Scan Class (Courtesy of Rockwell Automation, Inc.)

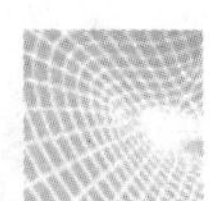

24. At the ***Scan Class*** pop up window, highlight name ***A*** and then change the ***Foreground Period*** and ***Background Period*** to the values shown in Figure 5-2-23. Click ***B*** to record the changes and then ***A*** to ensure the changes were recorded. Click ***OK***.

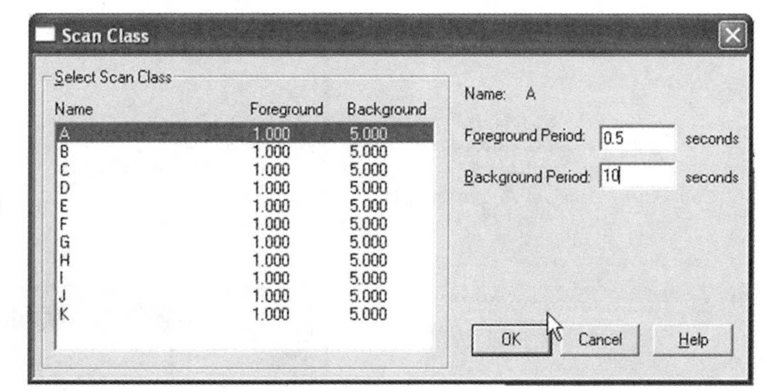

Figure 5-2-23: Configure Scan Class (Courtesy of Rockwell Automation, Inc.)

5.2.6 Creating Tags in the Tag Database

25. The next step is to set up the ***Tag Database.*** Once the graphics are in place, the *tags* link the PLC ladder diagram in the PLC to the graphic icons.

26. To begin this step, double click ***System*** and then ***Tag Database*** (Figure 5-2-24).

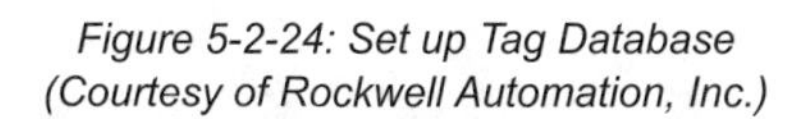
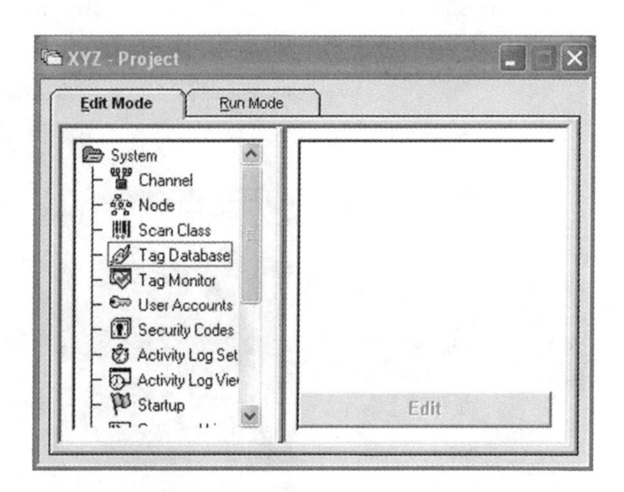

Figure 5-2-24: Set up Tag Database (Courtesy of Rockwell Automation, Inc.)

27. The screen shown in Figure 5-2-25 will appear.

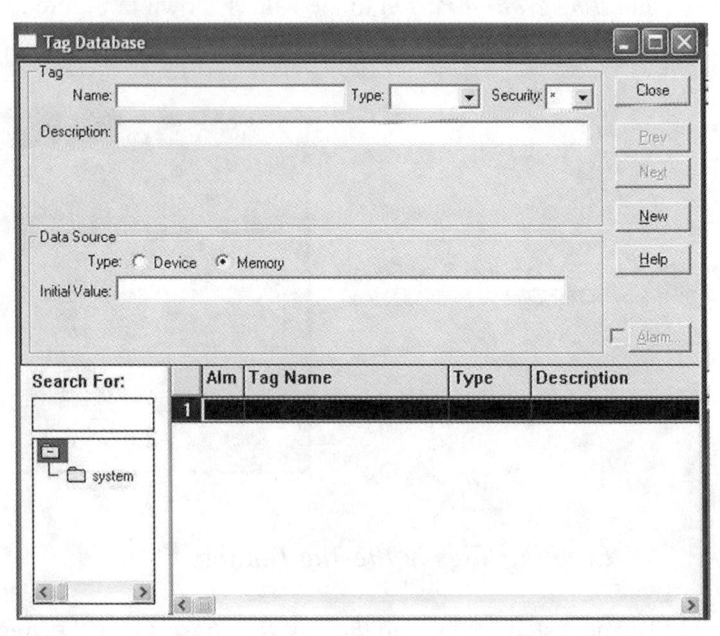

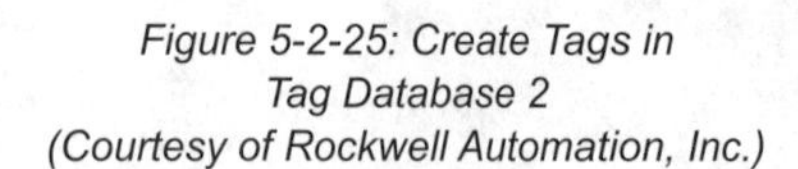
Figure 5-2-25: Create Tags in Tag Database 2 (Courtesy of Rockwell Automation, Inc.)

28. The ***Tag Database*** folder organizes the type of *tags* such as ***inputs, outputs, alarms, etc***., that will be used along with the PLC program.

    a. As shown in Figure 5-2-26, click **Edit** and then ***New Folder***

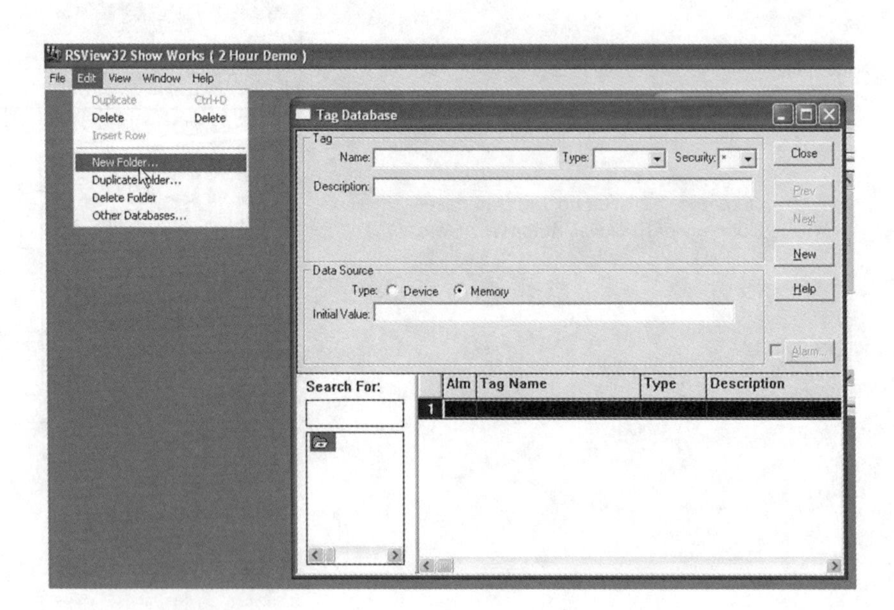

Figure 5-2-26 Tag Database 2 (Courtesy of Rockwell Automation, Inc.)

b. The small pop-up screen in Figure 5-2-27 will now appear.

Figure 5-2-27: Create New Folder
(Courtesy of Rockwell Automation, Inc.)

29. Type the word ***Inputs*** in the ***Name:*** dialog box and then click ***OK*** to close the ***Inputs*** window.

    a. Notice that an ***Inputs*** folder has been added underneath ***Search For:*** on the ***Tag Database*** screen.

30. Now click ***Edit and New Folder*** once again. Type ***Outputs*** into the ***Name:*** dialog box of the ***New Folder*** pop-up screen and then click ***OK***. Close the window if it does not happen automatically.

31. Double click ***Inputs*** in the ***Search For:*** menu shown in the lower right of Figure 5-2-28. Enter the data shown by typing the information into the screen dialog boxes.

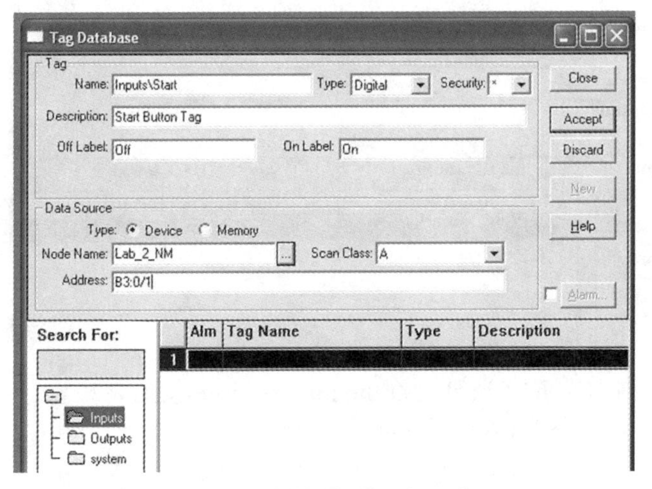

Figure 5-2-28: Tag Database 3
(Courtesy of Rockwell Automation, Inc.)

a. Double click the □ button to input the ***Node Name***.

b. Next, click ***Accept*** to ensure the information appears at the high lighted line ***1*** at the bottom of the screen; then click ***Close***.

32. Next, click the ***2,*** underneath line ***1***, as shown in Figure 5-2-29.

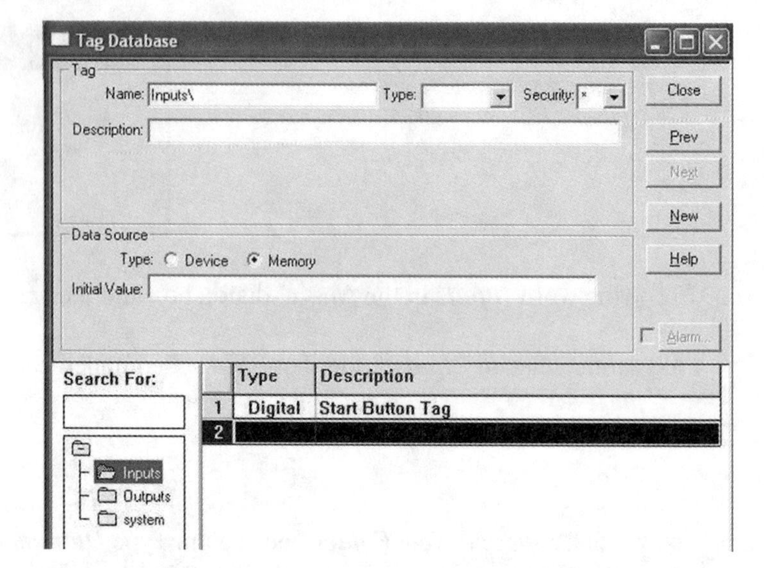

Figure 5-2-29: Tag Database 4
(Courtesy of Rockwell Automation, Inc.)

a. Enter the information onto the ***Tag Database*** screen, as shown in Figure 5-2-30.

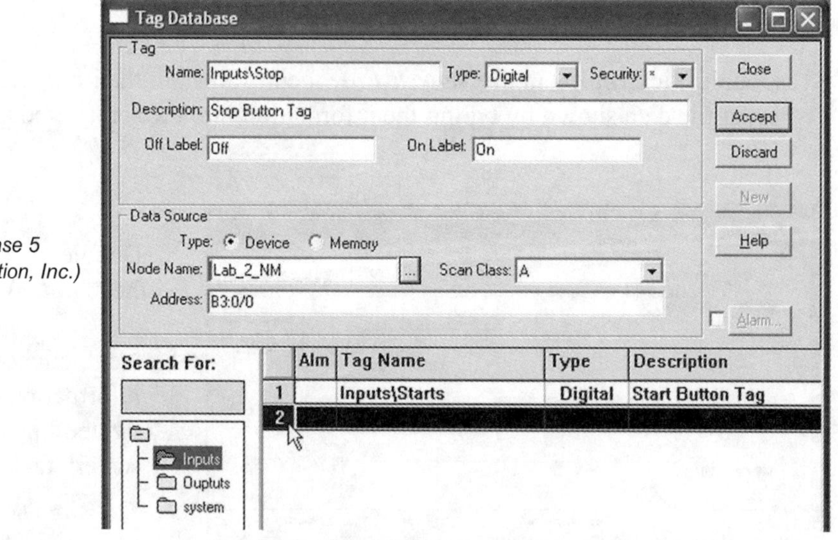

Figure 5-2-30: Tag Database 5
(Courtesy of Rockwell Automation, Inc.)

b. After inputting the information, click ***Accept.*** Check that the inputted information has been added to the second line, and then click ***Close*** (Figure 5-2-31).

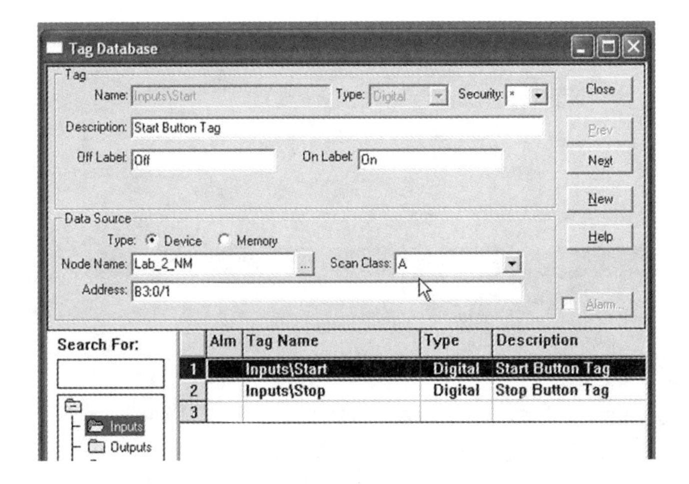

Figure 5-2-31: Tag Database 6
(Courtesy of Rockwell Automation, Inc.)

33. Double click ***Outputs*** in the ***Search For:*** menu. Enter the information onto the ***Tag Database*** screen, as shown in Figure 5-2-32. After clicking ***Accept***, notice that the inputted information has been added to line ***1.***

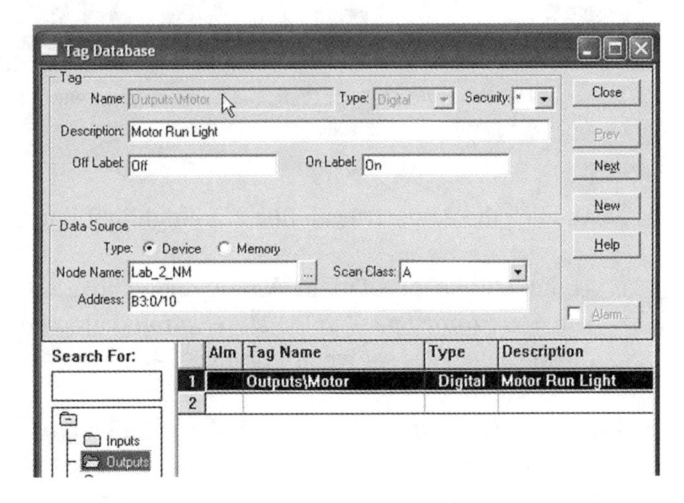

Figure 5-2-32: Tag Database 7
(Courtesy of Rockwell Automation, Inc.)

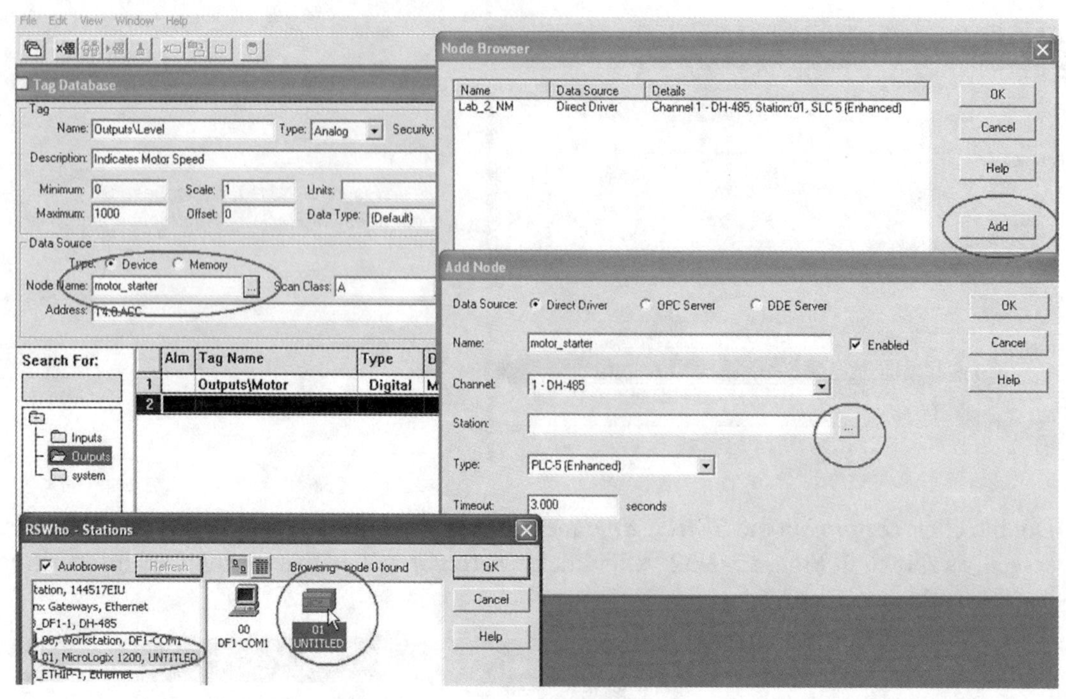

Figure 5-2-33: Complete Screens (Courtesy of Rockwell Automation, Inc.)

34. Next, click the ***2,*** underneath line 1, to highlight line 2, as shown in Figure 5-2-33.

    a. Enter ***Output\Level*** in the ***Name:*** dialog box. Select ***Analog*** for ***Type***. Add the words ***Indicates Motor Speed*** to the ***Description*** dialog box. Set the ***Minimum, Maximum,*** *and* ***Scale*** values, as shown in Figure 5-2-33.

35. Type in ***motor_starter*** at the ***Node Name*** dialog box and then click the □ button. Because there is no node named motor_starter, it will have to be added through the ***Node Browser***, so click the ***Add*** button.

36. At the ***Add Node*** pop-up screen, click the ***Stations*** dialog box □ button. At the ***RSWho – Station*** pop-up screen, make the appropriate ***PLC*** and ***Browser Node*** selections. Click ***OK*** twice to close out successive screens.

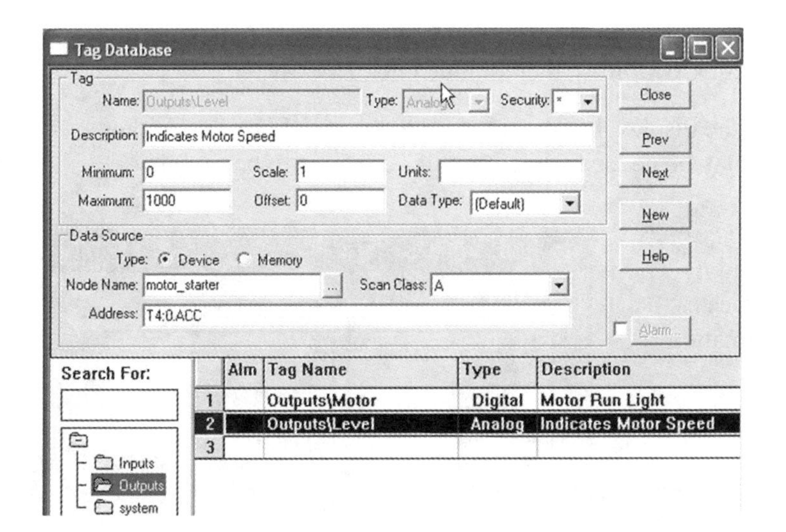

Figure 5-2-34: Tag Database 8 (Courtesy of Rockwell Automation, Inc.)

37. Next, type ***T4:0.ACC*** into the ***Address*** dialog box. Click ***Accept***. The ***Tag Database*** screen should appear as shown in Figure 5-2-34, and then click ***Close***.

5.2.7 Creating Graphic Images for the HMI

38. The next step is to create the graphics that will operate the outputs that were just tagged.

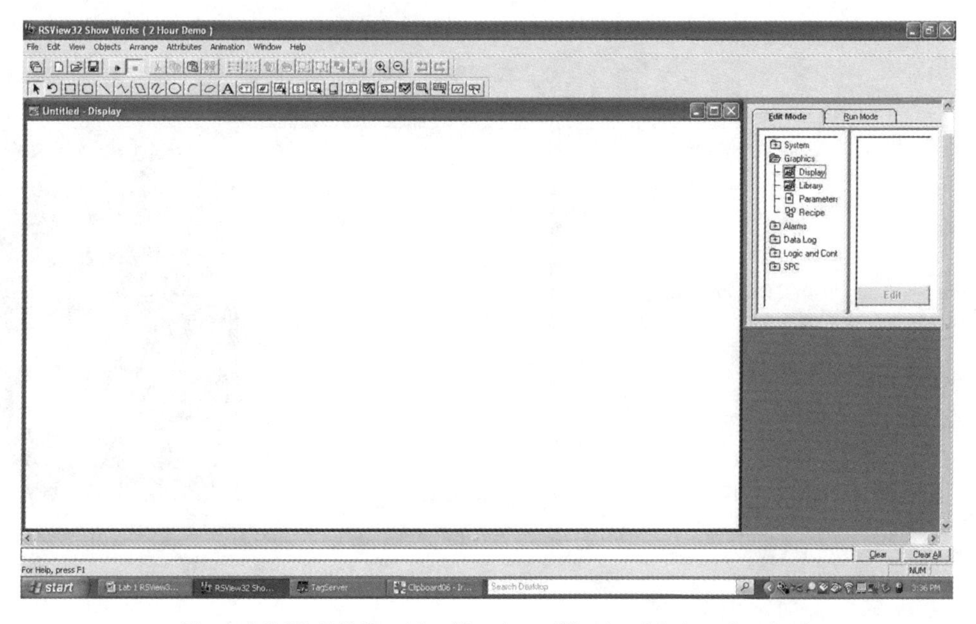

Figure 5-2-35: Edit Graphics (Courtesy of Rockwell Automation, Inc.)

a. In the ***Edit Mode*** screen (the right side of Figure 5-2-35), the **Systems** folder has a ***Graphics*** selection. Double click ***Graphics***, then double click ***Display***.

39. Resize this screen to fill most of the display, but do not overlap the small project window. Resizing of the project window may be needed while graphics continue to be set up.

40. In ***Edit Mode, System***, click ***Library*** once (only ONCE!!) in the project window (Figure 5-2-36). A list of graphic categories preinstalled to use for animation simulation purposes appears.

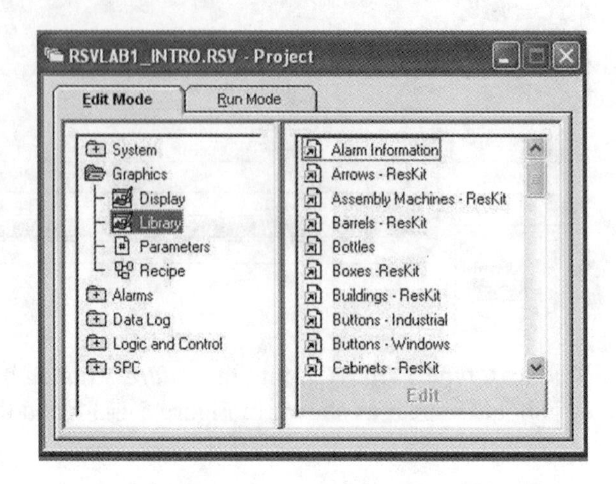

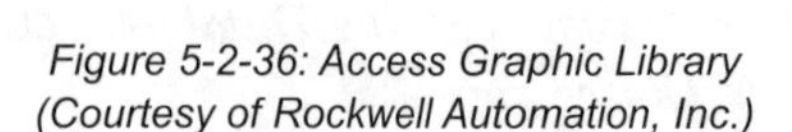
Figure 5-2-36: Access Graphic Library (Courtesy of Rockwell Automation, Inc.)

5.2.8 Creating Buttons and Motor Graphics

41. Double click the ***Buttons – Industrial*** selection. The graphics menu in Figure 5-2-37 will appear.

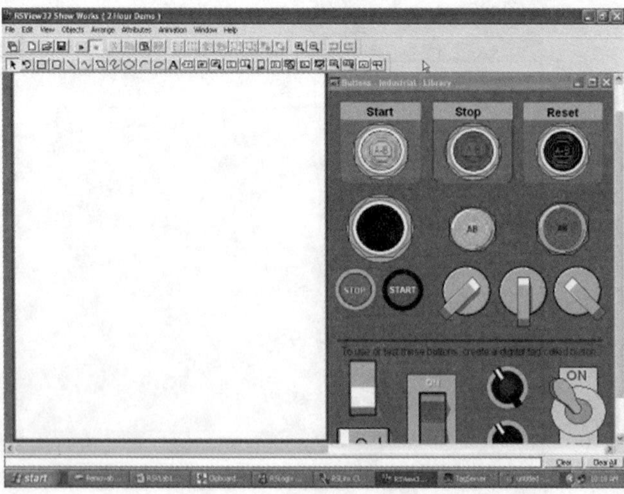

Figure 5-2-37: Industrial Graphic Icons (Courtesy of Rockwell Automation, Inc.)

42. Click, drag, and drop an Allen-Bradley ***Start*** button icon and a ***Stop*** button icon from the top row onto the ***Display*** screen. Close the ***Buttons – Library*** menu. If prompted, ***DO NOT SAVE*** changes to the library. Saving may change permanent icons in this library.

43. Similar to the previous step, open the ***Motors*** selection. Click, drag, and drop a motor icon onto the ***Display*** screen. Close the ***Motors – Library*** menu. If prompted, ***DO NOT SAVE*** changes to the library. Saving may change permanent icons in this library.

44. The ***Untitled – Display*** screen should appear similar to what is shown in Figure 5-2-38.

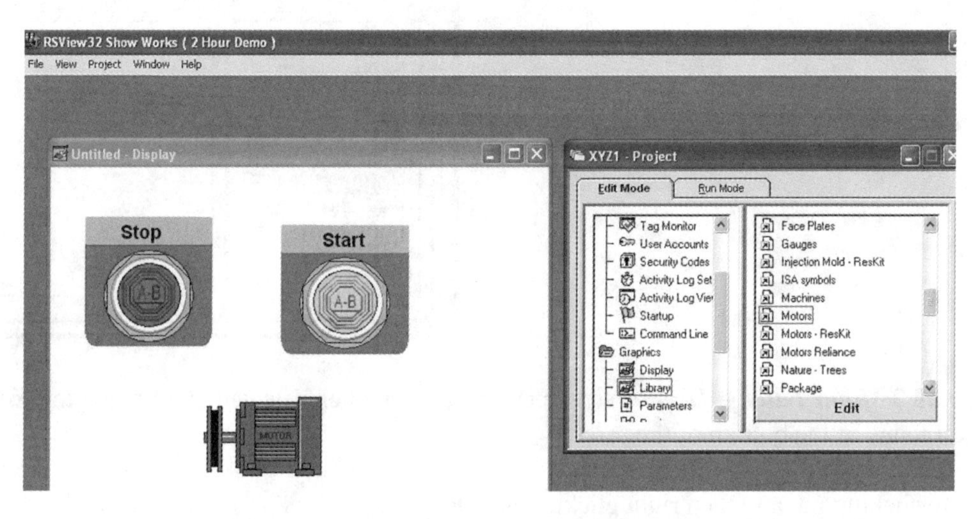

Figure 5-2-38: Three Icons Selected (Courtesy of Rockwell Automation, Inc.)

45. In the graphic in Figure 5-2-39, click the **red X** at the top right of the ***Untitled – Display*** screen to close this screen.

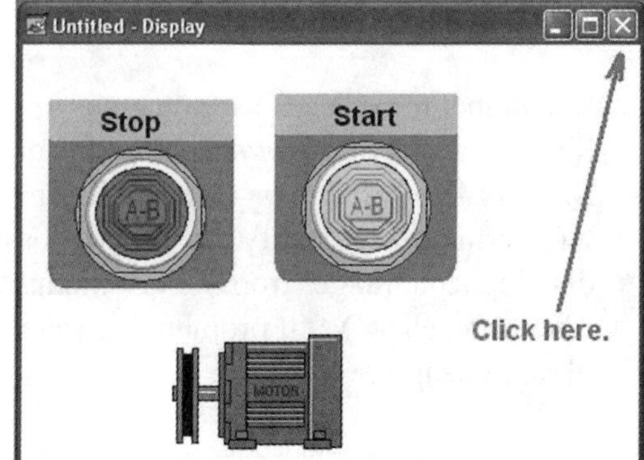

Figure 5-2-39: Magnified Screen of Figure 5-2-38 (Courtesy of Rockwell Automation, Inc.)

46. When prompted by the pop-up screen that appears, click ***Yes.*** In the **Save-As** screen**,** save the display graphic in the ***RSLinx_RSLogix500 Programs*** storage folder. Save the ***Gfx file*** with a file name like ***Lab 2 Motor Latch*** similar to the file name used to save the RSLogix 500 PLC program.

47. Close the ***Motors – Library*** screen. However, ***DO NOT SAVE CHANGES*** to this icon library if prompted to do so.

48. Once again, click the ***Display*** tab in the ***Graphics*** folder, which is in the ***Systems*** folder.

49. Considering the previous step, the graphic representation file, titled with a name like ***Lab 2 Motor Latch,*** will appear as shown in Figure 5-2-40.

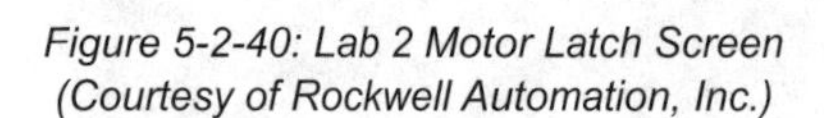

Figure 5-2-40: Lab 2 Motor Latch Screen (Courtesy of Rockwell Automation, Inc.)

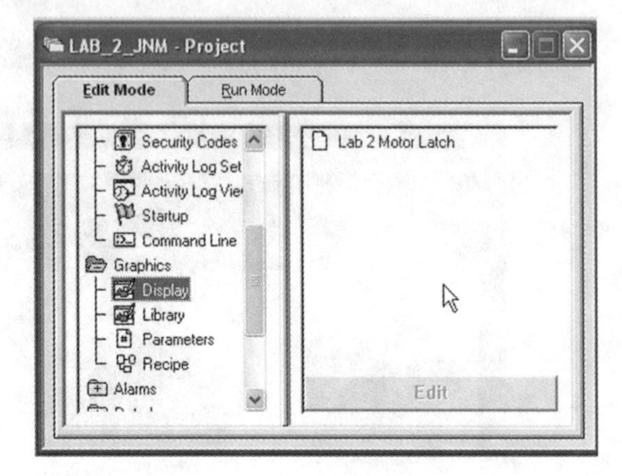

50. The ***Lab 2 Motor Latch – Display*** screen can be edited when selected by clicking the ***Edit*** button found at the bottom of the screen.

    a. By clicking on, and then right clicking, single graphic icons in that screen can be deleted, duplicated, and so forth.

51. Practice resizing and moving the buttons. For example, move the ***Stop*** button next to the ***Start*** button and try to resize them.

52. Remember to **NEVER** save changes to library screens like ***Motors – Library,*** or ***Buttons – Industrial****, or* ***Gauges***. Saving will permanently modify the library, possibly deleting icon images from library menus. Therefore, click ***NO*** if prompted to save library changes.

53. To close the screen in Figure 5-2-41, click the **red X** at the top right of the screen. To close Figure 5-40, also click its red X to save the screen.

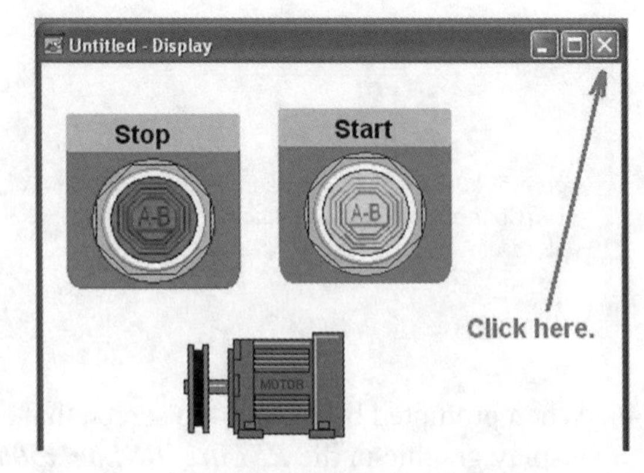

Figure 5-2-41: Lab 2 Motor Latch Display (Courtesy of Rockwell Automation, Inc.)

54. The next step is to add text and time to the ***Motor Latch – Display*** screen. Re-open the ***Motor Latch – Display*** screen with the same procedure used in Step 53. Click ***Objects*** and then ***Text***, as shown in Figure 5-2-42, from the ***RSView32*** screen.

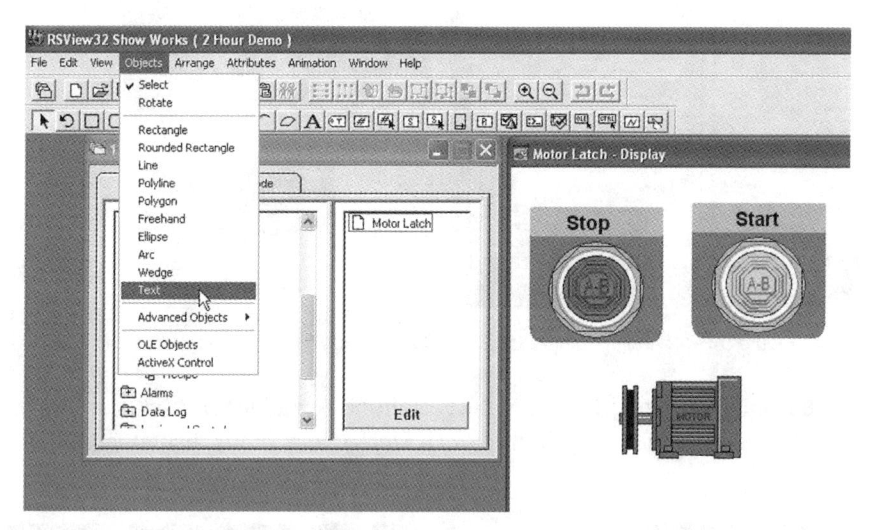

Figure 5-2-42: Adding text to Motor Latch Display (Courtesy of Rockwell Automation, Inc.)

55. Place the computer cursor somewhere in the display screen other than on the icon graphic representations in the ***Motor Latch – Display*** screen place. As shown in Figure 5-2-43, the text ***Motor Latching Circuit*** was added to the display area.

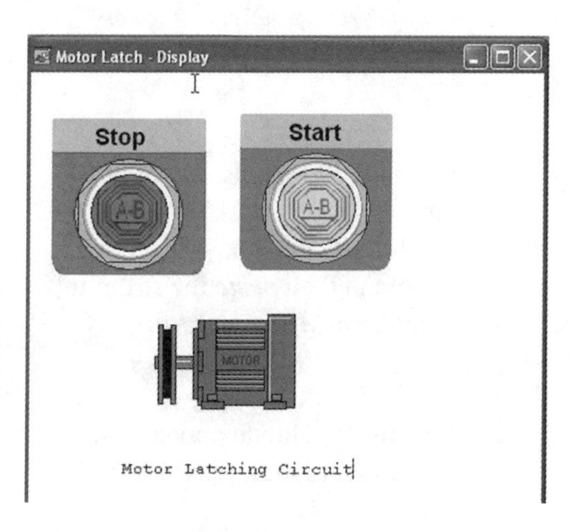

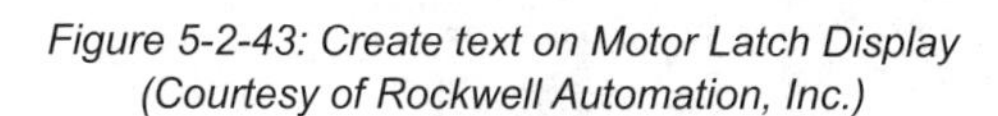

Figure 5-2-43: Create text on Motor Latch Display (Courtesy of Rockwell Automation, Inc.)

5.2.9 Creating Digital Clock Display

56. Next, place both the **date** and the **time** in the display area. Double click ***Graphics***, then single click ***Library***, and then double click ***Clocks*** from the ***Project*** screen (Figure 5-2-44).

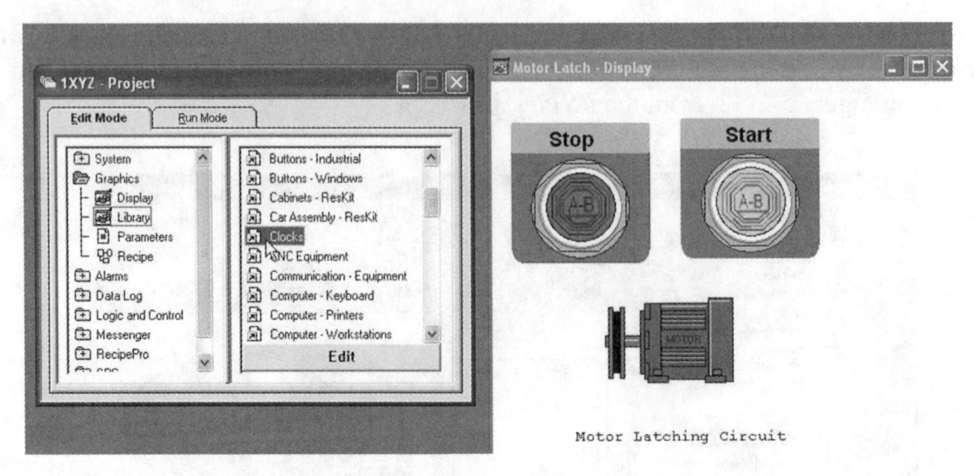

Figure 5-2-44: Adding Other Graphics (Courtesy of Rockwell Automation, Inc.)

a. From the library shown in Figure 5-2-45, right click the ***boxed "ssssssss" string***, click copy, and then paste the string in the ***Start-Stop-Motor*** icons display area.

Figure 5-2-45: Create Time string box
(Courtesy of Rockwell Automation, Inc.)

Figure 5-2-46: Create Date string box
(Courtesy of Rockwell Automation, Inc.)

b. From the ***library,*** as shown in Figure 5-2-46, left click then right click the ***4-digit date*** box string, click ***copy***, and then ***paste*** the string in the ***Start-Stop-Motor*** icons display area.

c. The display should appear similar to the representation provided in Figure 5-2-47.

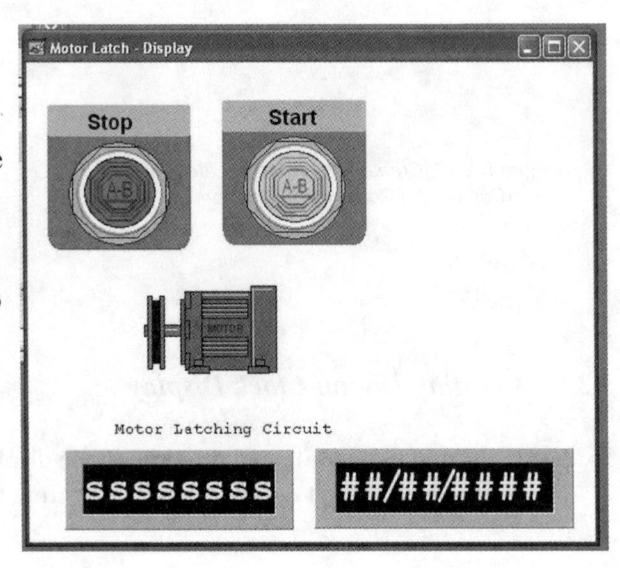

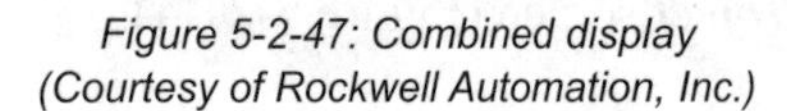

Figure 5-2-47: Combined display
(Courtesy of Rockwell Automation, Inc.)

d. Click the ***red X*** at the top right of the ***Display*** screen to save and close the screen. Remember: DO **NOT SAVE** any changes to ***Library*** screen. Therefore, click the ***red X*** click at the top of the ***Library*** display screen, but select ***NO*** if prompted to save changes to it.

5.2.10 Animating Input and Output Graphics

57. The next step is to animate display inputs and outputs. After opening up the ***Motor Latch – Display*** screen again, right click the ***Start*** button, highlight ***Animation***, and then click ***Touch***, as demonstrated in Figure 5-2-48.

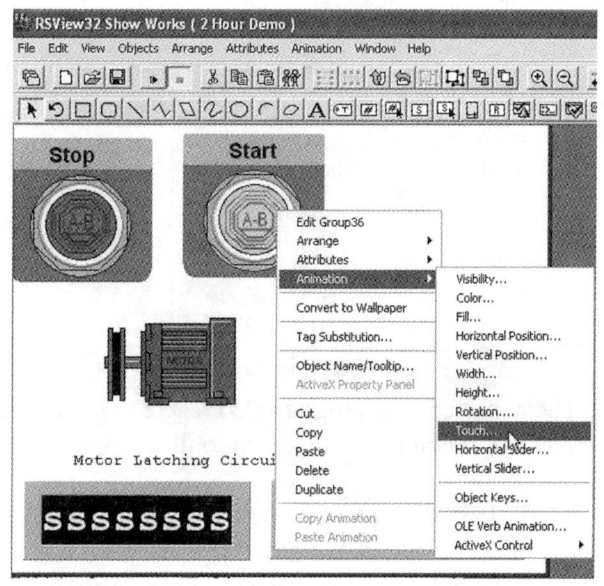

Figure 5-2-48: Adding Animation (Courtesy of Rockwell Automation, Inc.)

a. As an ***Animation*** pop-up screen appears, input information to the dialog boxes, as shown in Figure 5-2-49. Be careful to input the information exactly as it is shown. Incorrect spaces and characters will cause the animation to function improperly.

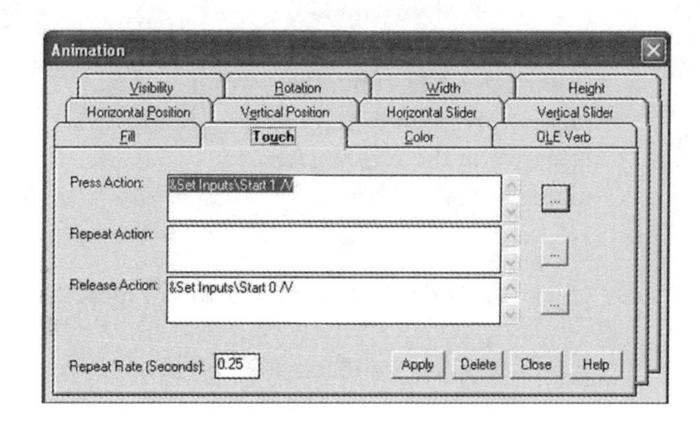

Figure 5-2-49: Add Touch Animation (Courtesy of Rockwell Automation, Inc.)

b. Click ***Apply*** and ***Close*** to close this screen. The ***Start*** button has just been programmed to go to a high state of '***1***' or ***on*** when touched (depressed) and a low state of '***0***' or ***off*** when the button is released.

58. Right click the ***Stop*** button, highlight ***Animate***, and then click ***Touch***, inputting values and characters exactly as shown in Figure 5-2-50. Click ***Apply*** and ***Close*** to close this screen. The ***Stop*** button has just been programmed to go to a low state of ***0*** or ***off*** when touched (depressed) or a high state of ***1*** or ***on*** and when the button is released.

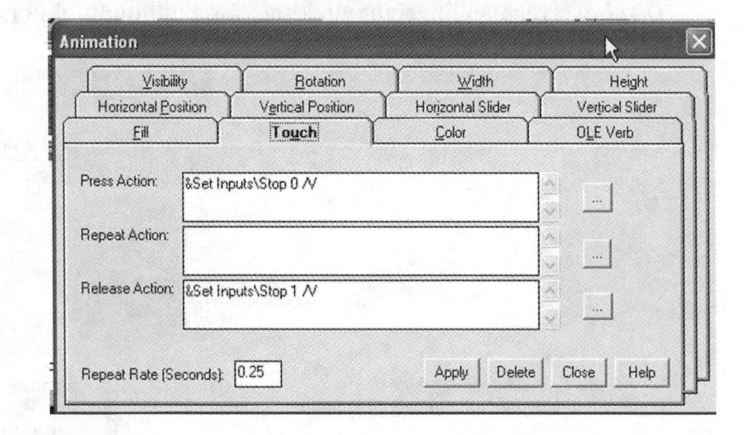

Figure 5-2-50: Completion of Animation (Courtesy of Rockwell Automation, Inc.)

59. The next step is to program ***RSView32*** to monitor inputs and outputs through color control. The use of color can add much to the animation of an input or output device.

a. To begin this process, right click the output ***Motor*** icon, highlight ***Animation***, then highlight and click ***Color***.

b. As an ***Animation pop-up screen*** appears, enter the expression as shown and change the colors as shown (Figure 5-2-51). Input the ***logic expression*** exactly as it is shown in the ***Expression*** dialog box.

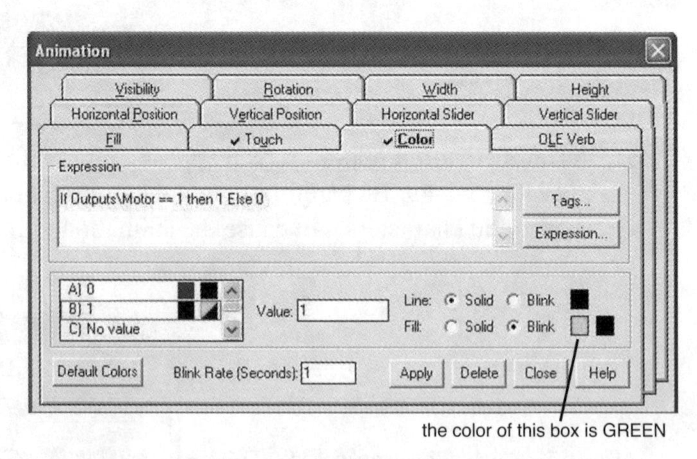

Figure 5-2-51: Add Color (Courtesy of Rockwell Automation, Inc.)

c. Click the ***A) 0*** selection as shown in Figure 5-2-51.

d. Next, left click the black box on the opposite side of the ***Animation*** screen to access the

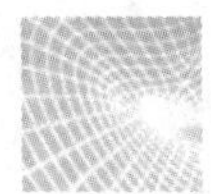

hidden color palette (Figure 5-2-52). Select a **color other than black** from the palette to determine how ***A) 0*** is affected.

e. Note too that if the ***Blink*** dialog were to be selected rather than the ***Solid*** dialog, the output (display screen motor icon) would blink black.

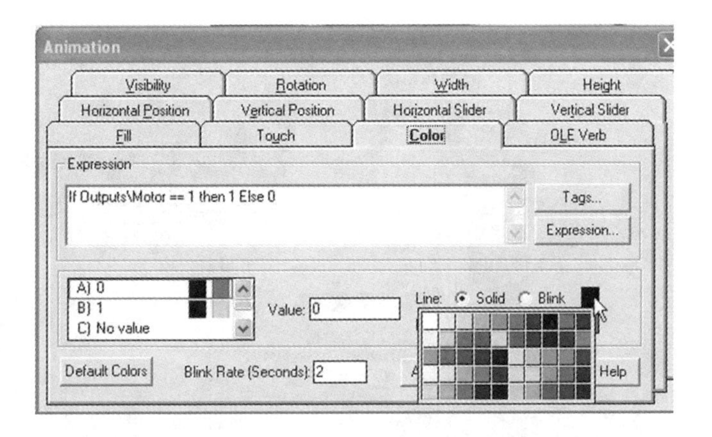

Figure 5-2-52: Select Color (Courtesy of Rockwell Automation, Inc.)

f. Now click ***B) 1***. Use the color pallet on the right side of the screen to ***Blink*** colors. Highlight the ***Green*** and ***Black*** colors as indicated in Figure 5-2-53. This means the ***motor icon*** will blink the colors of ***Green to Black*** when activated.

g. Click ***Apply*** and ***Close*** to save this color animation.

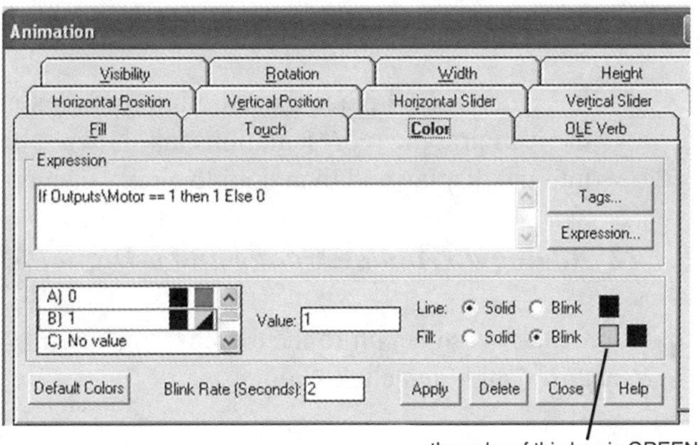

Figure 5-2-53: Color Animation (Courtesy of Rockwell Automation, Inc.)

5.2.11 Adding Visibility and Invisibility to Animation

60. The next step is to add ***invisibility*** to the animation. Right click the ***Motor*** icon in Figure 5-2-54 and highlight ***Animation*** in its task bar. On the drop down window, highlight and click ***Visibility***. Enter the tag information as shown in the ***Expressions*** dialog box in Figure 5-2-53.

   a. This animation choice will make the ***Motor invisible*** when it is stopped. Pressing the ***Start*** button in Figure 5-2-54 will make the motor ***visible*** and ***Color*** will blink the motor from ***Black to Green*** repeatedly. Click ***Apply*** and ***Close*** to close Figure 5-2-53.

61. Click ***Objects*** from Figure 5-2-54 task bar and then select ***Text*** near the bottom of the drop down window, as shown in Figure 5-2-54.

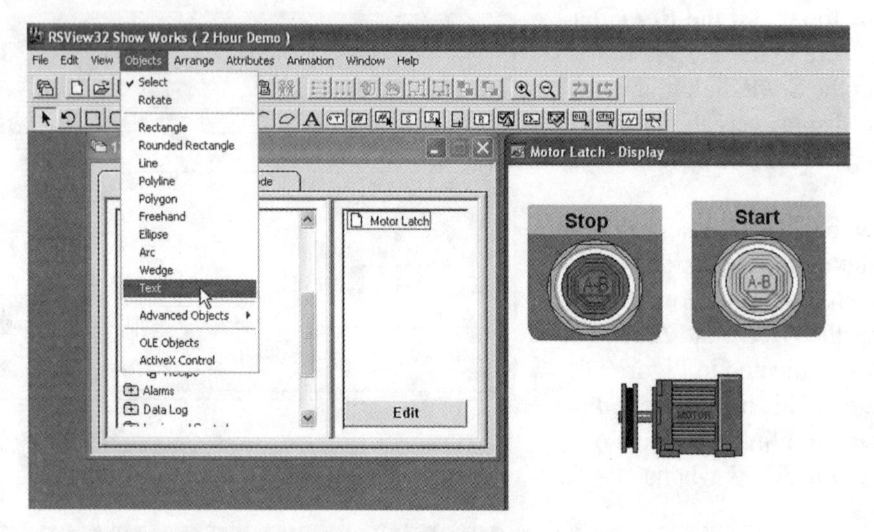

Figure 5-2-54: Adding Objects (Courtesy of Rockwell Automation, Inc.)

a. In Figure 5-2-54 place the computer cursor somewhere in the display screen other than on the icon graphic representations, as shown. Next, type some ***Text*** such as ***Motor Latch*** into the display and then double click.

5.2.12 Adding an Analog Gauge and a Digital Meter Graphic

62. Next, add a ***gauge icon*** to the display, as shown in Figure 5-2-55. Open ***Library*** beneath ***Graphics***, and then click ***Gauges***.

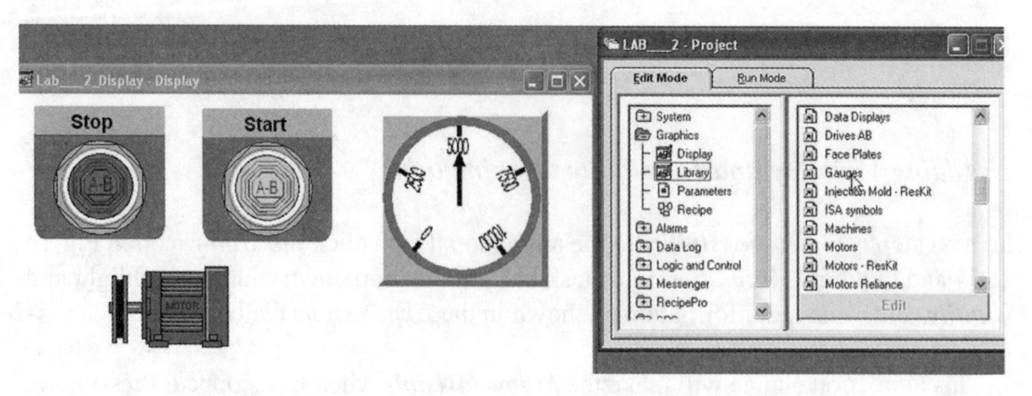

Figure 5-2-55: Adding Gauges (Courtesy of Rockwell Automation, Inc.)

63. Next, add a digital display from Figure 5-2-56 ***Shapes and Borders*** located in the ***Graphics Library***.

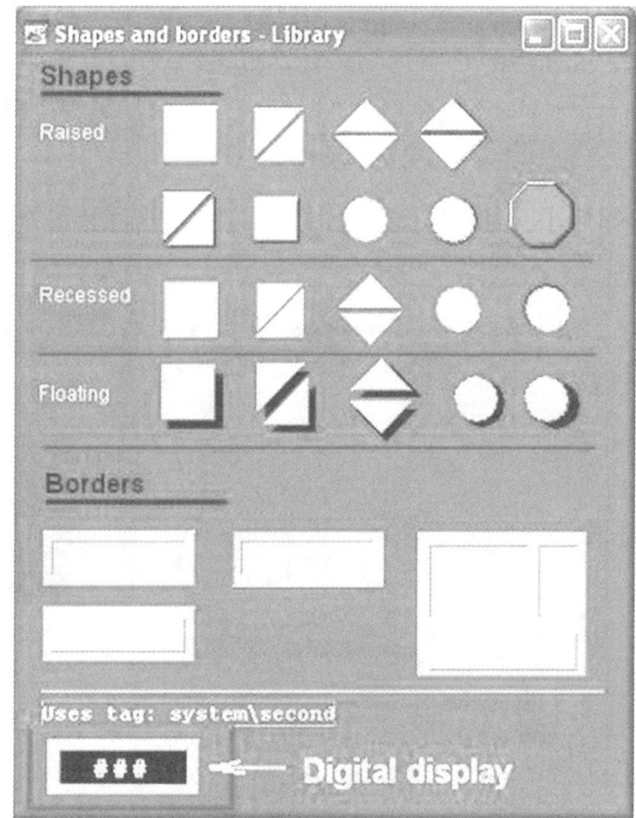

Figure 5-2-56: Shapes and Borders Icon Library (Courtesy of Rockwell Automation, Inc.)

64. After placing the digital display in the screen, click ***Edit,*** as shown in Figure 5-2-57, to bring up the menu bar at the top of the ***RSView32 Show Works*** page.

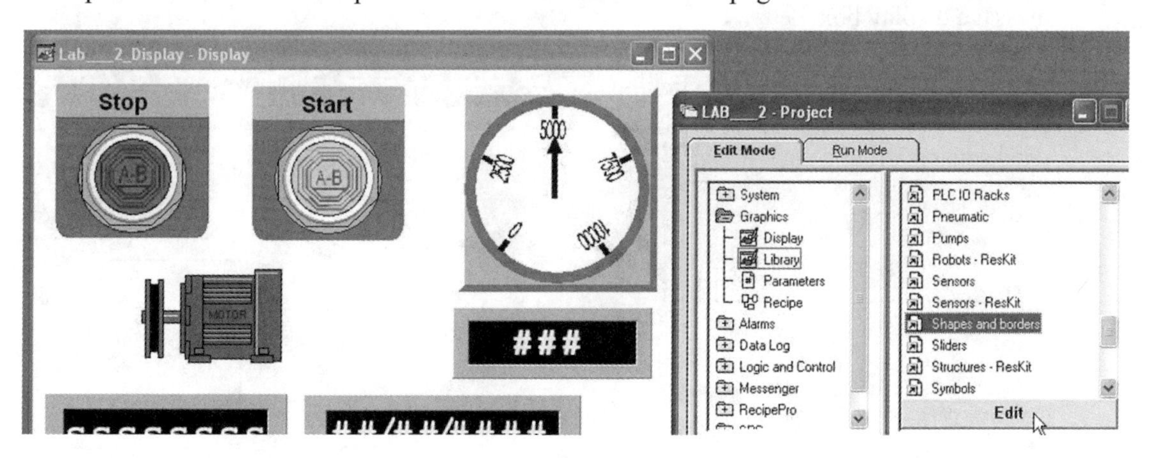

Figure 5-2-57: Adding Digital Values (Courtesy of Rockwell Automation, Inc.)

5.2.13 Adding a Warning Message Display with Color

65. Now add a **warning message** to the display. To do this, click the ***rectangle*** icon from the top menu bar, as shown in Figure 5-2-58.

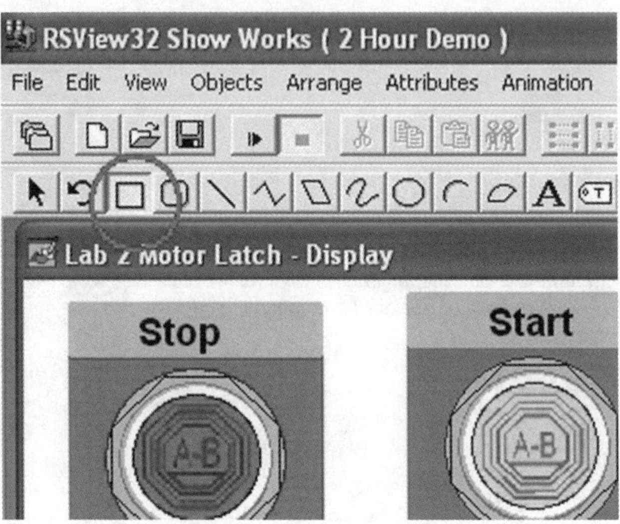

Figure 5-2-58: Add Warning Message (Courtesy of Rockwell Automation, Inc.)

a. Place the mouse cursor in an open area on the display. Then click to draw a box with the cursor, as shown in Figure 5-2-59.

b. Repeat this process to add another message display box.

c. These boxes can be resized by clicking them and then manipulating the blackened squares around each image.

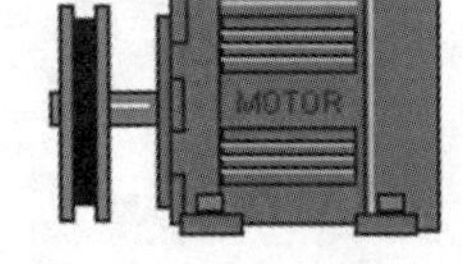

Figure 5-2-59: Add Message Box (Courtesy of Rockwell Automation, Inc.)

66. Right click just inside the bottom edge of one of the boxes to animate them. To animate one of the boxes, right click ***Animation***, and then ***Visibility***. Click the ***Tags*** button next to the ***Expression*** dialog box from the ***Animation*** pop-up screen. Click ***Outputs*** and the **Outputs\ Motor Tags**, as shown in Figure 5-2-60, and then click ***OK***. Click ***Apply*** and ***Close*** at the ***Animation*** pop-up screen to close it.

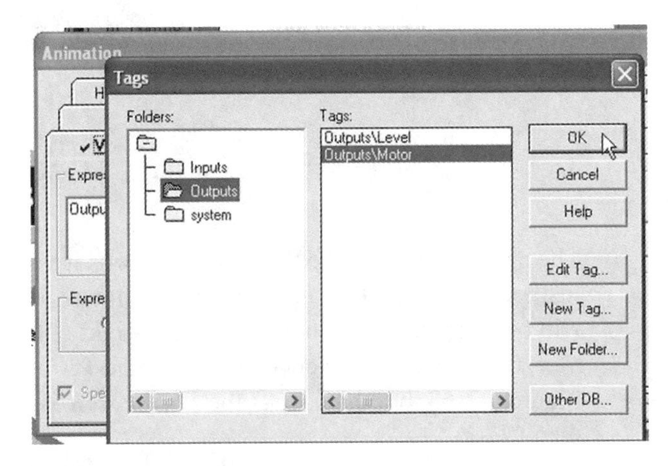

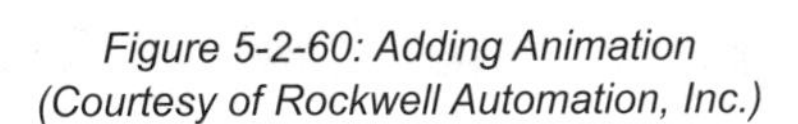
Figure 5-2-60: Adding Animation
(Courtesy of Rockwell Automation, Inc.)

67. Right click just inside the bottom edge of the same box again. A series of windows will open in the box. Click ***Animation*** and then ***Color***. Make the animation actuate with the colors, as shown in Figure 5-2-61. Click ***Apply*** and ***Close*** at the *Animation* pop-up screen to close it.

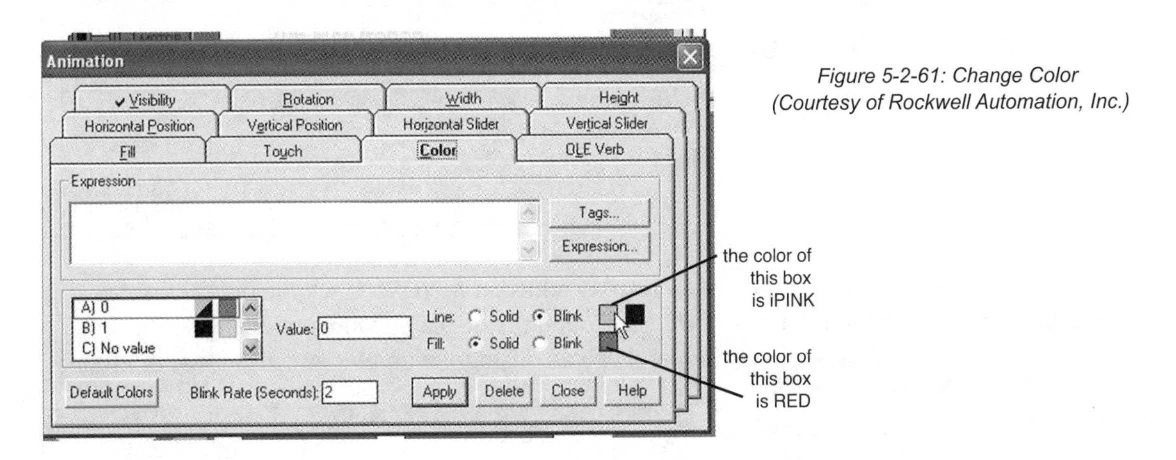

Figure 5-2-61: Change Color
(Courtesy of Rockwell Automation, Inc.)

68. Repeat Steps ***65 and 66*** for the **second warning box**.

69. The next step will be to create warning messages for these warning boxes. To do this, click the capital ***A*** text box icon from the top menu bar, as shown in Figure 5-2-62.

a. Place the cursor within one of the boxes that was just created. Type the text words ***Motor Running*** in one of the boxes; then right click. Select ***Attributes,*** then ***Font.*** Then notice the screen in Figure 5-2-63. Change the ***Font*** to ***Arial*** and make the ***Color*** of the font ***Yellow***. Double click to end this command.

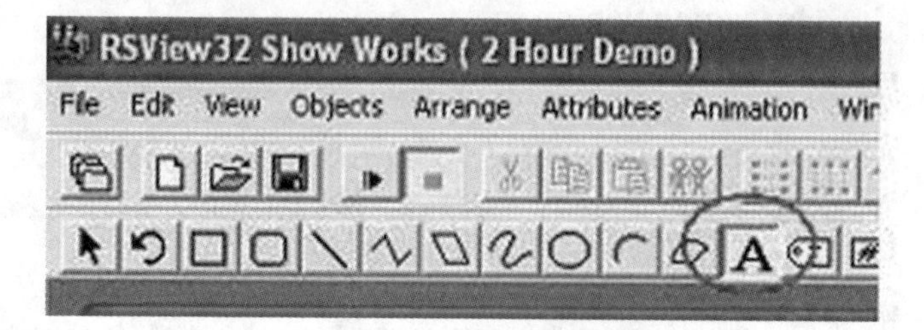

Figure 5-2-62: Create Warning Message (Courtesy of Rockwell Automation, Inc.)

b. The next step is to make the warning words invisible for the two warning boxes that were just created. To do this, click the words in one of the warning boxes. A box of pink dots will become visible around the words. Right click the box of pink dots. Select ***Animation***, if it is not already check marked, and ***Visibility***. Tag the ***Expression*** box as shown in Figure 5-2-63, but then select the ***Expression True State*** to be ***Invisible***. Click ***Apply*** and then ***Close***.

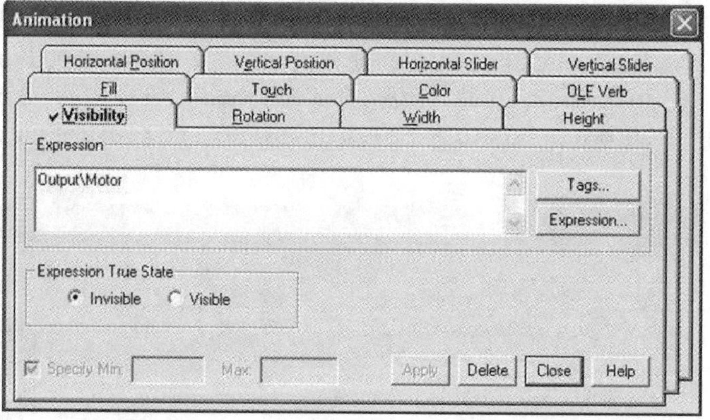

Figure 5-2-63: Adding Visibility Animation (Courtesy of Rockwell Automation, Inc.)

70. The next step is to configure the digital display added at ***Step 38***. Click the digital display number symbols, ###, until a box of ***dots*** surround the display. Double click inside this box until the ***Numeric Display pop-up screen*** appears. Using the graphic image shown in Figure 5-2-64, click the ***Tags*** button to input the proper expression. Enter the rest of the settings into the screen, as shown in the figure, and then click ***OK.***

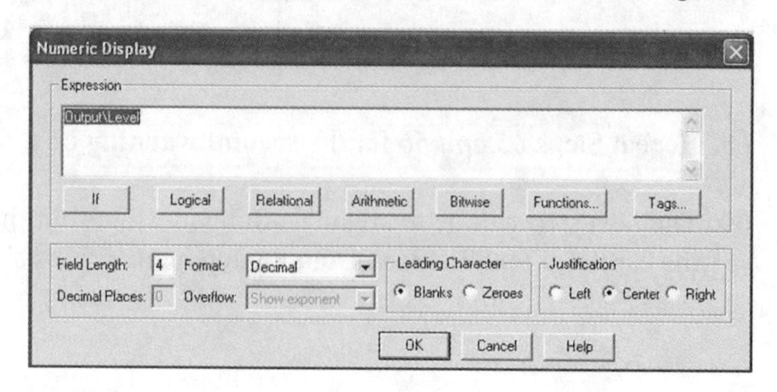

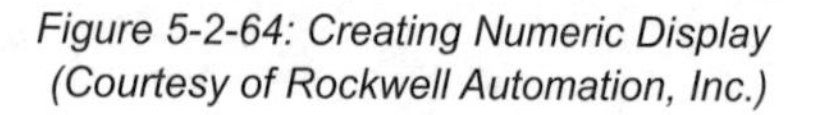

Figure 5-2-64: Creating Numeric Display (Courtesy of Rockwell Automation, Inc.)

71. The final step prior to running the animation is to configure the gauge display added at ***Step 37***.

    a. Double click the gauge graphic until a box of dots is surrounding it.

    b. Select ***Animation*** and then ***Rotation***. Tag the expression properly with the appropriate settings, as shown in Figure 5-2-65. Close the ***Motor Latch – Display*** screen to save the latest programming changes.

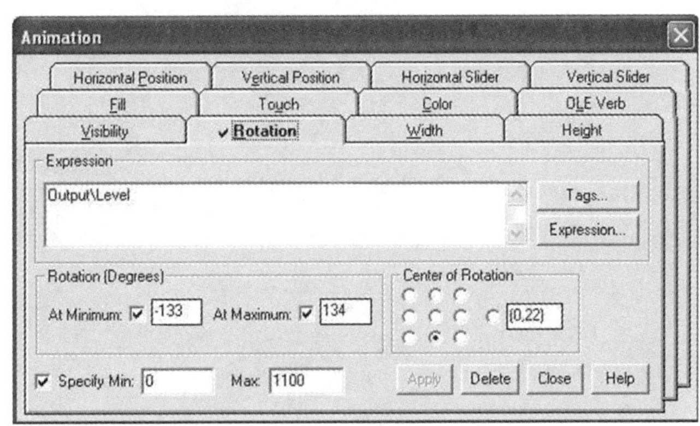

Figure 5-2-65: Add Rotation (Courtesy of Rockwell Automation, Inc.)

72. Pull up the ***RSLogix 500*** ladder program. To ensure that the RSView animation is reading the latest ladder diagram, go ***Offline*** with the ladder diagram program. Next, download the ladder diagram again into RSLogix 500. Do this by clicking ***Comms***, then ***Download***, and at successive screens, click ***Yes*** until the screen displays "go back online".

5.2.14 Running the Project

73. Open the Lab 2 file in ***RSView***. Click ***Graphics*** and ***Display***, and the Lab 2 selection to the pop-up screen's right side.

    a. With the ***Run Mode*** tab selected, click the ***Start*** button at the bottom of the pop-up screen.

    b. Depress the green start button on the ***Motor Latch – Display*** to run the simulation.

74. This completes this lab.

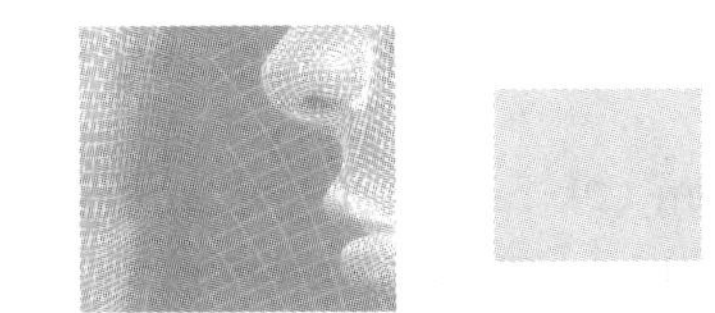

RSView Lab 5-3: Pump and Tank

5.3.1 Introduction

A common process in the chemical industry and other industries involves the use of tanks that contain liquids and use motor propelled pumps. This laboratory will incorporate knowledge gained from the previous two laboratories to simulate and control the operation of a motorized-pump, tank-filling operation with level alarms.

5.3.2 Creating the PLC Program and Configuring RSView

1. Enter the PLC program into the ***MicroLogix 1200 PLC***, as shown in Figure 5-3-1. Modifications may be made (using the same ***Node*** labeling for ***Tag Database*** Inputs/Outputs tags). Use the saved program from Laboratory 5-2 to program this ladder diagram.

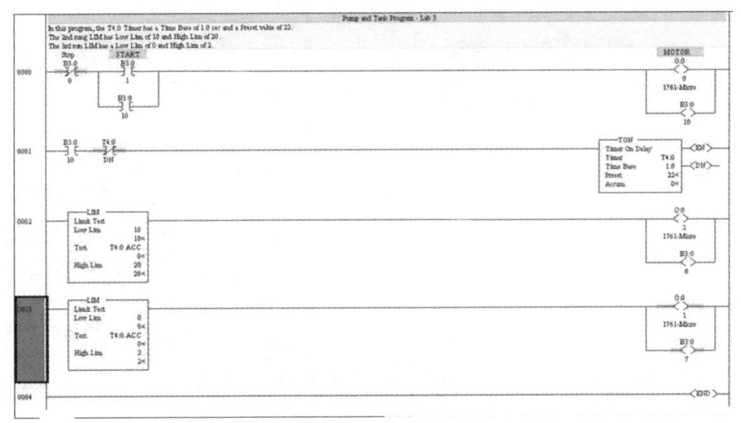

Figure 5-3-1: PLC program for Lab 5-3 based on Lab 5-2 PLC program (Courtesy of Rockwell Automation, Inc.)

2. Download this ladder into the PLC and be sure to place the program into the ***RUN*** mode. Also, be sure the ***B3:0/0*** Stop Button Bit is toggled properly to the high state in ***RSLogix 500*** when stopping and starting the simulation.

3. Next, make sure the ***RSView*** HMI monitoring and control software is configured properly, similar to previous labs. Refer to Laboratory 5-2, or modify the settings and use, regarding ***Channel, Node, Scan,*** *and* ***Tag Database*** settings.

5.3.3 *Creating Existing Tags and Two New Tags*

4. In addition to the tags from Lab 5-2, two new tags will need to be added to the database: a ***HighLevel*** indicator tag and a ***LowLevel*** indicator tag, as shown in Figures 5-3-2 and 5-3-3. Notice that there is no space between these words in the ***Name:*** *dialog box.*

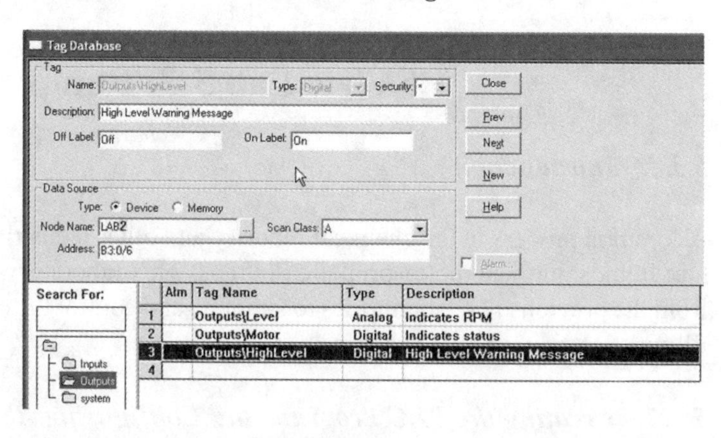

Figure 5-3-2: Enter High Level tags in the Tags Database (Courtesy of Rockwell Automation, Inc.)

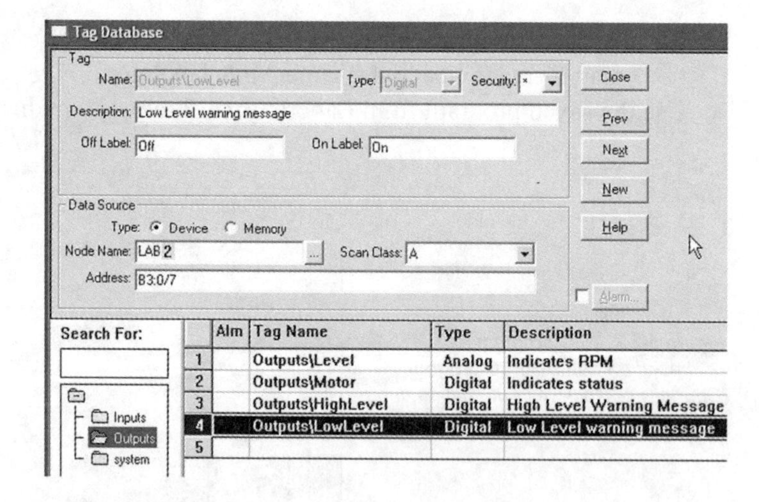

Figure 5-3-3: Enter Low Level tags in the Tags Database (Courtesy of Rockwell Automation, Inc.)

5. Be sure to click ***Accept***, then ***Close*** each new tag.

5.3.4 *Creating the Pump and Tank Graphics*

6. The next step is to add a tank and pump to the existing HMI program from Lab 5-2. The ***Pump and Tank HMI graphic*** screen should look like the screen shown in Figure 5-3-5.

7. To add the tank, click ***Graphics,*** then ***Library*** (Figure 5-3-4). The tank is located in the ***Tanks*** library.

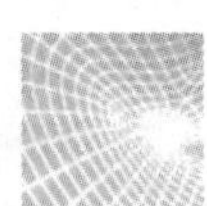

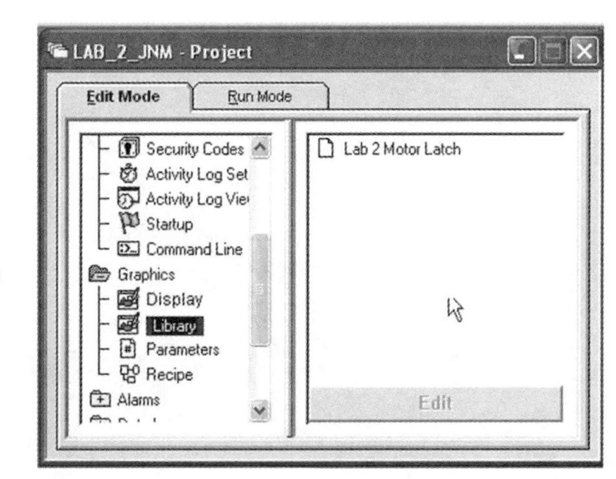

Figure 5-3-4: Select Library to create Pump and Tanks HMI icons (Courtesy of Rockwell Automation, Inc.)

8. The piping is in the ***Pipe3*** library menu (don't forget the flanges!). The small pump can be added from the ***Pumps*** library menu.

9. Save the screen as ***Lab_3*** (Figure 5-3-5).

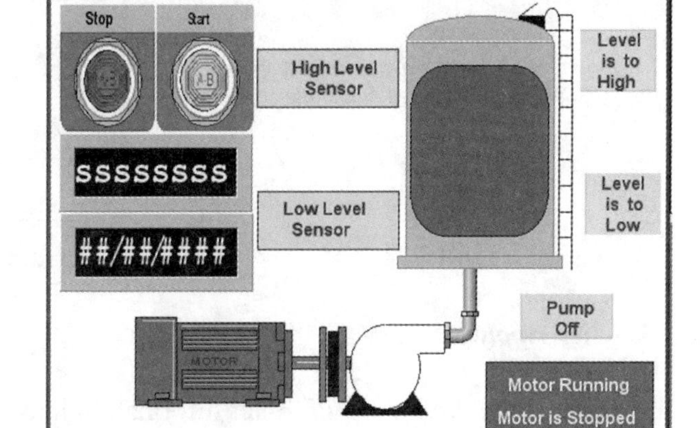

Figure 5-3-5: Pump and Tank HMI (Courtesy of Rockwell Automation, Inc.)

10. Be aware that the Lab 2 ***Node*** is attached to the Lab 2 ***Tag Database***.

5.3.5 Animating the Pump and Tank and Displays

11. Now that the tank with pump graphics has been assembled on the display screen, the next step is to animate components. Place the mouse pointer in the middle of the tank and click until a ***box with dots*** opens, surrounding the tank cutaway, as shown in Figure 5-3-6.

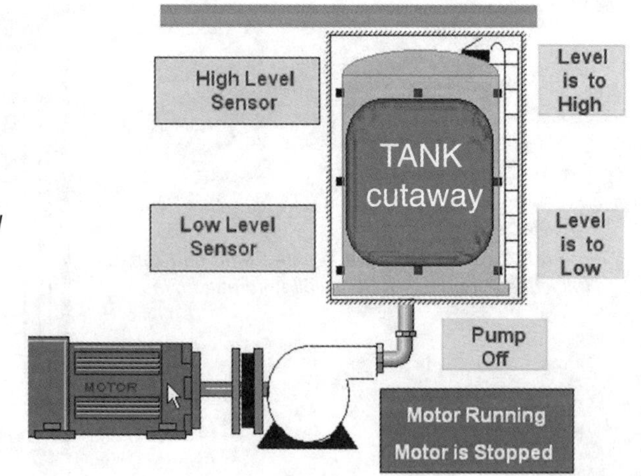

Figure 5-3-6: Adding Animation to Pump and Tank HMI (Courtesy of Rockwell Automation, Inc.)

12. For the tank, right click in the box surrounded by dots; then click ***Animation*** and ***Color***. Enter information, as shown in Figure 5-3-7, into the ***Animation*** pop-up screen.

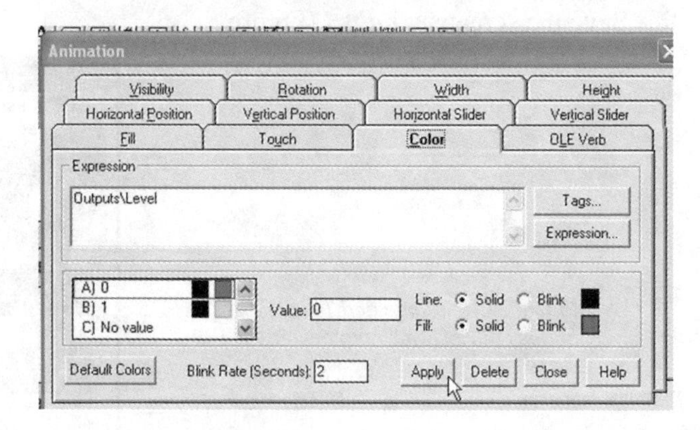

Figure 5-3-7: Enter Animation Data Values (Courtesy of Rockwell Automation, Inc.)

13. Click ***Apply.***

14. Click ***Animation*** *and* ***Fill***. Enter information into the ***Animation*** pop-up screen, as shown in Figure 5-3-8.

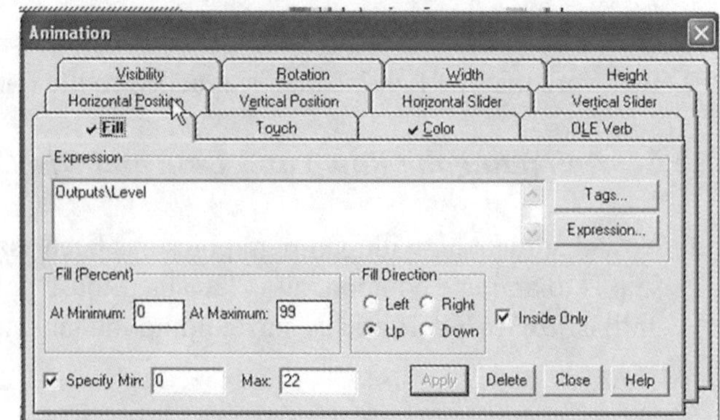

Figure 5-3-8: Animation of Filling using Color (Courtesy of Rockwell Automation, Inc.)

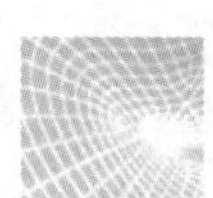

15. Click ***Apply** and **Close. Save the Lab_3 – Display*** screen and then re-open it as necessary.

16. Test the tank filling process using the ***Test Run*** mode. The display screen should appear similar to the graphic representations shown in Figures 5-3-9 and 5-3-10, with the tank filling up and emptying.

17. Figure 5-3-9 shows the process starting and filling.

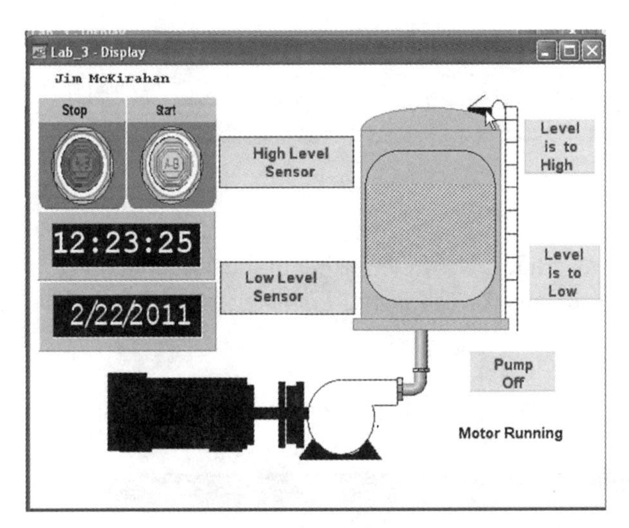

Figure 5-3-9: Final HMI for Starting, Running, and Filling Mode (Courtesy of Rockwell Automation, Inc.)

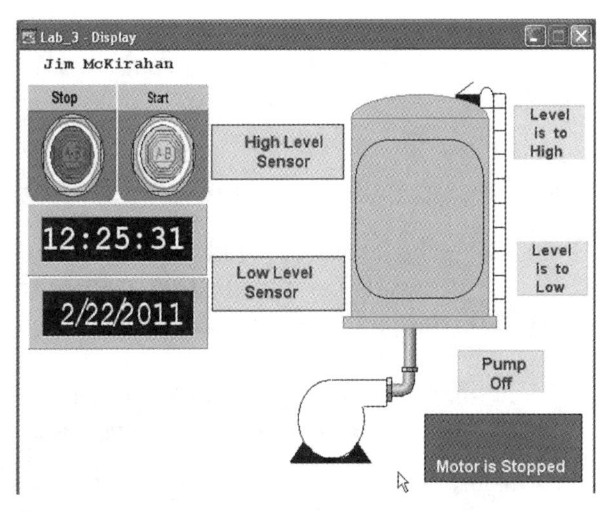

Figure 5-3-10: HMI stopped filling mode (Courtesy of Rockwell Automation, Inc.)

18. Figure 5-3-10 shows after the process has stopped from the filling mode.

5.3.6 Creating Warning Messages and Display Features

19. Looking at the screens in Figures 5-3-11 and 5-3-12, recreate Steps 6 and 7 for ***Color** and **Visibility***, but this time for the ***Level is too High*** warning message.

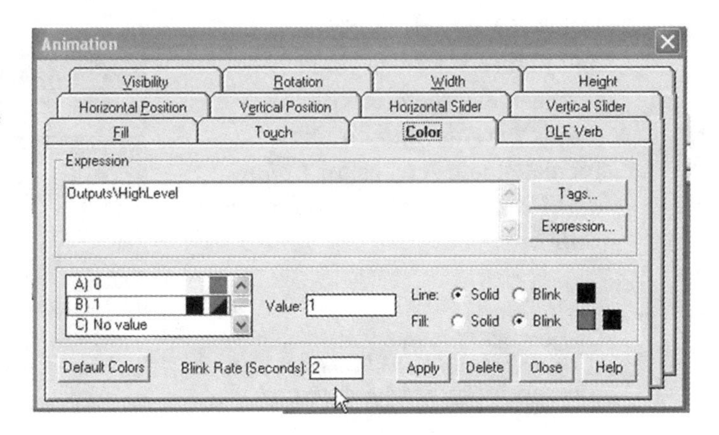

Figure 5-3-11: Add Color for High Level warning (Courtesy of Rockwell Automation, Inc.)

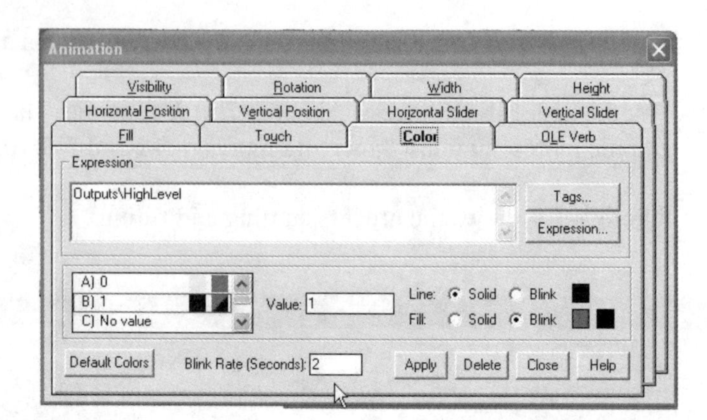

Figure 5-3-12: Add Visibility and Color for High Level warning (Courtesy of Rockwell Automation, Inc.)

20. Using the screens in Figures 5-3-11 and 5-3-12 as guides, click just inside of the ***box*** containing the words ***Level is too High.*** Recreate these screens for ***Color and Visibility***, but for the ***box*** containing the words ***Level is too High.***

21. Using the screens in Figures 5-3-13 and 5-3-14, recreate Step 9 for ***Color and Visibility***, but for the ***Level is too Low*** warning message.

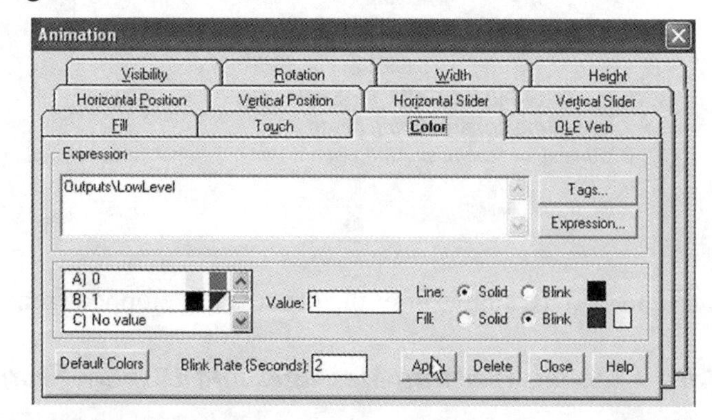

Figure 5-3-13: Adding Color for Low Level warning (Courtesy of Rockwell Automation, Inc.)

22. Using the screens in Figure 5-3-13 and Figure 5-3-14 as guides, click just inside of the ***box*** with the words ***Level is too Low.*** Now recreate these screens for ***Color and Visibility***, for the ***Level is too Low box.***

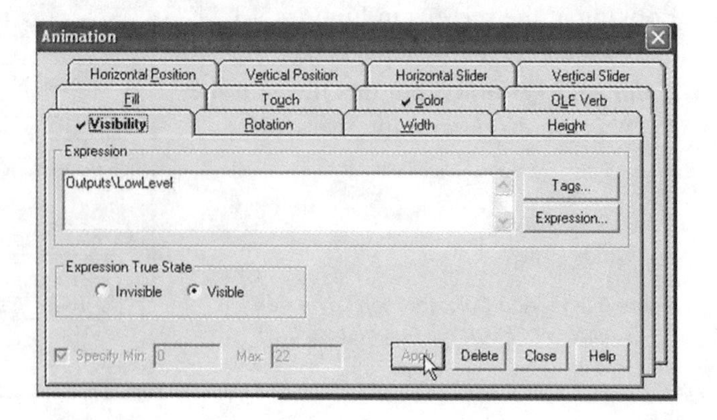

Figure 5-3-14: Adding Visibility and Color for Low Level warning (Courtesy of Rockwell Automation, Inc.)

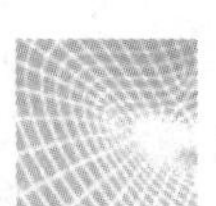

a. As a result of Steps 9 and 10, both the words and the warning words will appear and not appear, in unison, relative to the filling of the tank.

5.3.7 Creating Pump Animation and Display Messages

23. With the warning labels animated, a warning message that the pump is not on (i.e., off) needs to be created. Click just inside of the ***box*** containing the warning words ***Pump Off***.

24. Click just inside the bottom edge of the box to get a ring of dots. Right click inside the ring of dots, click ***Animation***, and then click ***Visibility***. Enter the information shown in Figure 5-3-15.

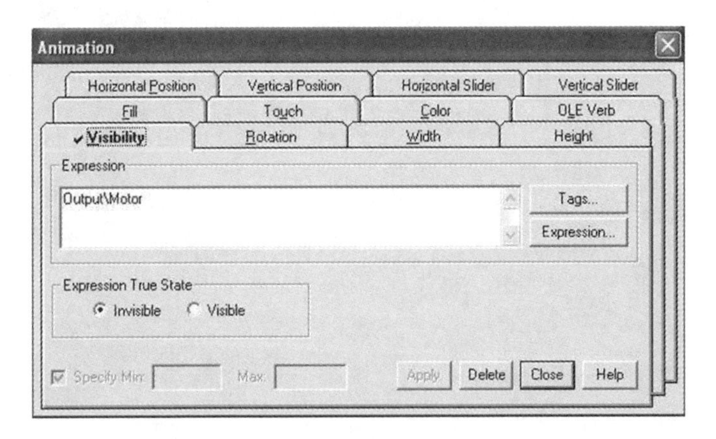

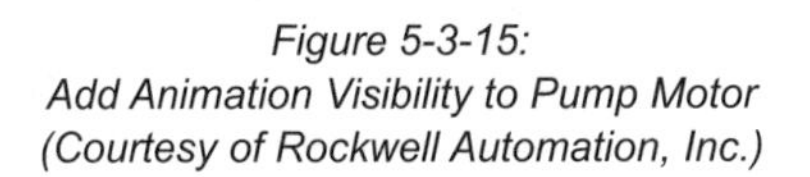

Figure 5-3-15: Add Animation Visibility to Pump Motor (Courtesy of Rockwell Automation, Inc.)

25. Click the words **Pump Off** to get a box of dots surrounding the words. Click **Animation**, then click **Visibility.** Again enter the information shown in Figure 5-3-15.

26. The words **Pump Off** have now been programmed as well as the box containing the word **invisible**, shown when the motor is running.

27. Test the tank filling process using the ***Test Run*** mode.

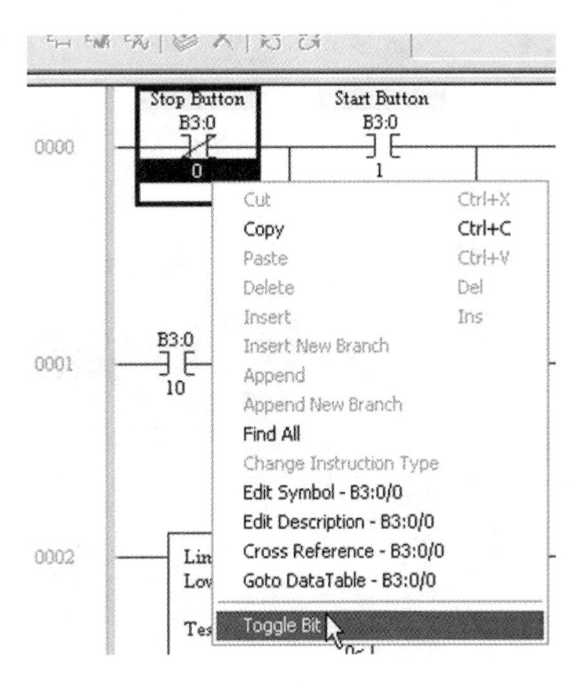

Figure 5-3-16: Example Test Run (Courtesy of Rockwell Automation, Inc.)

28. Be sure to toggle the **B3:0/0** Stop Button Bit to the high state in **RSLogix 500** when stopping and starting the PLC program, as shown in Figure 5-3-16.

29. This completes this lab.

RSView Lab 5-4: Alarms

5.4.1 Introduction

Use Labs 5-1, 5-2, and 5-3 as guides for this laboratory when creating the HMI display screen to include alarms.

In Allen-Bradley's RSView software, an alarm is a special screen that identifies a crucial event or events that need to be displayed and logged. In addition, the criticality of the alarmed event and the day and time of occurrence are recorded.

To begin this laboratory, enter the PLC diagram, as shown in Figure 5-4-1 in RSLogix 500 software. This PLC program simulates the generation of varying temperatures that can go up or down. When the temperature exceeds 700°, a light activates, indicating a high temperature condition has been equaled or exceeded. A low temperature condition is also displayed.

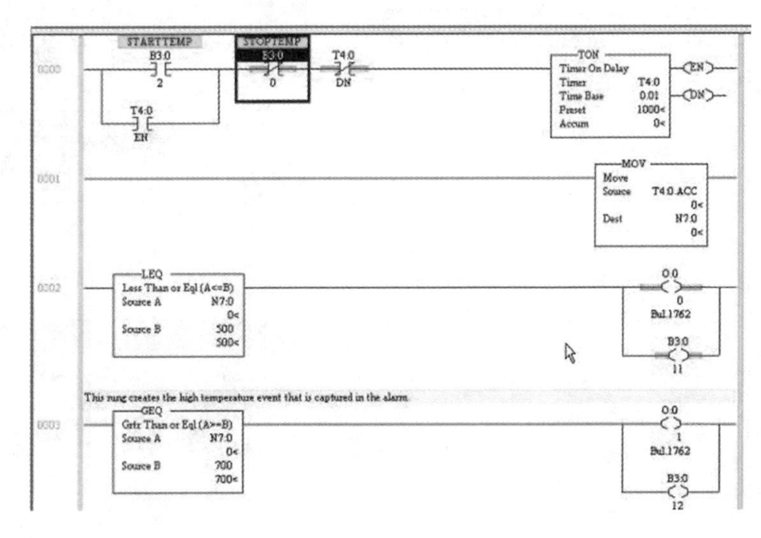

Figure 5-4-1:
PLC program for Lab 5-4
(Courtesy of Rockwell Automation, Inc.)

5.4.2 Configuring RSLogix, RSLinx, and RSView

1. Program the ladder diagram from Figure 5-4-1 into ***RSLogix***. ***Save*** the program and then download it into the PLC. Place it in ***RUN*** mode, and then toggle the ***B3:0/2*** contact between high and low states to ensure the ladder program operates correctly.

2. Make sure the ***RSLinx*** communication software is configured properly. Refer to any of the three prior laboratories for proper configurations settings and properties.

3. Using ***RSView*** ensures ***Channel**, **Node, and Scan Class*** settings are configured similarly to settings used in any of the three prior laboratories.

5.4.3 Creating Tags in the Tags Database

4. Use ***RSView*** to create the following ***Input** and **Output*** tags for the ***Tag Database*** (Figures 5-4-2 through 5-4-8) .

5. Remember to click ***Accept*** for each tag; then click ***Close*** after creating all tags. Note there are more Inputs and Outputs than will be used in this activity.

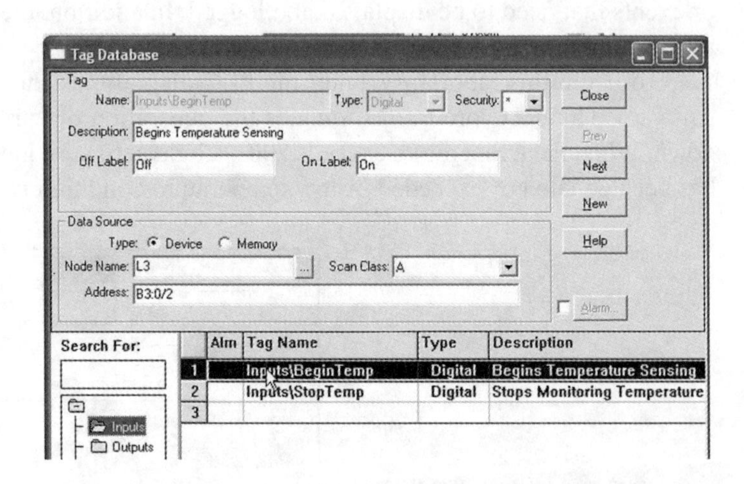

Figure 5-4-2: Creating Input tags in the Tag Database (Courtesy of Rockwell Automation, Inc.)

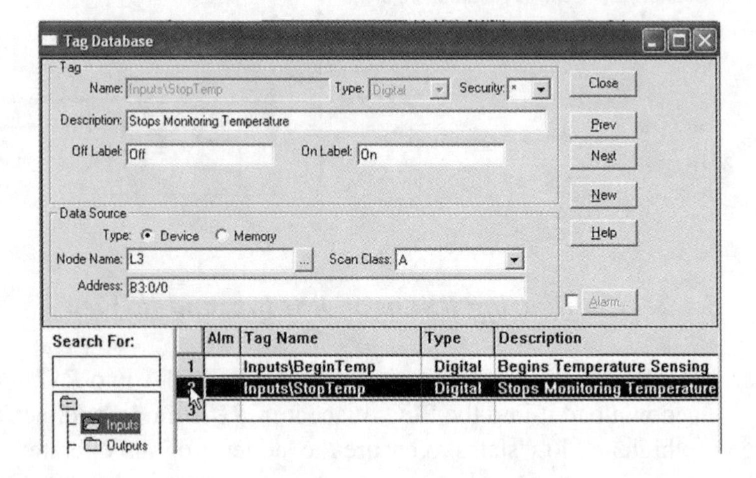

Figure 5-4-3: Continue creating Input tags in the Tag Database (Courtesy of Rockwell Automation, Inc.)

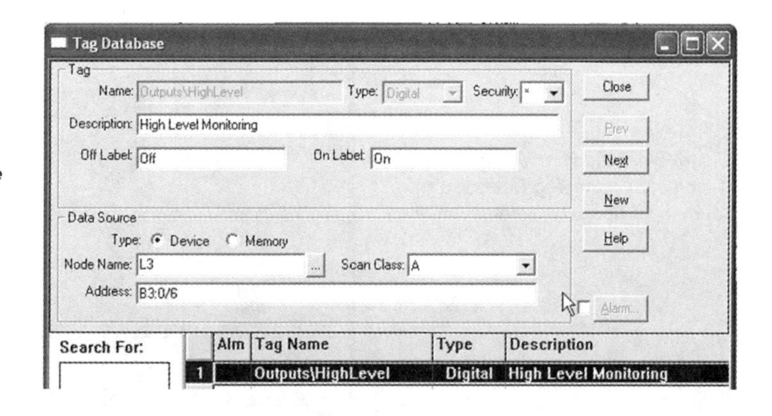

Figure 5-4-4: Creating Output tags in the Tag Database (Courtesy of Rockwell Automation, Inc.)

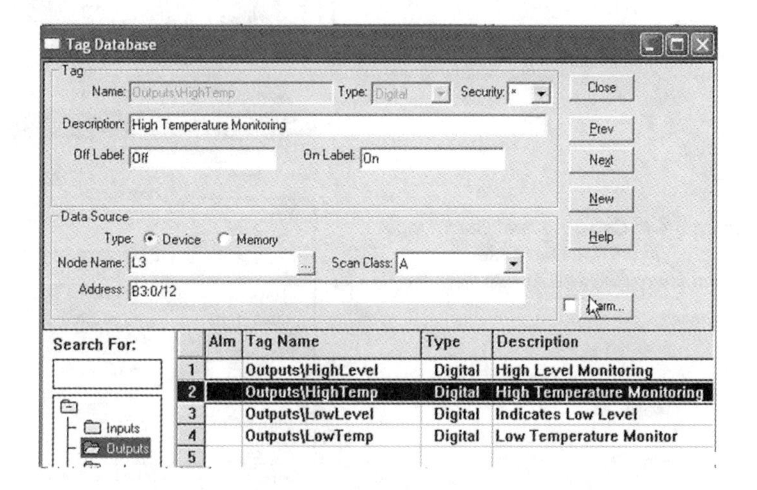

Figure 5-4-5: Continue creating Output tags in the Tag Database (Courtesy of Rockwell Automation, Inc.)

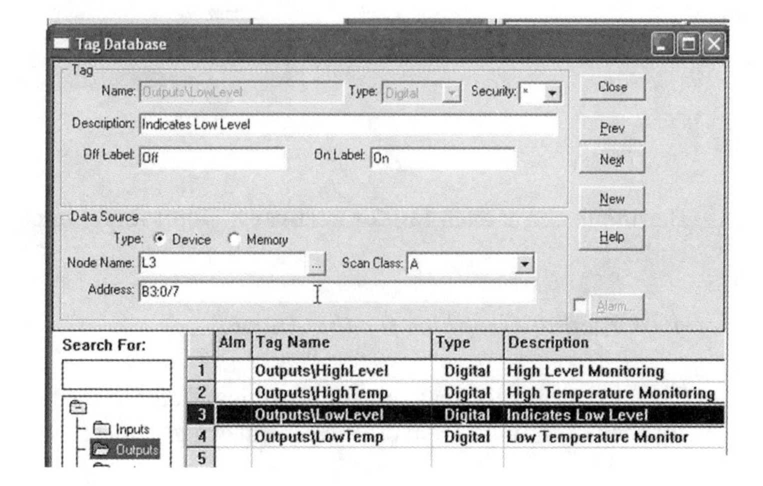

Figure 5-4-6: Continue again creating Output tags in the Tag Database (Courtesy of Rockwell Automation, Inc.)

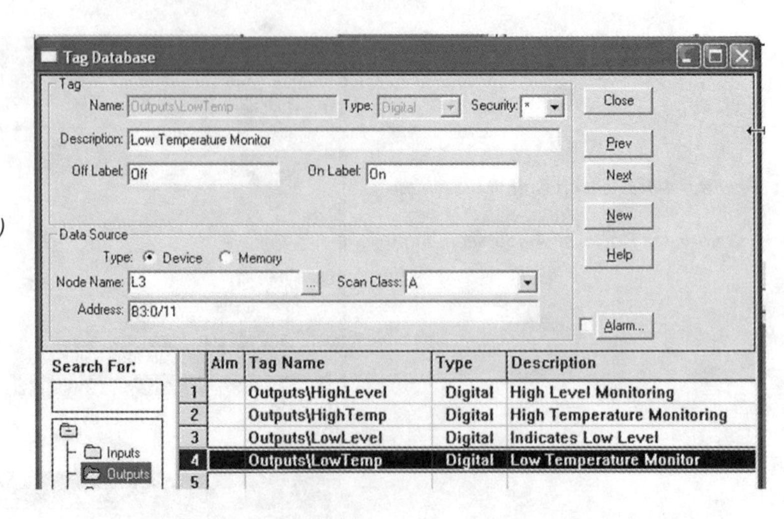

Figure 5-4-7: Additional Output tags created in the Tag Database (Courtesy of Rockwell Automation, Inc.)

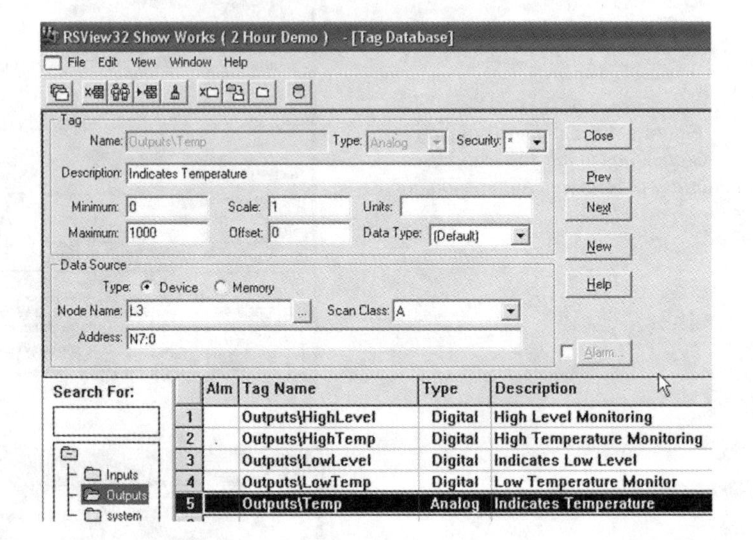

Figure 5-4-8: Complete final Output tags in the Tag Database (Courtesy of Rockwell Automation, Inc.)

6. Be sure to check **each tag for accuracy**. Don't forget to click ***Accept** and **Close*** for all tags entered.

5.4.4 Creating Graphics for the Alarms

7. The next step is to create **alarm graphics**. Refer to the representation shown in Figure 5-4-9 for Steps 8 through 13 in creating the alarm graphics.

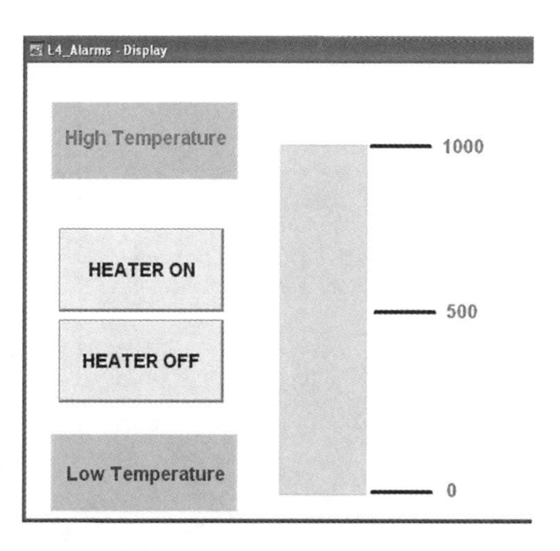

Figure 5-4-9: Alarm Graphics to be created (Courtesy of Rockwell Automation, Inc.)

8. Using the screen in Figure 5-4-10, under the ***Edit Mode*** folder, click the ***Graphics*** folder. Then click ***Display*** and create a screen for this lab as something like ***L4_Alarms***. Be sure to ***save*** the screen.

9. Now click ***Objects*** and then ***Advanced Objects*** *and* ***Buttons***.

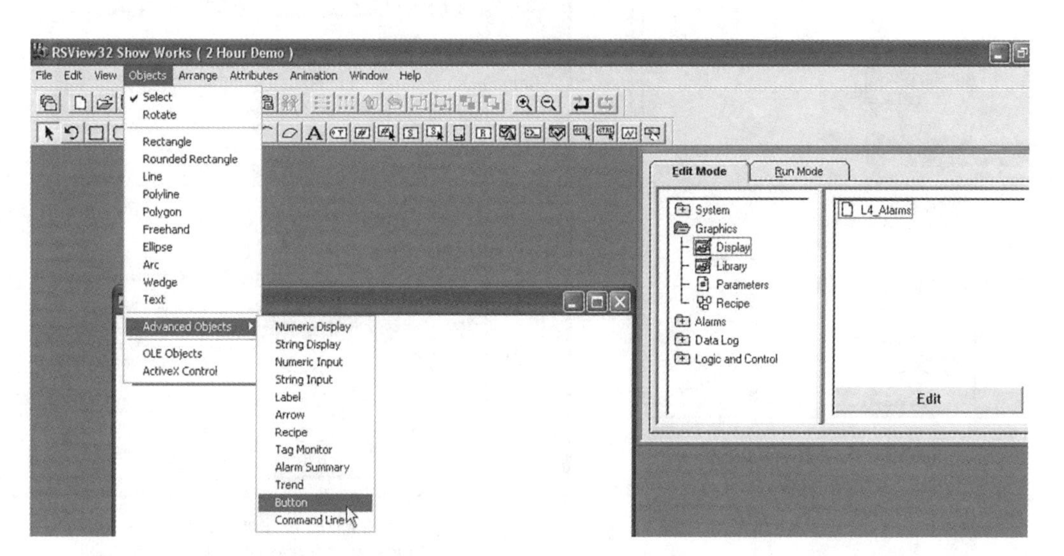

Figure 5-4-10: Editor screen to add Objects, Buttons, etc. (Courtesy of Rockwell Automation, Inc.)

10. Create an ***alarm box*** by clicking and then dragging the box, as shown in Figure 5-4-11. Note that the ***configuration box*** will appear as shown in the figure.

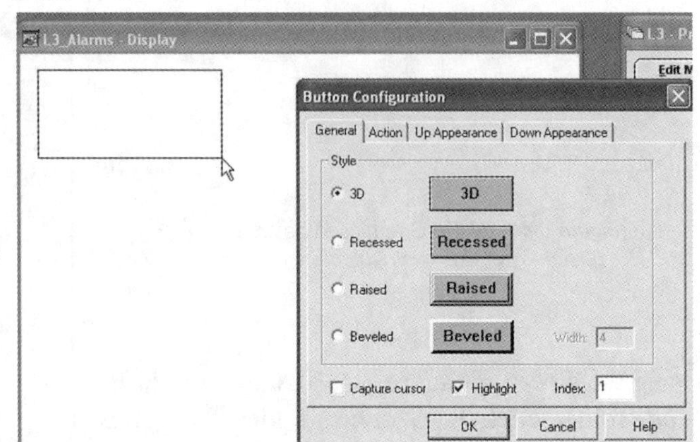

Figure 5-4-11: Alarm Box
(Courtesy of Rockwell Automation, Inc.)

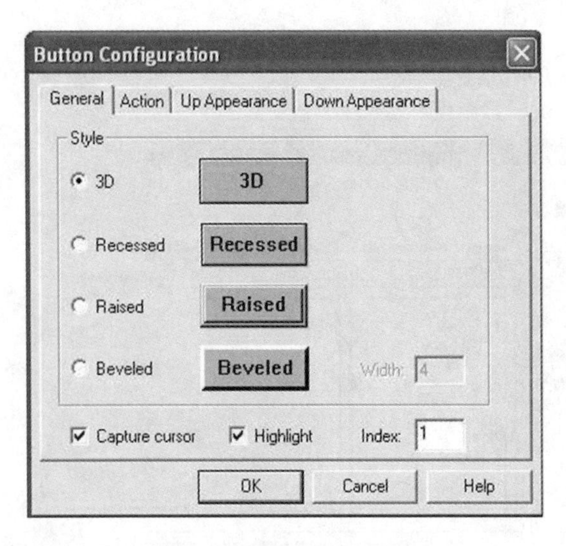

Figure 5-4-12: Create settings for Alarm Box buttons
(Courtesy of Rockwell Automation, Inc.)

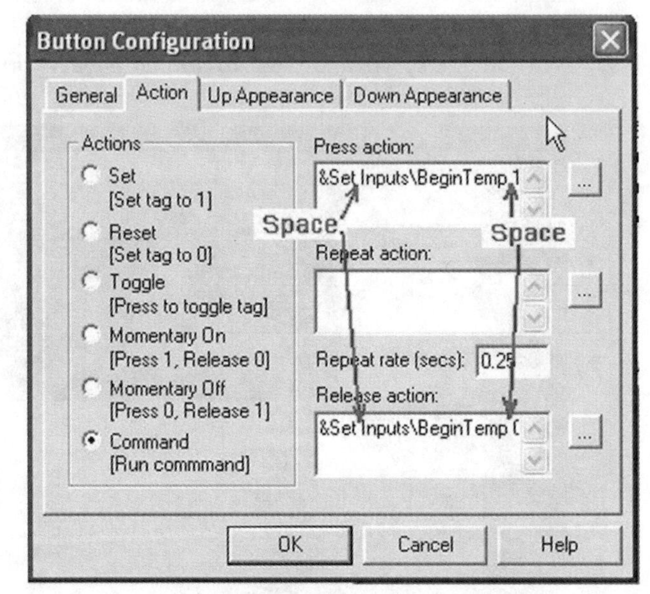

Figure 5-4-13: Enter actions in Press Action
(Courtesy of Rockwell Automation, Inc.)

11. As shown in Figure 5-4-12, re-create the settings for each tab and for the box just created.

    a. At the ***Button Configuration – Action*** tab, enter the command ***&Set Inputs\BeginTemp 1*** in ***Press action*** and ***&Set Inputs\BeginTemp 0*** in ***Release action*** dialog boxes, as shown in Figure 5-4-13.

    b. Type ***HEATER ON*** into the dialog box for the ***Up Appearance*** tab, as shown in Figure 5-4-14.

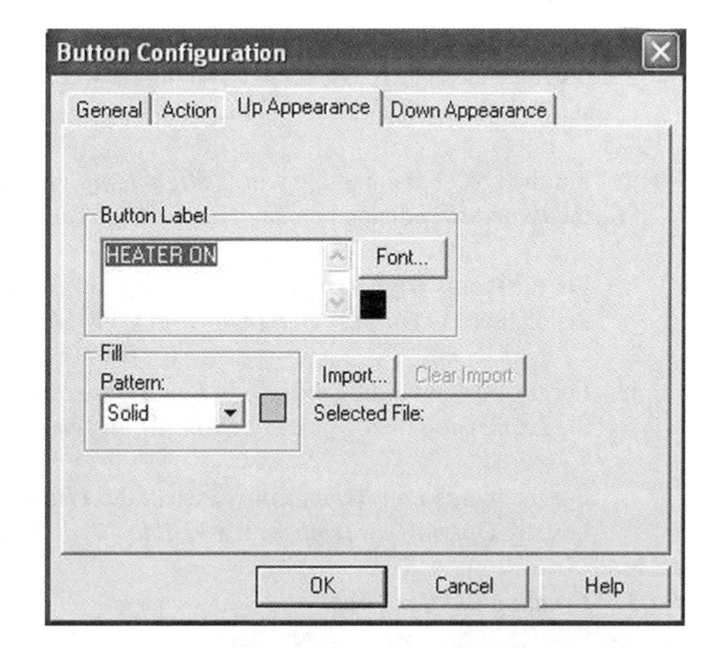

Figure 5-4-14: Create HEATER ON action (Courtesy of Rockwell Automation, Inc.)

c. As shown in Figure 5-4-15, the words ***HEATER ON*** should appear in the warning box.

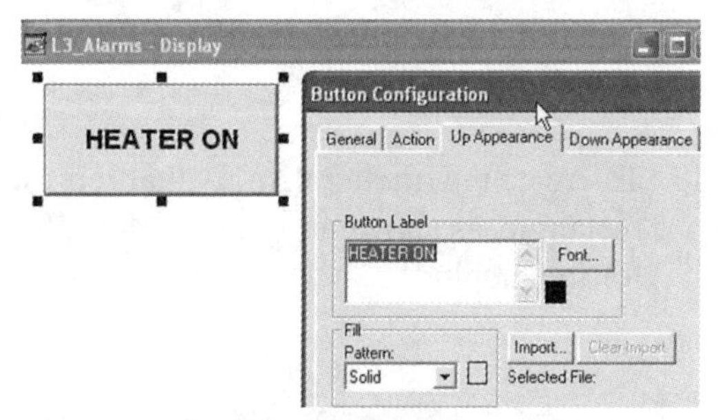

Figure 5-4-15: Warning Box with HEATER message (Courtesy of Rockwell Automation, Inc.)

d. Follow the same procedure to create a ***HEATER OFF*** button. In the ***Action*** tab, change the word Begin to the word ***Stop (StopTemp),*** and in the ***Up Appearance*** tab, change **Button Label** dialog box to ***HEATER OFF***. See Figure 5-4-16.

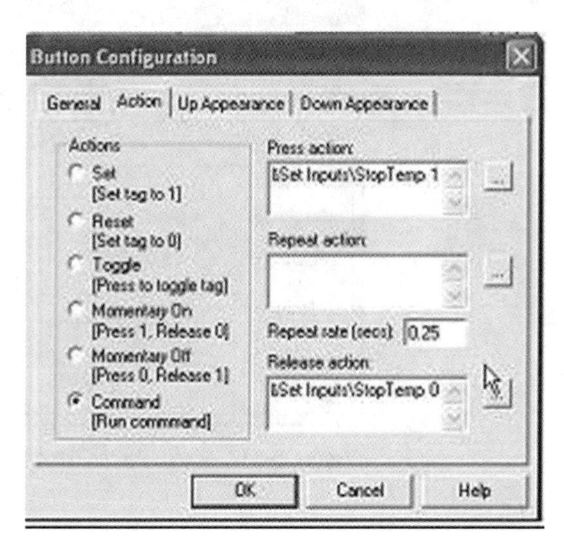

Figure 5-4-16: Create HEATER OFF and related actions (Courtesy of Rockwell Automation, Inc.)

e. You may want to review Section 5.5.4 to help create the actions listed in the following steps.

f. Next, ***High Temperature*** and ***Low Temperature*** alarm displays need to be created. Use previous labs where ***Object*** *and* ***Rectangle*** were used to create these two display message screens, for example, Lab 4-3 and Lab 4-4.

g. For the box_containing the words ***High Temperature***, select the ***Visibility*** **tag** and make the ***Expression*** dialog box say ***Output\HighTemp*** with a ***Visible - True State Expression***.

h. For the words ***High Temperature***, select the ***Visibility*** **tag** and make the ***Expression*** dialog box say ***Output\HighTemp*** with a ***Visible - True State Expression***.

i. For the box_containing the words ***Low Temperature***, select the ***Visibility*** **tag** and make the ***Expression*** dialog box say ***Output\LowTemp*** with a ***Visible - True State Expression***.

j. For the words ***Low Temperature***, select the ***Visibility*** **tag** and make the ***Expression*** dialog box say ***Output\LowTemp*** with a ***Visible - True State Expression***.

5.4.5 Creating the Bar Graph

12. Create a **vertical bar graph** for the analog ***Outputs\Temp*** (Indicates Temperature) tag in the ***Tag Database***.

13. To create the **analog display** that represents low and high temperature levels, create a vertical rectangle, as shown in Figure 5-4-17. Select ***Animation*** *and* ***Fill*** to enter the values as shown for the graphic.

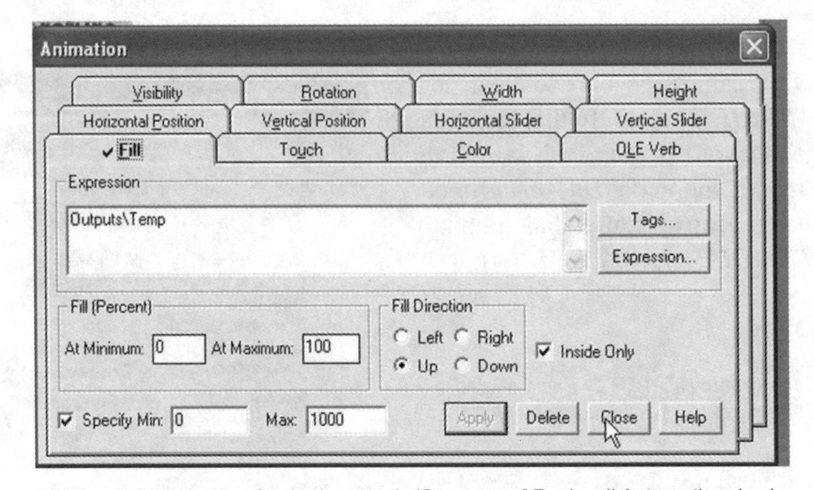

Figure 5-4-17: Actions for the bar graph (Courtesy of Rockwell Automation, Inc.)

14. Create the scale (0, 500, and 1000) next to the ***Outputs\Temp*** graphic using the line tool to make the tic lines and the text tool **A** to create the numbers (Figures 5-4-18 and 5-4-19). Estimate the half-way point to place the ***500*** gradient. Experiment with the ***Attributes*** of the lines to thicken them.

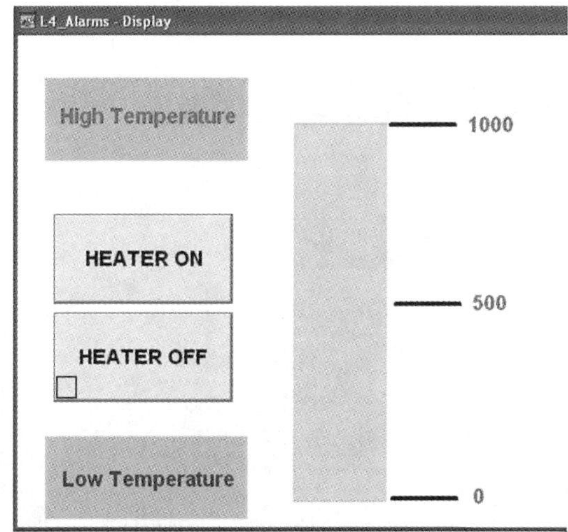

Figure 5-4-18: Bar graph
(Courtesy of Rockwell Automation, Inc.)

5.4.6 Creating an Alarm Button and Alarm String

15. Create the ***Alarm*** button by clicking ***Object,*** then ***Advanced Objects****, and* ***Button*** from the top bar menu in ***RSView*** to then place the button onto the display, as shown in Figure 5-4-19.

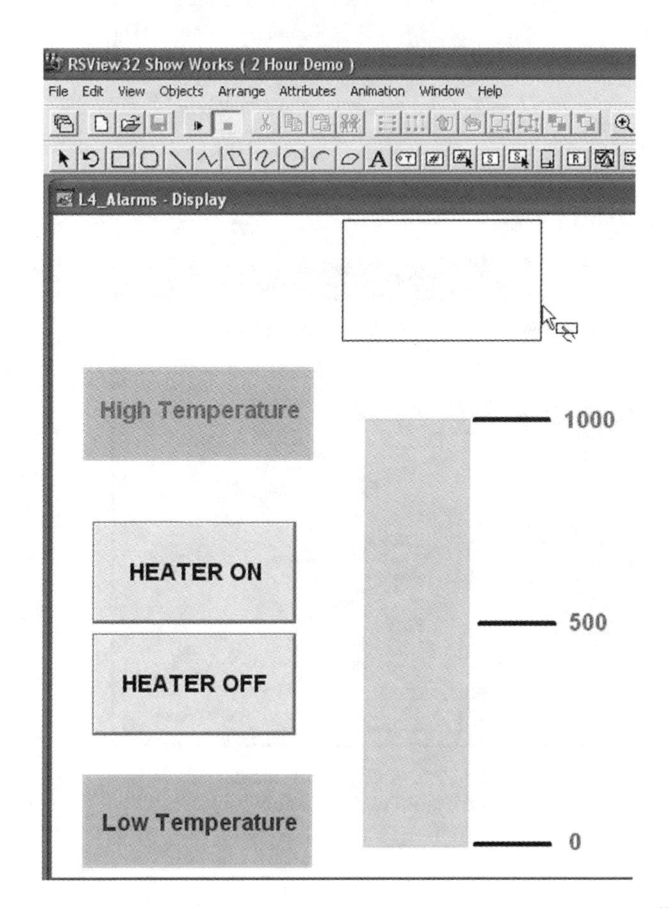

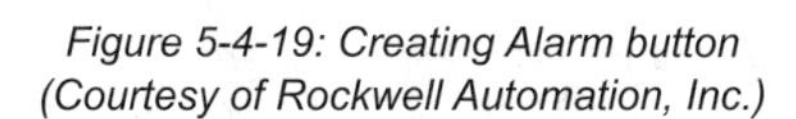
Figure 5-4-19: Creating Alarm button
(Courtesy of Rockwell Automation, Inc.)

16. After clicking to place the box, make the information for the ***General, Action, Up Appearance***, and ***Down Appearance*** tabs appear as shown in Figures 5-4-20 through 5-4-23: ***General, Action, UpAppearance,*** and ***Down Appearance.***

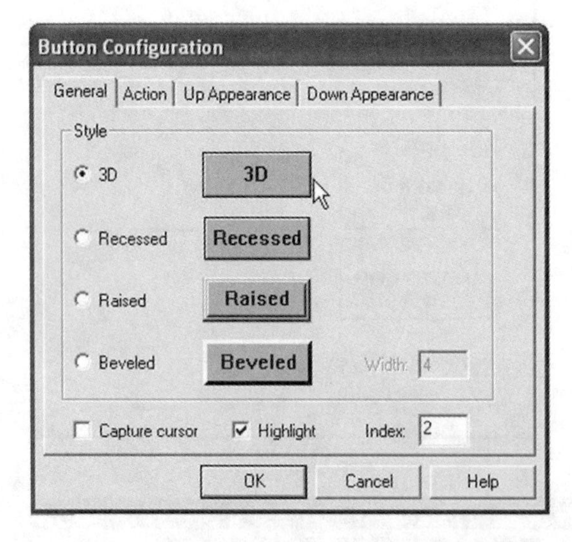

Figure 5-4-20: Create actions for buttons listed (Courtesy of Rockwell Automation, Inc.)

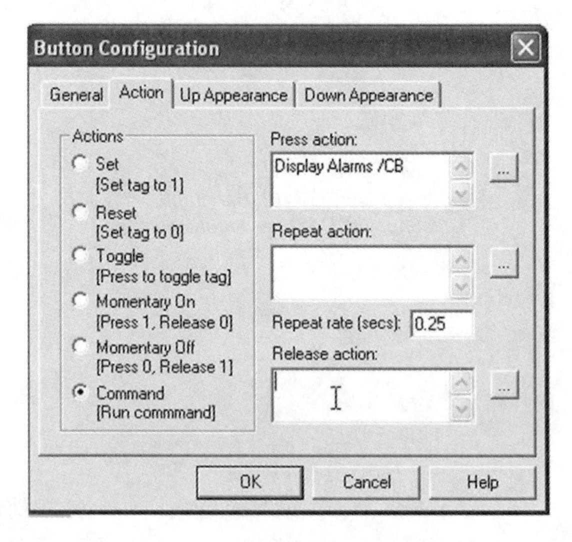

Figure 5-4-21: Continue from Figure 5-4-20 (Courtesy of Rockwell Automation, Inc.)

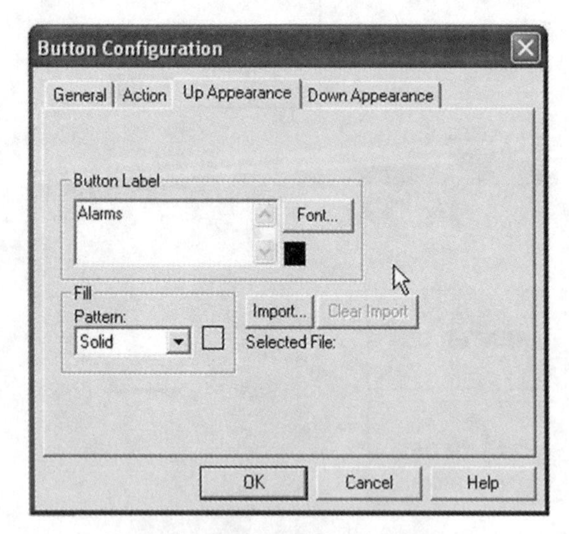

Figure 5-4-22: Creating Alarm Button (Courtesy of Rockwell Automation, Inc.)

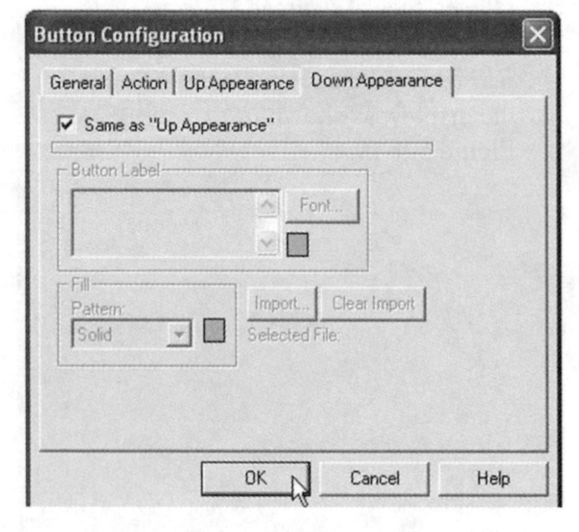

Figure 5-4-23: Completion of Alarm Button (Courtesy of Rockwell Automation, Inc.)

17. With the ***Alarms Button*** created, an ***alarm string (ssssssss)*** needs to be created underneath the ***Alarms*** button, as shown in Figures 5-4-24 and 5-4-25.

18. Click ***Object,*** then ***Advanced Objects***, and ***String Display*** from the top bar menu in ***RSView***; then place the rectangle where the string will appear.

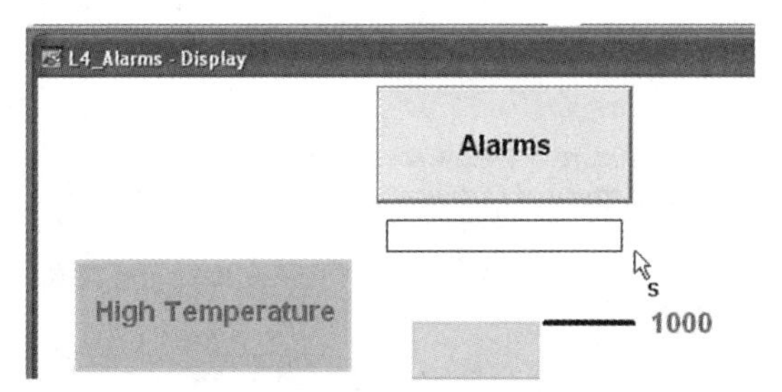

Figure 5-4-24 Creation of String Display (Courtesy of Rockwell Automation, Inc.)

19. The ***Alarm string*** screen, as shown in Figure 5-4-25, will appear. Upon making the ***ssssssss string***, a pop-up screen will appear.

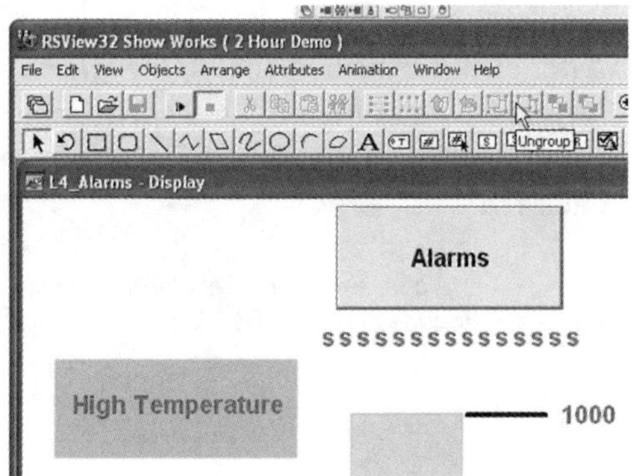

Figure 5-4-25: String of s's will appear (Courtesy of Rockwell Automation, Inc.)

20. **Enter the values** as shown into the pop-up screen. Create and enter the ***Expression** and* ***values*** as shown in Figure 5-4-26.

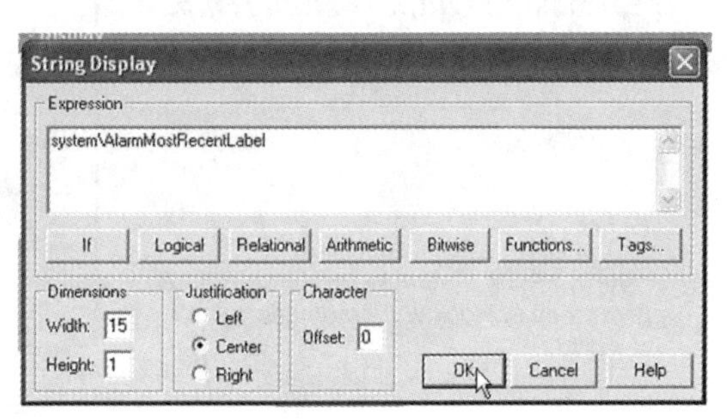

Figure 5-4-26: Enter Expression (Courtesy of Rockwell Automation, Inc.)

21. Then, to set the ***Alarm string*** messages, click ***System*** and then ***Tag Database***. Select ***Outputs\HighTemp***, as shown in Figure 5-4-27.

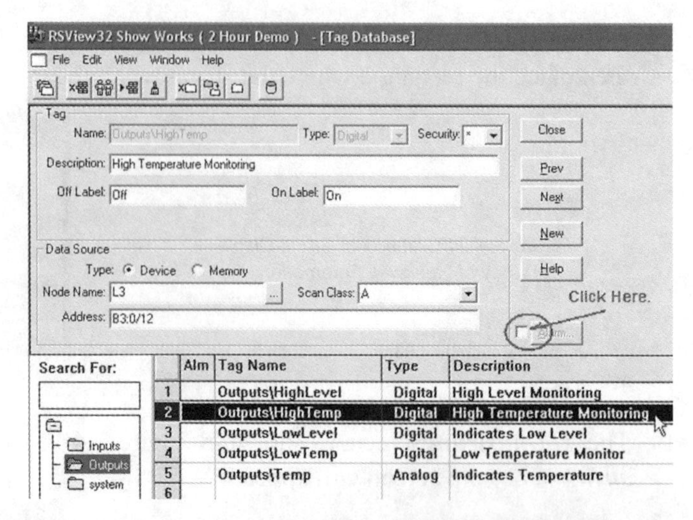

Figure 5-4-27: Click Alarm button
(Courtesy of Rockwell Automation, Inc.)

22. After clicking the box next to the ***Alarm*** button, the screen shown in Figure 5-4-28 will appear. Enter the values, as shown, for the ***Digital Alarm*** screen; then click ***OK***.

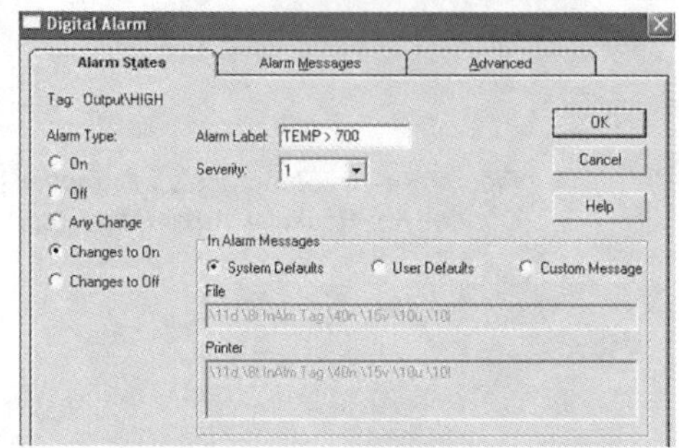

Figure 5-4-28: Digital Alarm Screen
(Courtesy of Rockwell Automation, Inc.)

23. In the ***Tag Database*** screen (Figure 5-4-29), click ***Accept*** to see an ***X*** appear next to ***Outputs\HighTemp.***

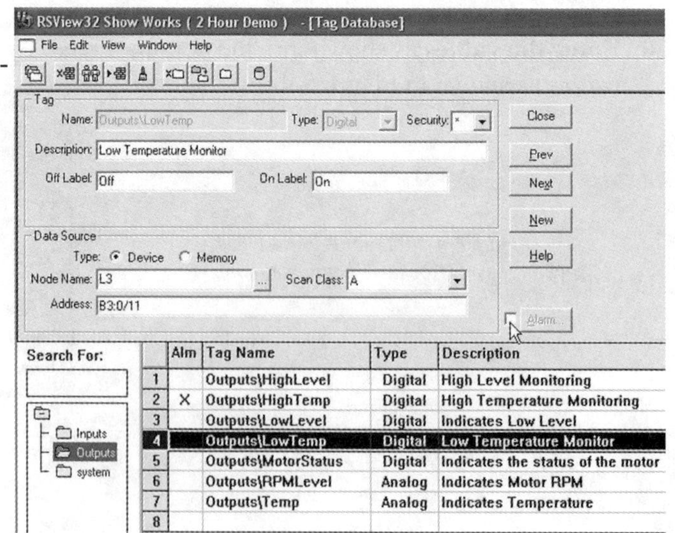

Figure 5-4-29: Outputs\Low Temp Alarm
(Courtesy of Rockwell Automation, Inc.)

24. Repeat the process just followed for the ***Outputs\LowTemp*** tag. As shown in Figure 5-4-30, input information and change values to make the ***Digital Alarm*** tag represent the pop-up screen.

25. Click ***OK***, and then click ***Accept.*** An ***X*** should appear next to ***Outputs\LowTemp.*** An alarm action will be activated that is associated with this tag. This completes the Alarms Button and Alarms string.

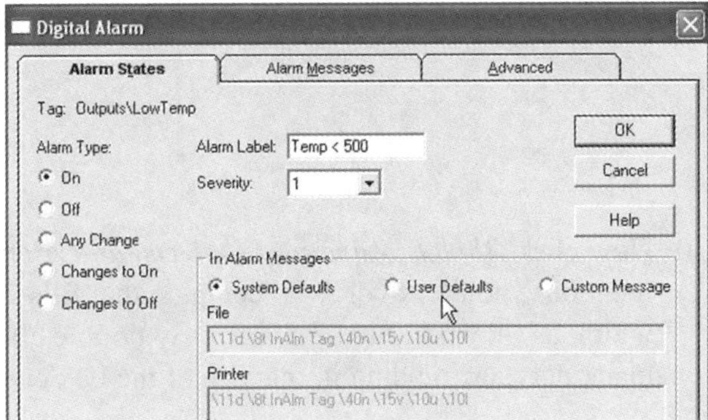

Figure 5-4-30: Alarm States (Courtesy of Rockwell Automation, Inc.)

5.4.7 Creating the Alarm Display Screen

26. Select ***Graphics*** and then double click ***Display***. An ***Untitled-Display*** should appear as shown in Figure 5-4-31.

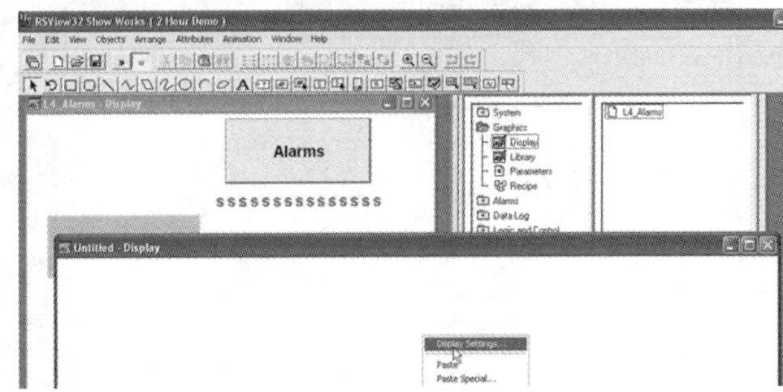

Figure 5-4-31: Alarm Display Screen (Courtesy of Rockwell Automation, Inc.)

27. Enlarge this screen, but make sure the ***Alarms-Display*** and ***Project/Edit-Run*** screens are still visible as shown in Figure 5-4-32.

28. Right click inside the ***Untitled–Display*** and then select ***Display Settings***.

29. Enter the values into the ***Display Setting*** screen, as shown in Figure 5-4-32, with a background color of yellow, and then click ***OK***.

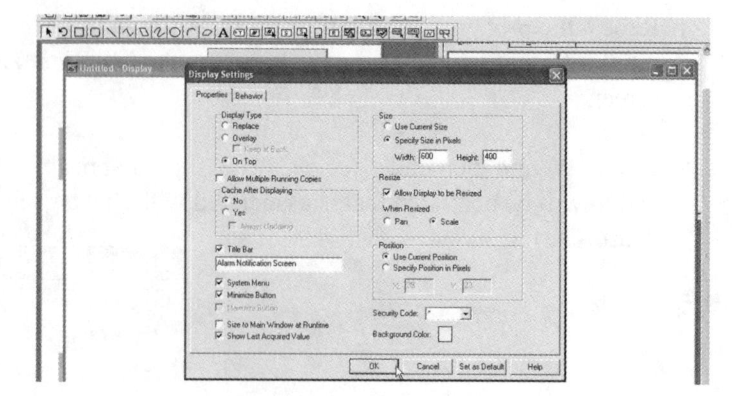

Figure 5-4-32: Checking Screens for Alarms
(Courtesy of Rockwell Automation, Inc.)

30. Now click ***Objects, Advanced Objects, and Alarm Summary***. As an array icon appears, move the mouse pointer to draw a rectangle that fills *2/3* of the height of the screen and nearly full width, as shown in Figure 5-4-33. By double clicking the ***Alarm Summary*** box to get the black dots surrounding it, the size of the box can be increased or decreased.

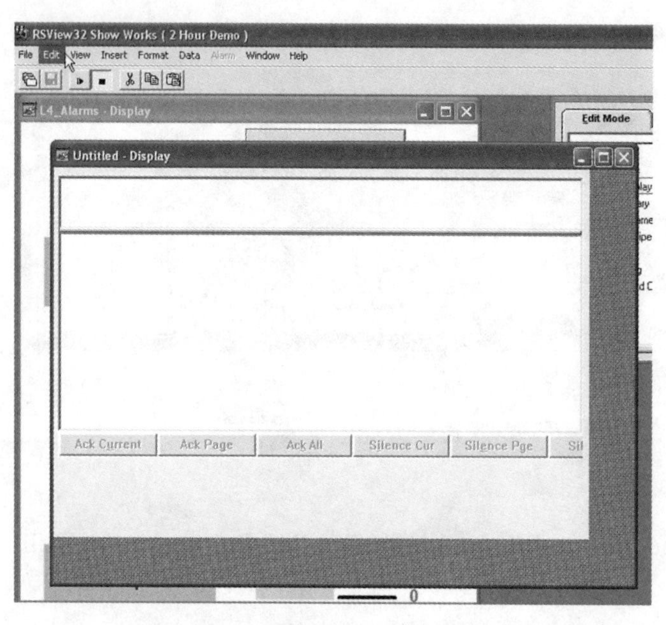

Figure 5-4-33: Create Alarm Screen
(Courtesy of Rockwell Automation, Inc.)

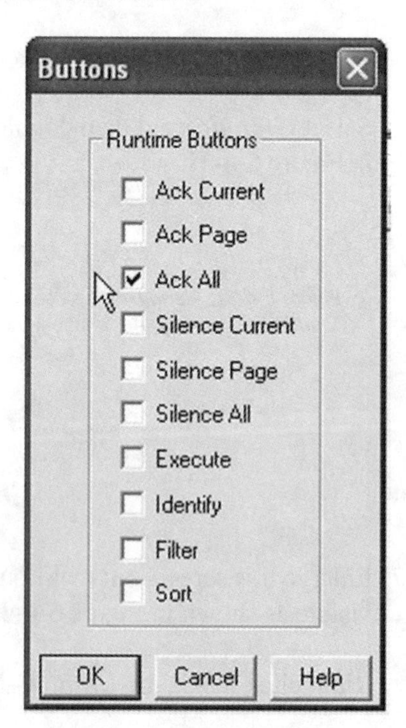

Figure 5-4-34: List of Buttons
(Courtesy of Rockwell Automation, Inc.)

31. Now click ***Format** and **Buttons*** from the top bar menu. A pop-up screen will appear, similar to the one shown in Figure 5-4-34. Leave the ***Ack All*** selection checked, but uncheck everything else; then click ***OK***.

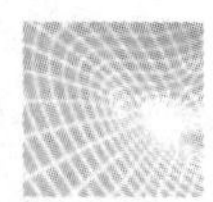

32. Note that an ***Ack All*** button has been placed on the ***Untitled – Display*** screen, as shown in Figure 5-4-35.

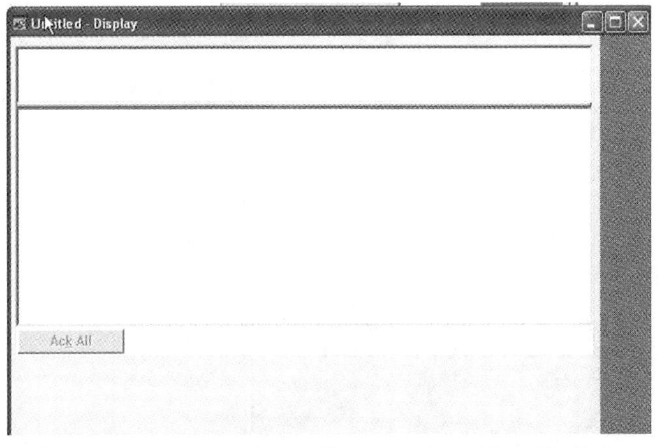

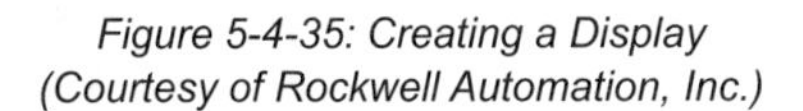
Figure 5-4-35: Creating a Display (Courtesy of Rockwell Automation, Inc.)

33. Click ***Insert** and **Tag Name*** from the top bar menu screen. A small rectangle with a black bar in the center will appear on the ***Untitled – Display*** screen.

34. Move this rectangle with a bar in it to the left side of the ***Alarm Summary*** box and click. The graphic in Figure 5-4-36 with the words ***Tagname*** will appear.

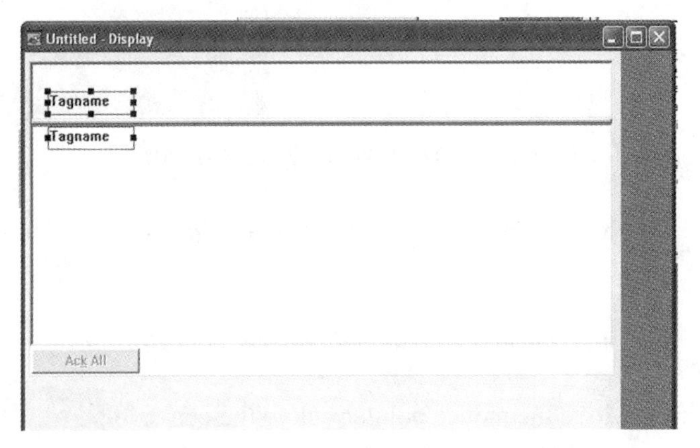

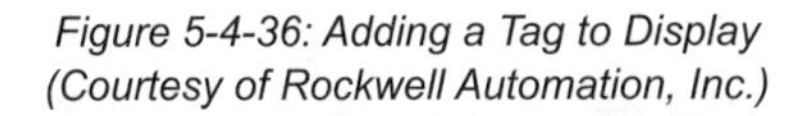
Figure 5-4-36: Adding a Tag to Display (Courtesy of Rockwell Automation, Inc.)

35. Left mouse click to leave this tag at the left side of the screen

36. Repeat Step 36 to insert ***Alarm Label, Alarm Time,** and **Alarm Severity*** tags into the screen, as shown in Figure 5-4-37.

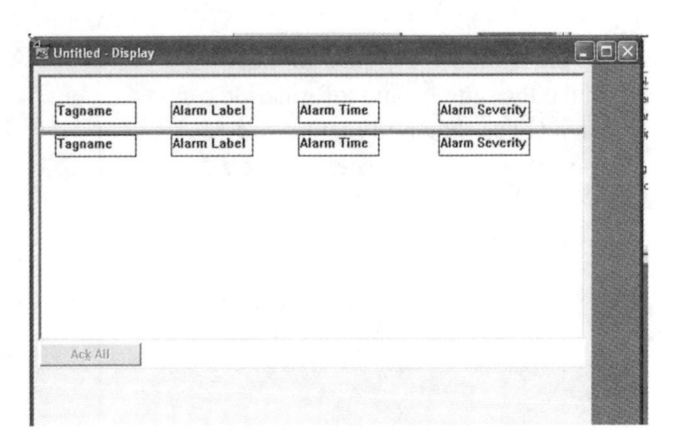

Figure 5-4-37: Alarm Tags for Alarm Display (Courtesy of Rockwell Automation, Inc.)

37. Upon inserting the ***Alarm Severity*** tag and clicking ***Close***, the screen in Figure 5-4-38 appears. By clicking this screen, the black dots surrounding the display, allow resizing and moving the display.

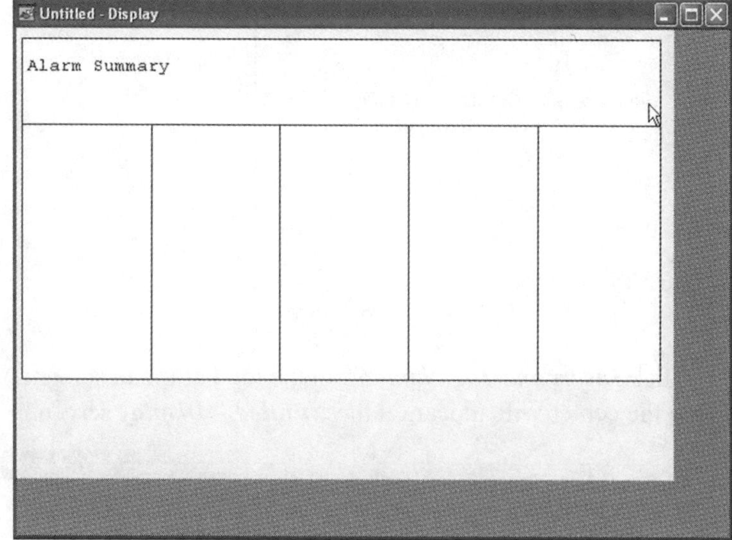

Figure 5-4-38: Alarm Summary Display (Courtesy of Rockwell Automation, Inc.)

5.4.8 Creating the Alarm Close Button

38. Now add a ***Close*** button to this ***Untitled–Display*** screen underneath the ***Alarm Summary*** screen.

39. Click ***Objects, Advanced Objects,*** *and* ***Button***.

40. Move the mouse pointer below the center of the ***Alarm Summary*** screen and place the box.

41. At the ***Button Configuration*** screen, make the settings and information for each tag as shown in Figures 5-4-39, 5-4-40, and 5-4-51. Then click ***OK***.

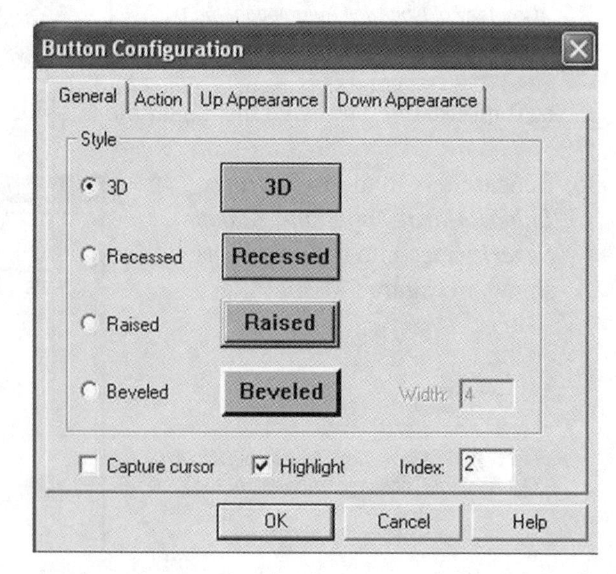

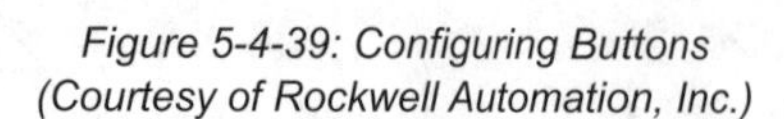

Figure 5-4-39: Configuring Buttons (Courtesy of Rockwell Automation, Inc.)

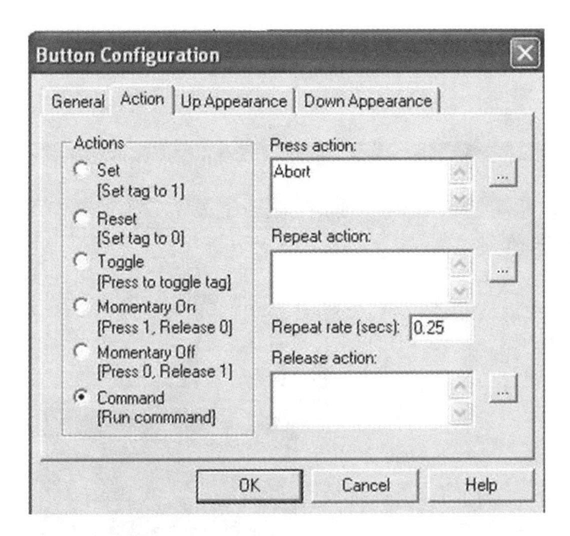

Figure 5-4-40: Additional Configuring Buttons
(Courtesy of Rockwell Automation, Inc.)

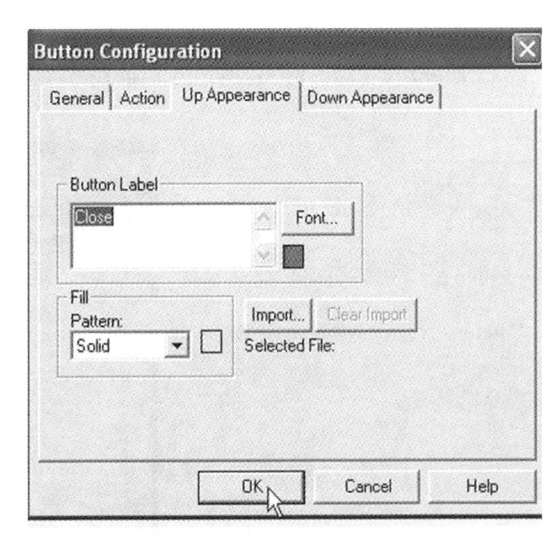

Figure 5-4-41: More Button Configurations
(Courtesy of Rockwell Automation, Inc.)

42. Left click somewhere besides the ***Alarm Summary*** and newly produced ***Close*** button to remove the black dots from around the ***Close*** button (Figure 5-4-42).

Figure 5-4-42: Close Button
(Courtesy of Rockwell Automation, Inc.)

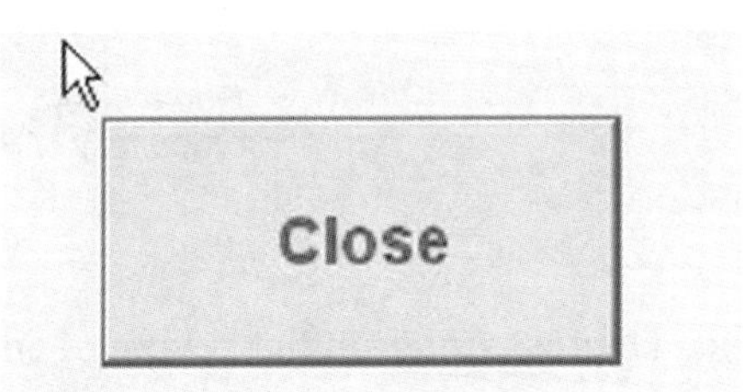

43. Close out this alarm screen by clicking the **red X**. When prompted, save this display with a name similar to the ***Lab4 Alarms – Display*** Screen, like ***L4_Alarms_AlarmsSummary*** shown in Figure 5-4-43.

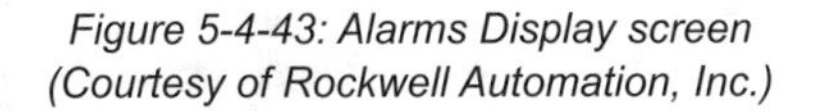

Figure 5-4-43: Alarms Display screen
(Courtesy of Rockwell Automation, Inc.)

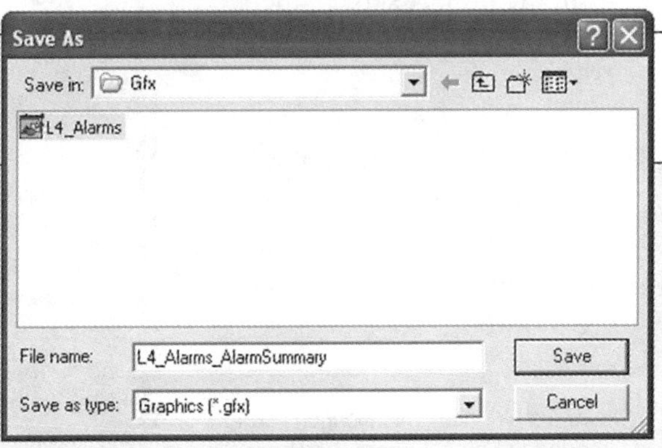

44. To put this screen into the ***Run mode***, ***Alarms*** must be activated. Click ***Systems***, then the ***Command Line***, as shown in Figure 5-4-44.

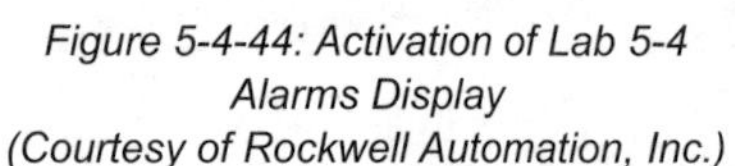

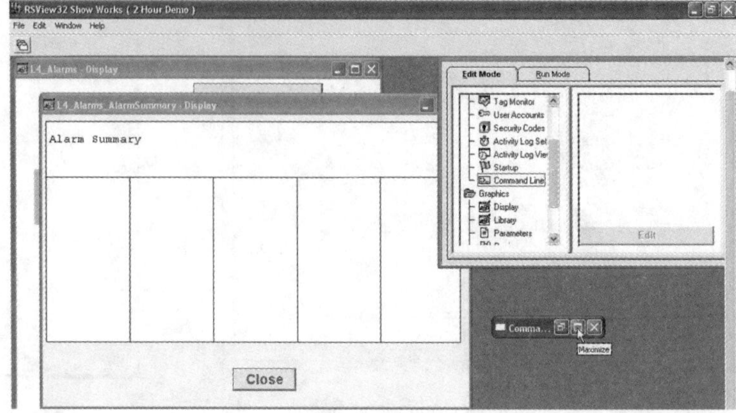

Figure 5-4-44: Activation of Lab 5-4 Alarms Display (Courtesy of Rockwell Automation, Inc.)

45. Be sure the ***Command Line screen*** is maximized before typing the words ***AlarmOn/H*** at the top of the screen, as shown in Figure 5-4-45.

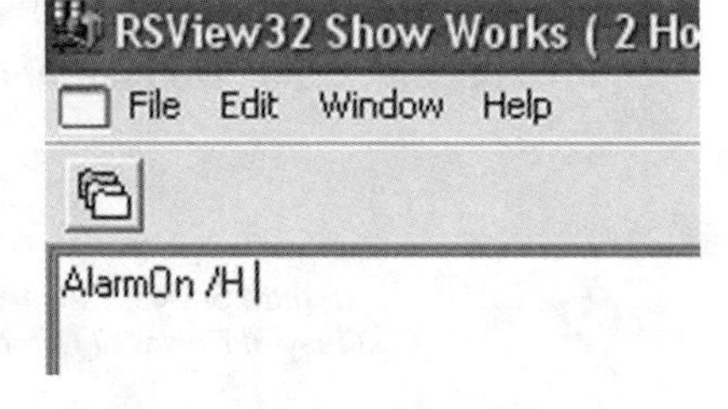

Figure 5-4-45: Enter AlarmOn/H to screen (Courtesy of Rockwell Automation, Inc.)

46. Depress ***Enter*** on the keyboard and note that the command disappears. Close the Command Line screen.

47. Run ***L4_Alarms_Alarms Summary.*** Next, click ***Run Mode*** from the project screen, then double click **Graphics,** and click ***Display.***

48. Now press the ***Start*** button located on the ***Run Mode*** screen. The ***Alarms Notification*** Screen should appear, as shown in Figure 5-4-46.

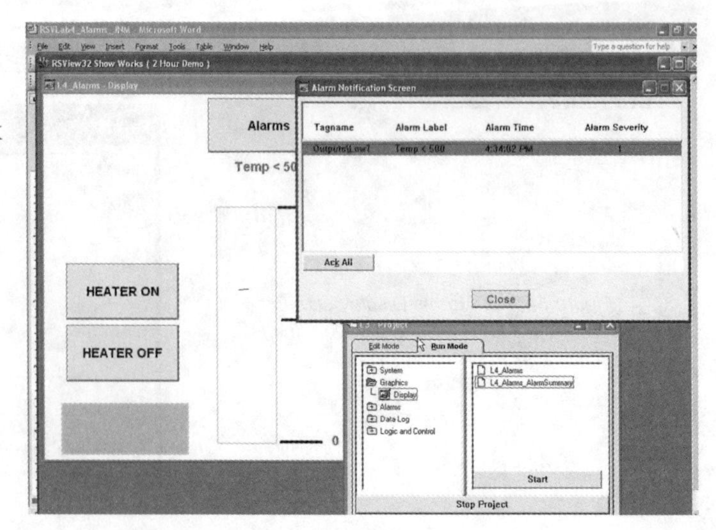

Figure 5-4-46: Alarm Notification screen added to Lab 5-4 Alarm Summary

49. Start the program by clicking ***HEATER ON*** to allow the program to cycle. The Alarm String display initially shows ***TEMP < 500***. As the heater heats up, the display will switch to ***TEMP > 700***.

50. Click ***HEATER ON*** and then click ***ALARM*** button. A screen similar to Figure 5-4-47 should appear.

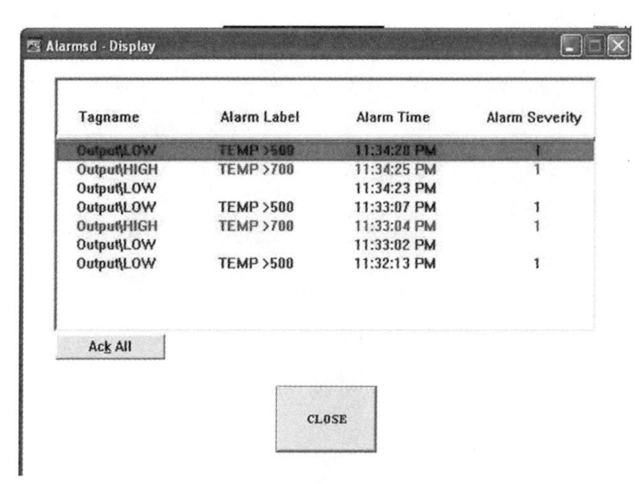

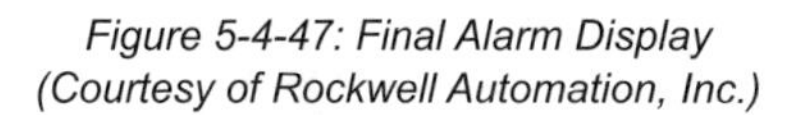
Figure 5-4-47: Final Alarm Display (Courtesy of Rockwell Automation, Inc.)

51. Press ***CLOSE*** to close out this screen. Explore running this project.

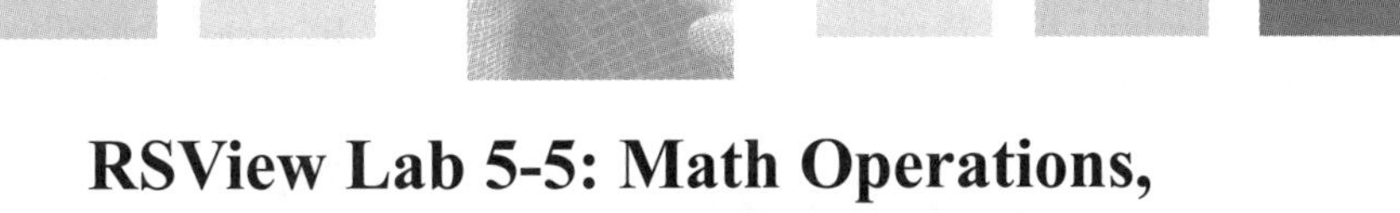

RSView Lab 5-5: Math Operations, Subprograms, and Alarms

5.5.1 *Introduction*

This lab illustrates the use of math operations and subprogramming in PLCs with HMI displays and alarms. It is based upon Lab 5-4.

Some PLCs have special inputs and outputs that are analog instead of digital. For example, analog inputs are used with sensors such as thermistors or thermocouples, which deliver a varying voltage versus temperature. This lab will show how these are used. Quite often mathematical operations are needed to "scale" or format the voltage value to display the correct physical quantity.

For this lab, the MicroLogix 1100 PLC has two input output types. The two types are analog and digital. The digital temperature sensors from PV Lab 5-4 will also be used. This sensor produces millivolt signals that are directly proportional to degrees Fahrenheit or Celsius.

5.5.2 *Programming the PLC*

1. Connect the ***1100 PLC*** through its serial connection to the computer and enter the ladder diagram shown in Figure 5.5.1. This diagram is the main ladder diagram called ***LAD 2 – MAIN PROG***.

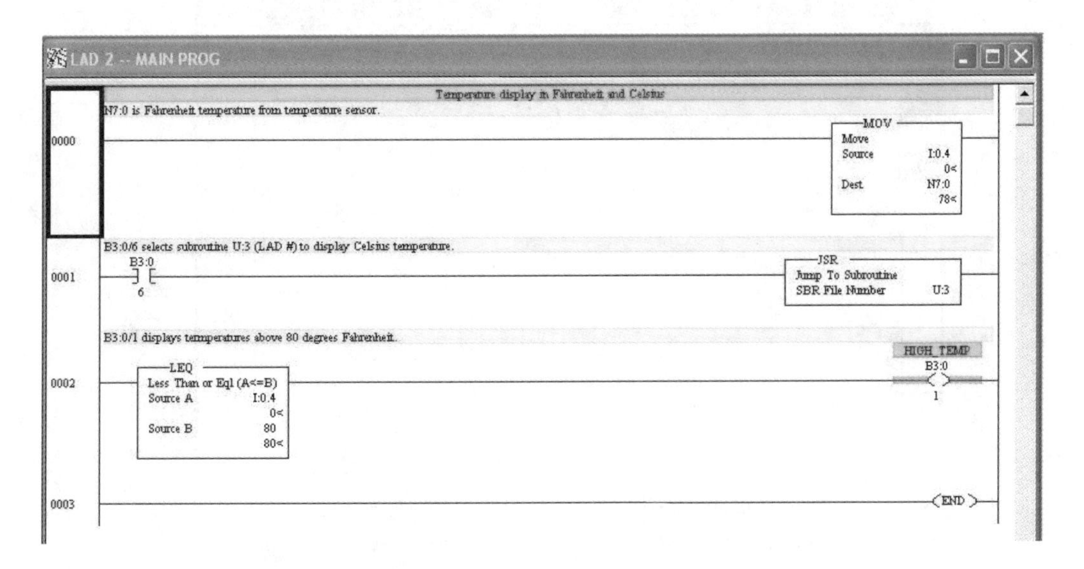

Figure 5.5.1: LAD 2 Main Program (Courtesy of Rockwell Automation, Inc.)

2. Note the use of ***Jump To Subroutine*** (JSR). A subroutine is another ladder diagram that is connected to ***LAD 2***. Figure 5-5-2 shows ***LAD 3 – SUB PROG***.

3. To make ***LAD 3***, right click ***Program Files***; then select ***New*** to create the subroutine, as shown in Figures 5-5-3 and 5-5-4.

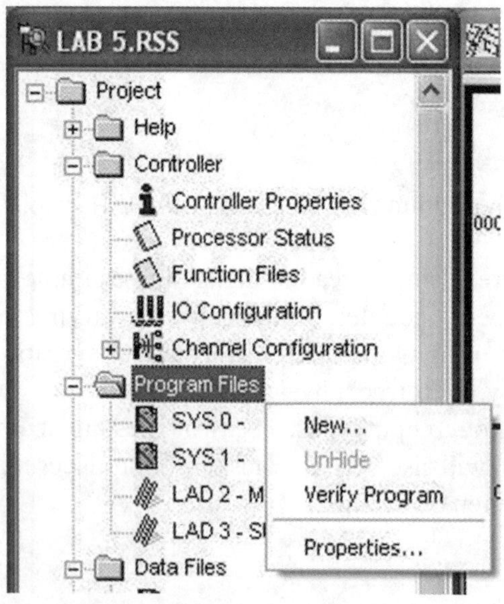

Figure 5-5-2: Selecting LAD 3
(Courtesy of Rockwell Automation, Inc.)

Figure 5-5-3: Creating LAD 3
(Courtesy of Rockwell Automation, Inc.)

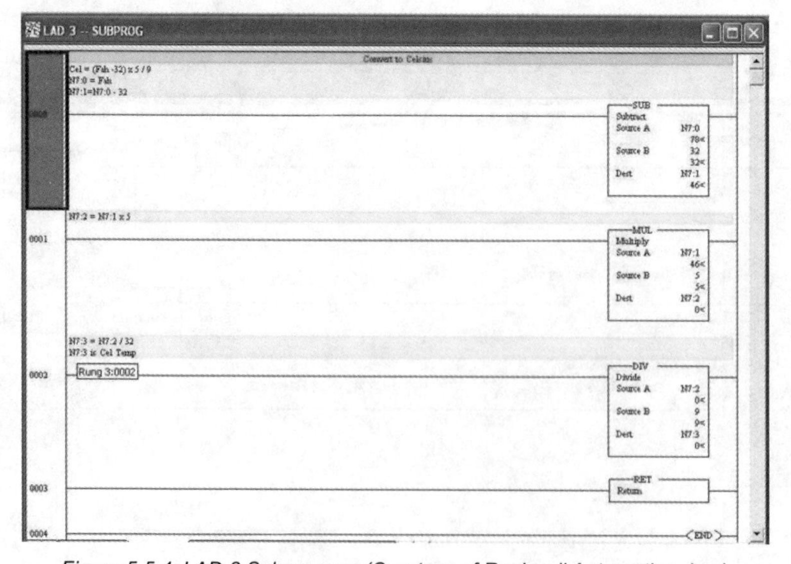

Figure 5-5-4: LAD 3 Subprogram (Courtesy of Rockwell Automation, Inc.)

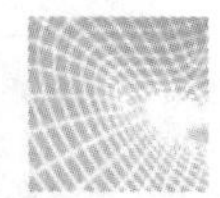

4. Connect the digital temperature sensor (see Appendix C) to input ***I:0/4***. Be sure to turn on the PLC in ***RUN*** mode.

5.5.3 Configuring RSView

5. Make sure the RSView is configured under ***Edit Mode, System***. Look at ***RSView Lab 1*** for the information on how to do this.

5.5.4 Creating the Tag Database

6. Begin by creating the tags in the tag database under the Tag Database (S***ystem > Tag Database)***. Make sure Input and Output folders have been created for the Tag Database.

7. Now create the following screens for ***Input and Output tags***, as shown in Figures 5-5-5 through 5-5-9.

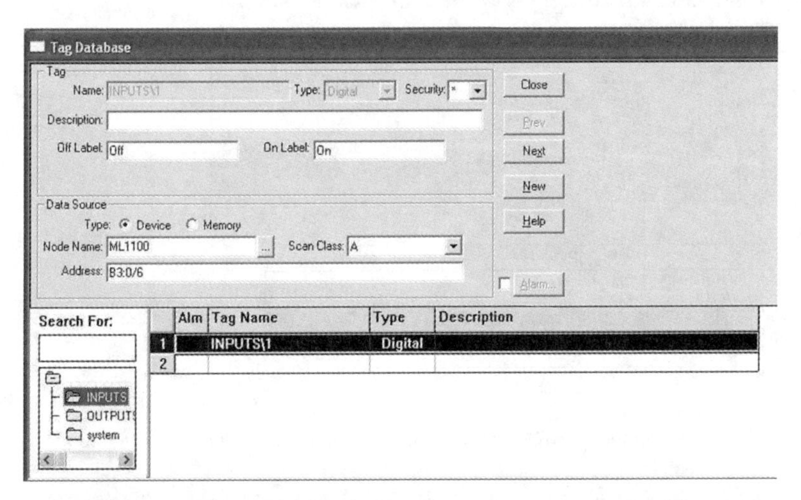

Figure 5-5-5: Screen 1: Entering Input Tag (Courtesy of Rockwell Automation, Inc.)

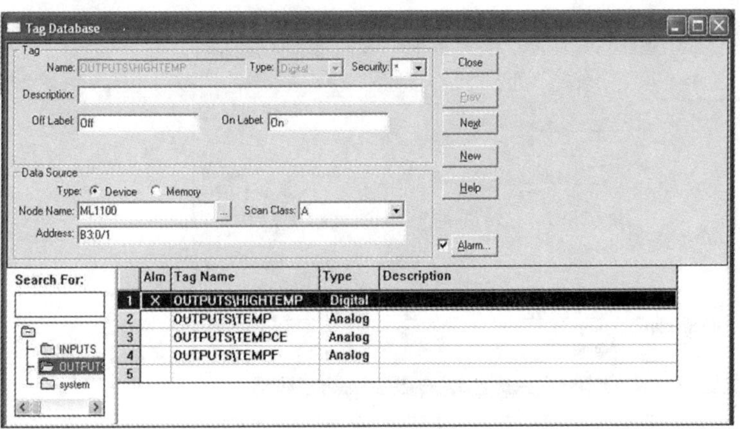

Figure 5-5-6: Screen 2: Entering Output Tags (Courtesy of Rockwell Automation, Inc.)

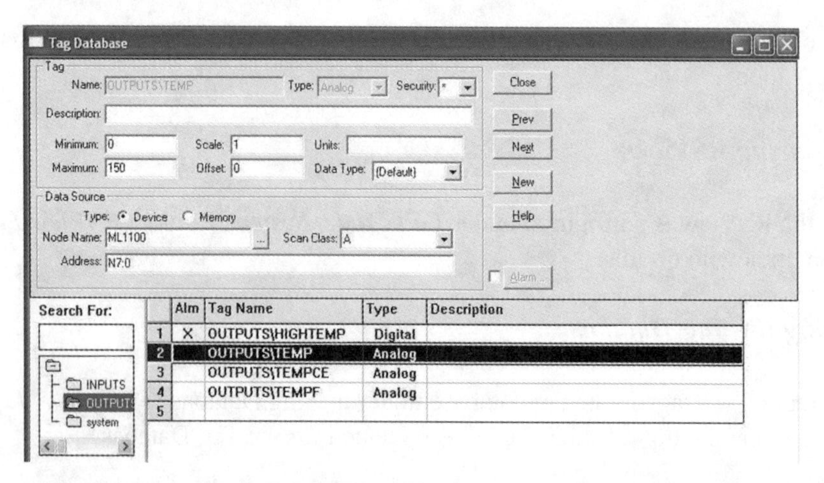

Figure 5-5-7: Screen 3: Entering Output Tags (Courtesy of Rockwell Automation, Inc.)

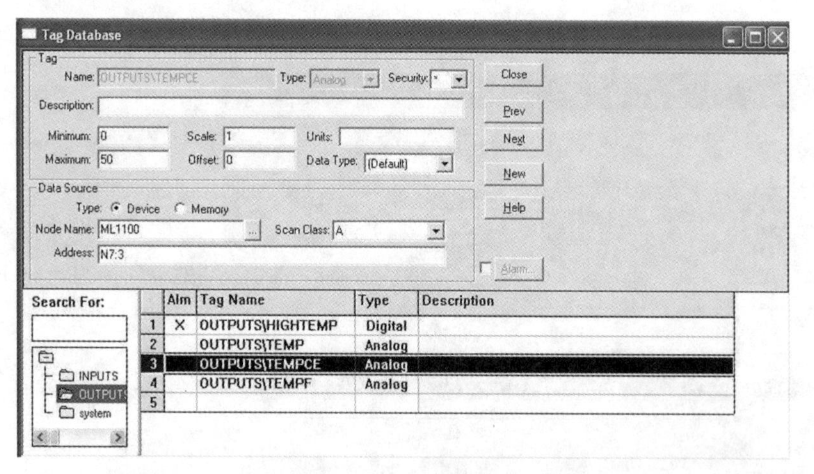

Figure 5-5-8: Screen 4: Entering Output Tags (Courtesy of Rockwell Automation, Inc.)

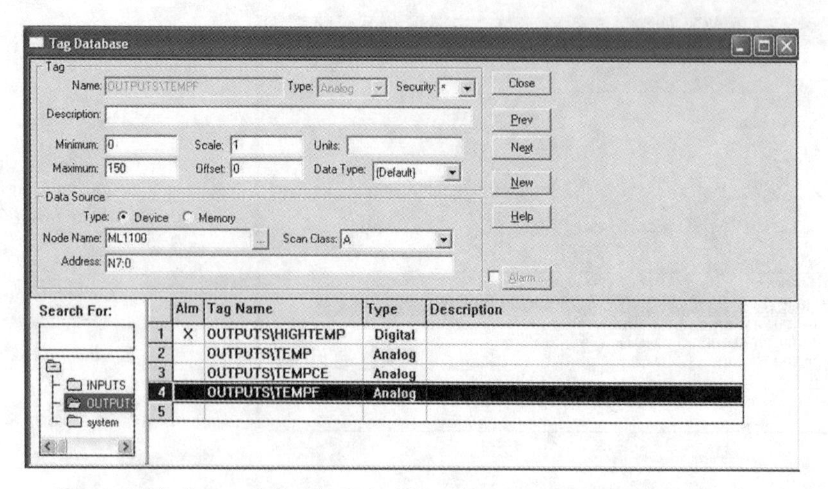

Figure 5-5-9: Screen 5: Create Output Tags (Courtesy of Rockwell Automation, Inc.)

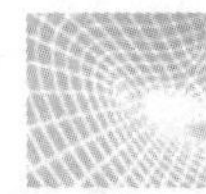

5.5.5 Creating the Graphic Screens

8. Noting Figure 5-5-10, click on the ***Display*** > ***Math SubProg & Alarm*** from the Graphics folder on the ***MATH_SUBPROGRAM.RSV*** screen.

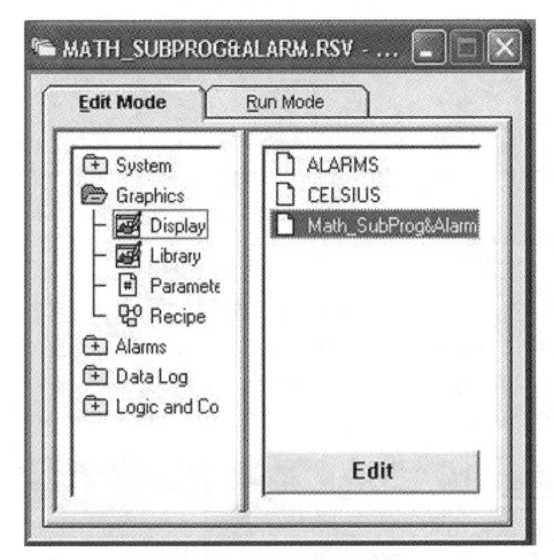

Figure 5-5-10: Screen 1:
Select Math SubProg & Alarm
(Courtesy of Rockwell Automation, Inc.)

Figure 5-5-11: Screen 2:
Main Screen
(Courtesy of Rockwell Automation, Inc.)

9. The main screen should look like the screen in Figure 5-5-11.

5.5.6 Creating a Digital Temperature Display

10. The next set of screens shows the graphical Fahrenheit analog scale, the digital temperature, and the Celsius button being created. You will also create the remaining items on the screen, including the Alarms. Look back at previous labs, including Labs 5-3 and 5-4, if you are unsure of what to do.

11. To create the digital display, click ***Objects*** > ***Advanced Objects*** > ***Numeric Display***. Place the mouse pointer where needed for the numeric display; then drag a small rectangle into the Math_SubProg & Alarm – Display screen. The display in Figure 5-5-12 will appear. Enter the values shown in the figure.

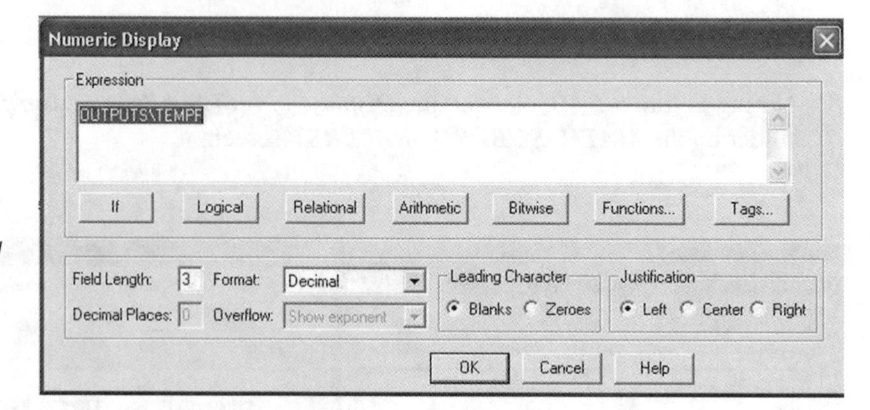

Figure 5-5-12: Screen 3: Creating a Digital Temperature Display (Courtesy of Rockwell Automation, Inc.)

5.5.7 *Creating an Analog Temperature Display*

12. Add a **rectangle** to create the analog temperature display seen in Figure 5-5-11. The 0, 75, and 225 graduated scale found in this figure illustrates the scale to be created. Add the **scale** as shown by drawing lines and adding **numbers**. Refer to Lab 5-4 and Sections 5.5.4 (*Creating Graphics for the Alarms*) and 5.5.5 (*Creating the Bar Graph*) to add the text and scale.

13. Right click **inside** the rectangle, select ***Animation > Fill*** and enter the values shown in Figure 5-5-13.

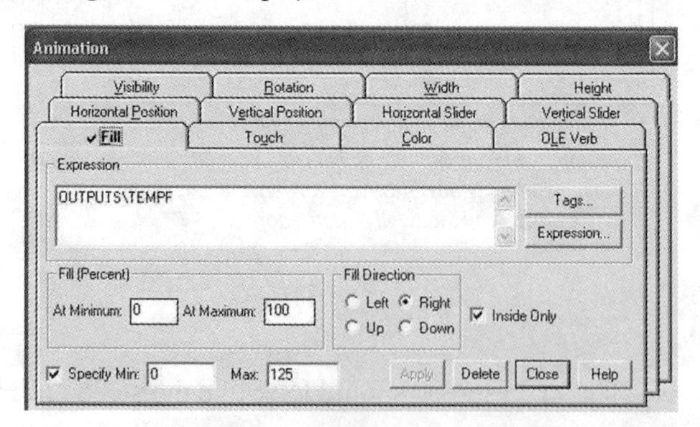

Figure 5-5-13: Screen 4: Creating an Analog Temperature Display (Courtesy of Rockwell Automation, Inc.)

5.5.8 *Creating the Celsius Button*

14. Make the ***Celsius Button*** by selecting ***Objects > Advanced Objects > Button***. Similar to the procedure in Step 12, draw the rectangle and enter the values shown in Figures 5-5-14 and 5-5-15.

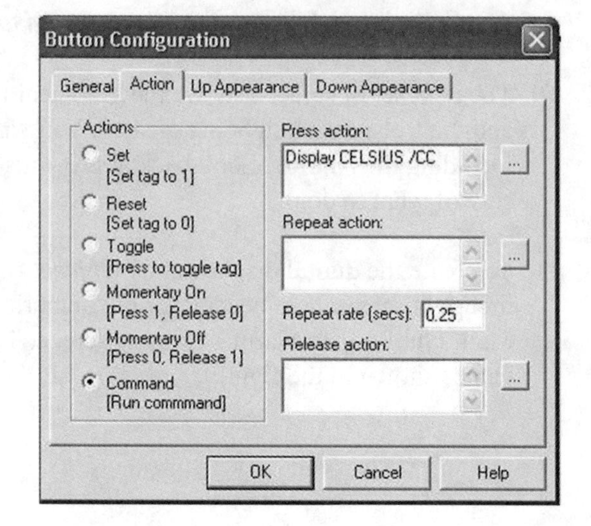

Figure 5-5-14: Creating the Celsius Button (Courtesy of Rockwell Automation, Inc.)

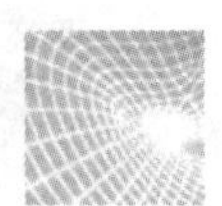

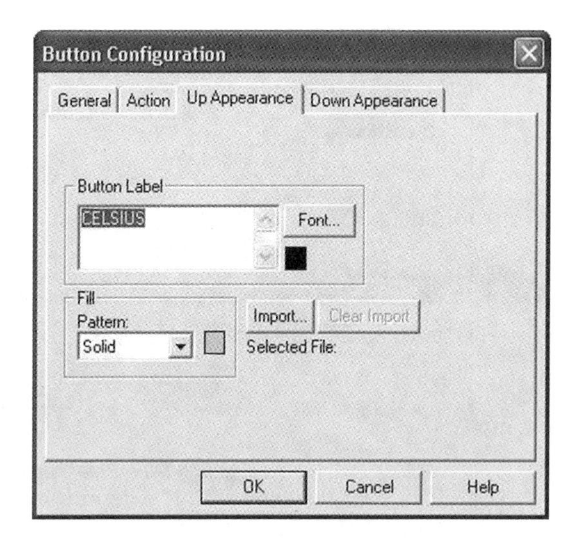

Figure 5-5-15: Configuring the Celsius Button (Courtesy of Rockwell Automation, Inc.)

15. Click "***Same as Up Appearance***" in the ***Down Appearance*** tab; then click ***OK***.

5.5.9 Creating the Celsius Screen

16. To create the ***Celsius*** screen, double click ***Display***; an **Untitled** blank display screen will appear. Enter the items shown in Figure 5-5-16 on the screen.

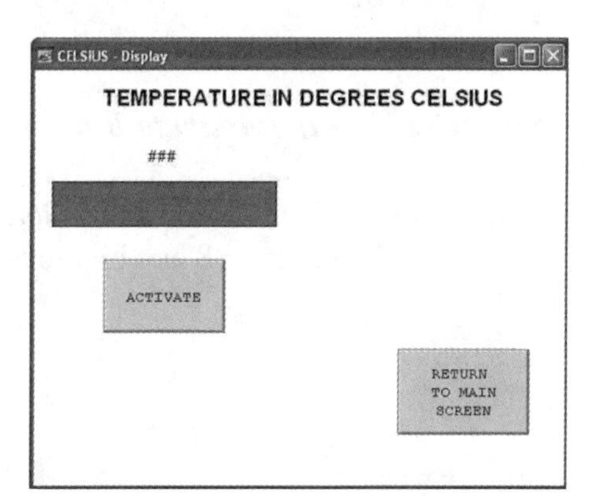

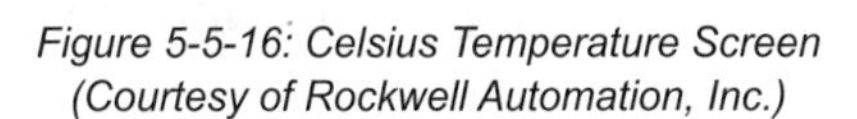
Figure 5-5-16: Celsius Temperature Screen (Courtesy of Rockwell Automation, Inc.)

17. Use the **text tool** to enter the title for the screen.

18. To enter the numeric display, **###,** click ***Objects > Advanced Objects > Numeric Display***. Place the asterisks on the screen and the window shown in Figure 5-5-17 opens. Enter the values shown; then click ***OK***.

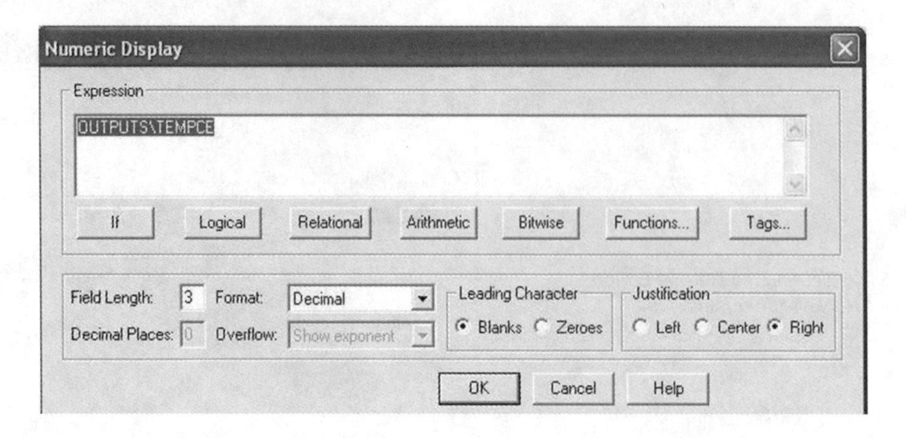

Figure 5-5-17: Enter Numeric Values for Celsius Screen (Courtesy of Rockwell Automation, Inc.)

5.5.10 Creating an Analog Display

19. Create the **Analog** display (horizontal rectangle) using the box tool. Then right click inside the rectangle and click ***Animation* > *Fill***. enter the values as before, shown in Figure 5-5-17 and click ***OK***.

5.5.11 Creating the Activate Button

20. Create the Activate button using ***Object* > *Advanced Objects* > *Button*** and enter the values shown in Figure 5-5-18 and Figure 5-5-19.

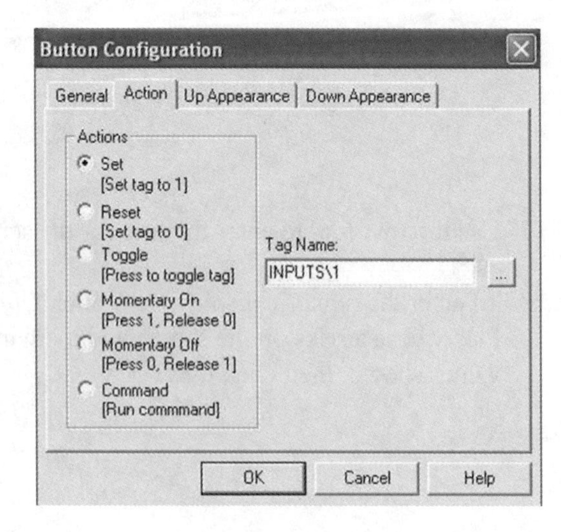

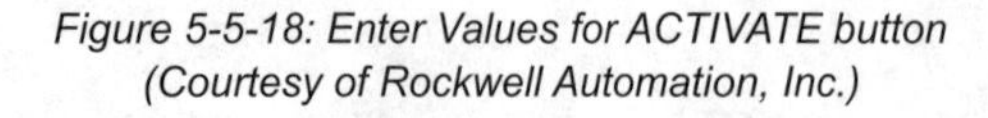
Figure 5-5-18: Enter Values for ACTIVATE button (Courtesy of Rockwell Automation, Inc.)

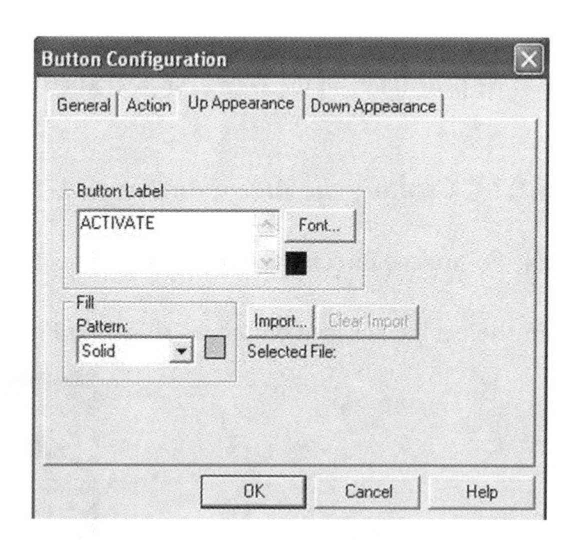

Figure 5-5-19: Create ACTIVATE button (Courtesy of Rockwell Automation, Inc.)

21. Click "***Same as Up Appearance***" in the ***Down Appearance*** tab; then click OK.

5.5.12 Return to Main Screen Button

22. Create the Return to Main Screen button using ***Object > Advanced Objects > Button*** and enter the values shown in Figures 5-5-20 and 5-5-21.

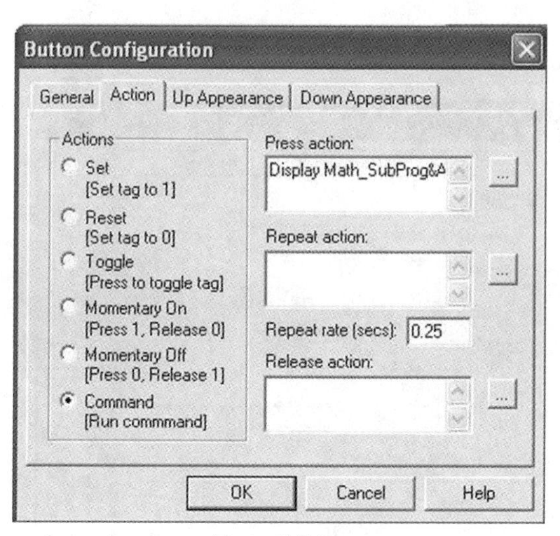

Figure 5-5-20: Enter Values for RETURN TO MAIN SCREEN button (Courtesy of Rockwell Automation, Inc.)

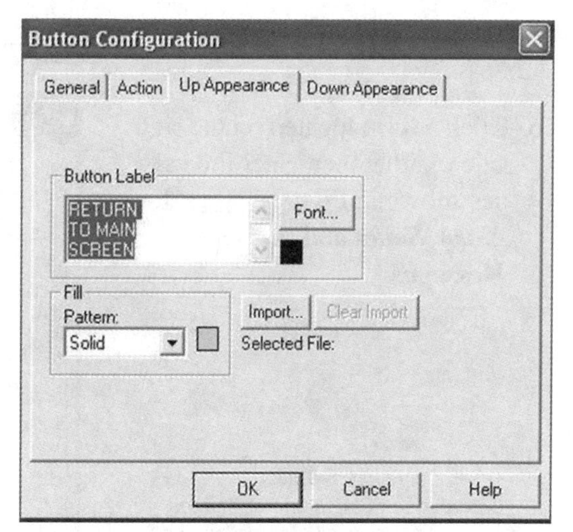

Figure 5-5-21: Create RETURN TO MAIN SCREEN button (Courtesy of Rockwell Automation, Inc.)

23. The **Press Action** should read ***Display Math_SubProg&Alarm /CC***. Click "**Same as Up Appearance**" in the **Down Appearance** tab; then click OK.

5.5.13 Creating the Alarm Button and Alarm Screen

24. Follow the directions in Lab 5-4 on how to create the ***Alarm Button and Alarm Screen***.

25. Select the alarmed output, as shown in Figure 5-5-22.

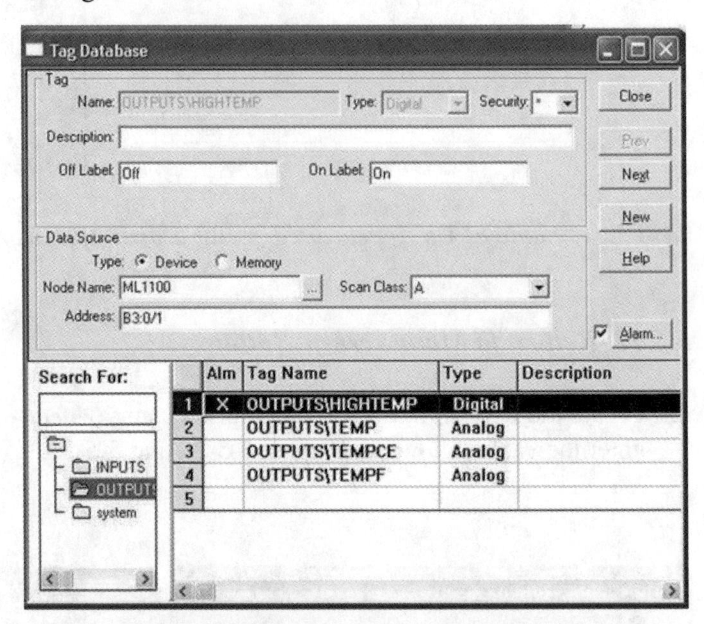

Figure 5-5-22: Create Alarmed Output (Courtesy of Rockwell Automation, Inc.)

26. Click ***Alarm*** located on the right side middle; then enter the values shown in Figure 5-5-23 for ***Alarm States*** and ***Alarm Messages***.

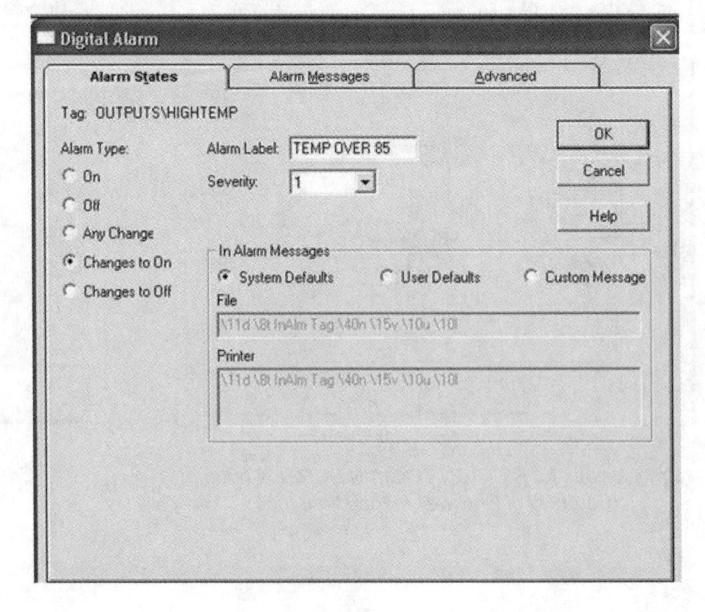

Figure 5-5-23: Create Alarm States and Messages (Courtesy of Rockwell Automation, Inc.)

27. **Close out** this screen and prepare to **run** the project.

5.5.14 Running the Project

28. Click ***Run Mode***, double click ***Graphics > Display > main screen name***. ***Start*** should be activated at the bottom of the ***Run Mode*** screen.

29. Don't forget to activate the ***Alarm mode*** using the ***Command Line***. **See Lab 5-4**.

30. When ***Run Mode*** begins, the temperature will appear in the digital and analog display. Click the ***Celsius*** button and **Activate** it to see the Celsius temperature display.

31. Hold the end of the temperature probe in the closed hand to start the temperature to rise. Click the ***ALARM*** button and note the temperature alarms, as shown in Figure 5-5-24.

32. Click ***CLOSE*** to close out this screen.
    Explore running the project.

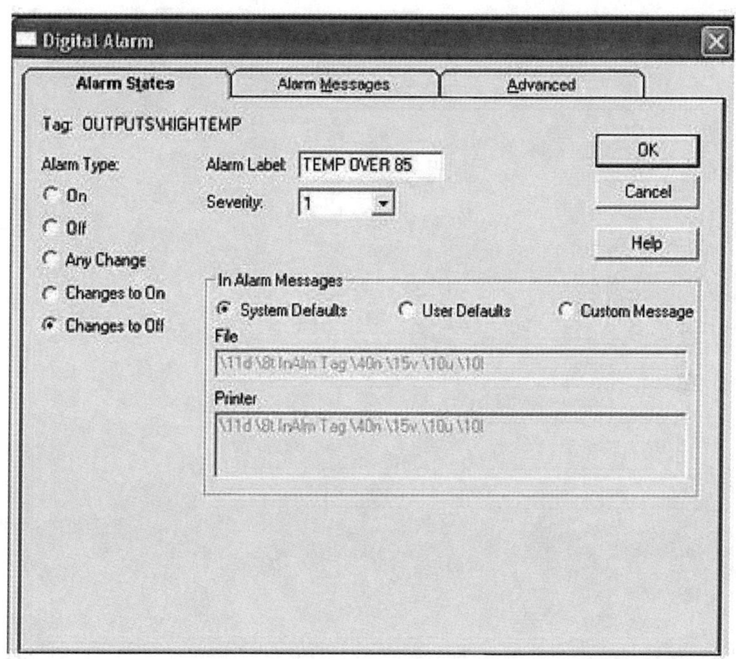

Figure 5-5-24: Alarm Screen for HIGH Temperature (Courtesy of Rockwell Automation, Inc.)

RSView Lab 5-6: Security

5.6.1 Introduction

This lab, which is based upon RSView Lab 5-2, introduces security accessibility that is implemented in HMI. Security is commonly used in industrial processes to prevent access by unauthorized personnel. **The PLC program shown in Figure 5-6-1 comes directly from Lab 5-2 Figure 5-2-1.** Copy and paste Figure 5-2-1 to Figure 5-6-1.

NOTE: When instructions such *as **Edit Mode, Objects>Advanced Objects>Numeric Display,*** and ***others*** appear and are not located on a figure, they are for use in RSView32 software that is running in the background.

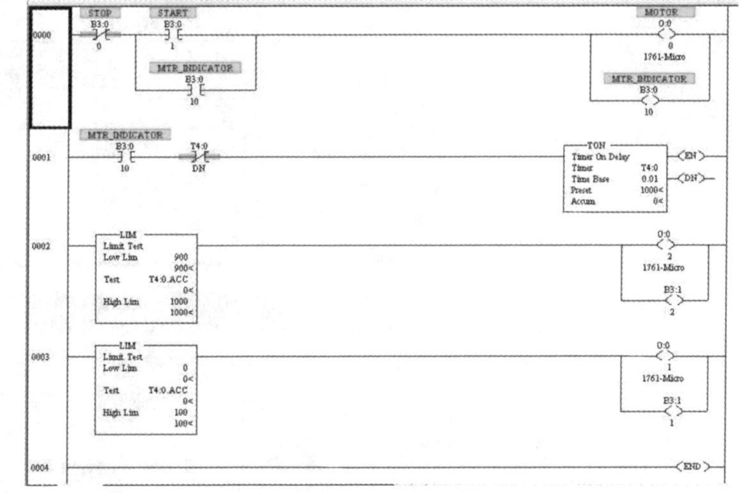

Figure 5-6-1:
Lab 5-2 PLC program used for Lab 5-6
(Courtesy of Rockwell Automation, Inc.)

1. Download the ladder diagram from Figure 5-6-1 into the ***PLC*** and be sure to place the PLC in ***RUN*** mode.

5.6.2 Configuring RSView

2. Make sure the ***RSView*** is configured under ***Edit Mode, System***. Look at RSView Lab 5-1 for the information on how to do this, if you have forgotten.

5.6.3 Creating the Tag Database

3. Load RSView Lab 5-2 and save it as Lab 5-6. Then modify the various parts of the project as described below.

4. Begin by looking at the tag database under ***System*** > ***Tag Database***. Make sure ***Input*** and ***Output*** folders are created for the ***Tag Database*** if they do not exist.

5. Create the ***Input and Output*** tags shown in Figure 5-6-2 if they do not already exist. Click ***New*** to create the tag. Make sure to click ***Accept*** for each tag, then ***Close*** when all tags have been created. Enter values shown in the image into Figure 5-6-2.

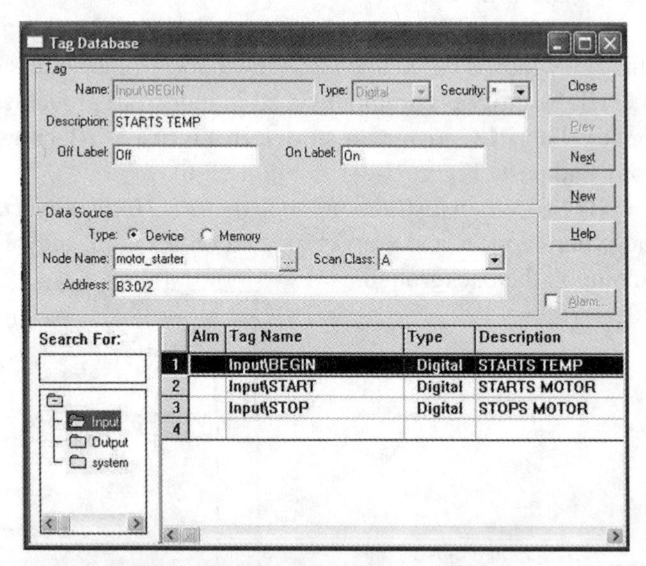

Figure 5-6-2: Begin entering Input tags to Tag Database (Courtesy of Rockwell Automation, Inc.)

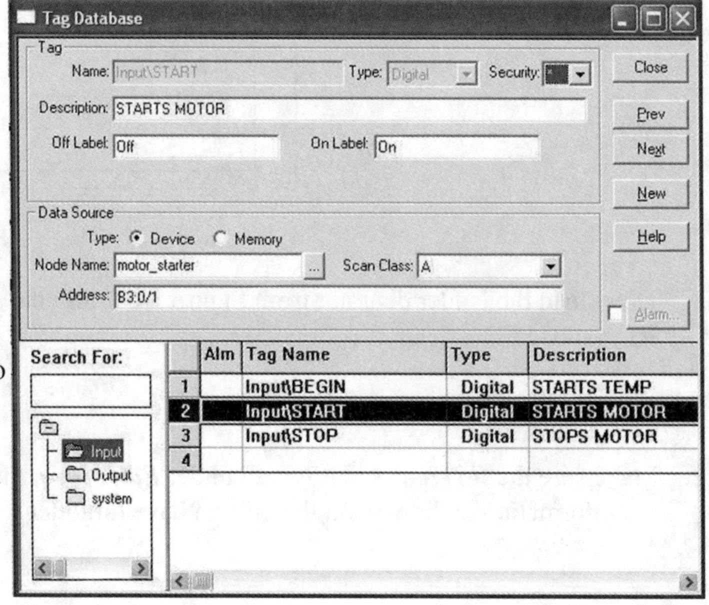

Figure 5-6-3: Enter Input Tags Screen #1 (Courtesy of Rockwell Automation, Inc.)

6. Create the nine ***Input / Output screens*** beginning in Figure 5-6-3 to Figure 5-6-11. If a screen with tags exists, go to the next screen.

7. Don't forget to click ***Accept*** and ***Close*** when all tags are entered.

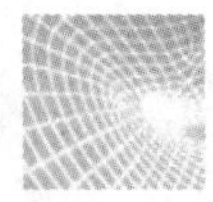

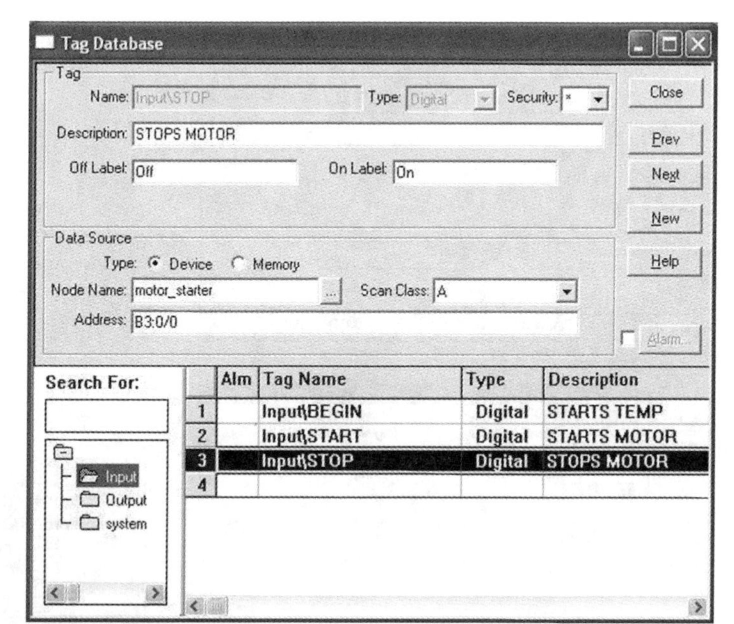

Figure 5-6-4: Enter Input Tags Screen #2 (Courtesy of Rockwell Automation, Inc.)

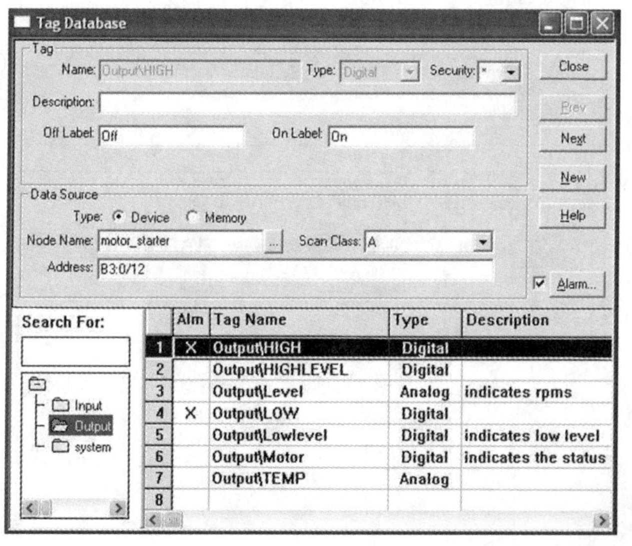

Figure 5-6-5: Enter Output Tags Screen #3 (Courtesy of Rockwell Automation, Inc.)

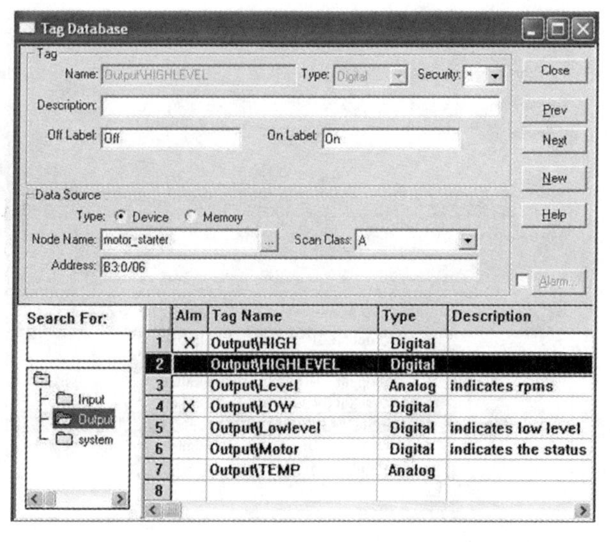

Figure 5-6-6: Enter Output Tags Screen #4 (Courtesy of Rockwell Automation, Inc.)

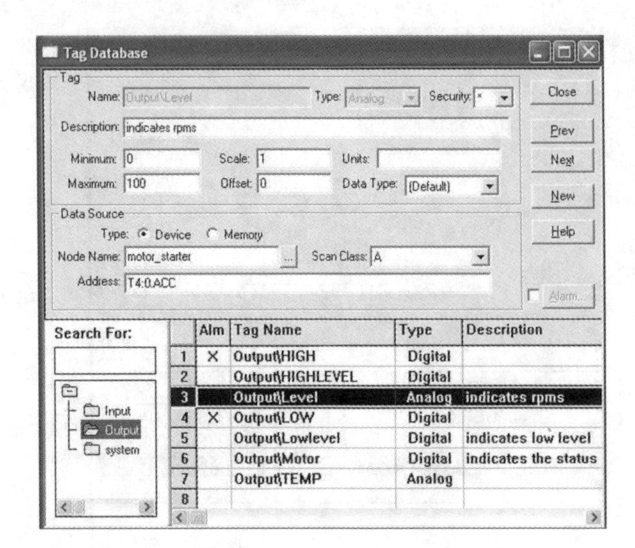

Figure 5-6-7: Enter Output Tags Screen #5
(Courtesy of Rockwell Automation, Inc.)

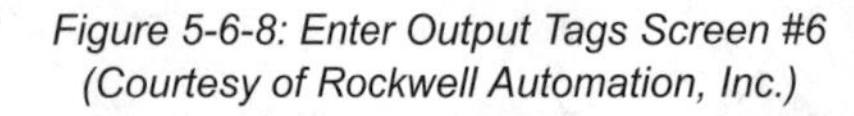

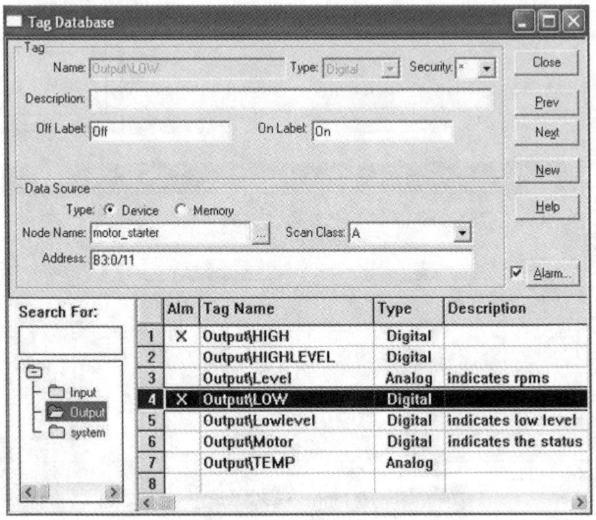

Figure 5-6-8: Enter Output Tags Screen #6
(Courtesy of Rockwell Automation, Inc.)

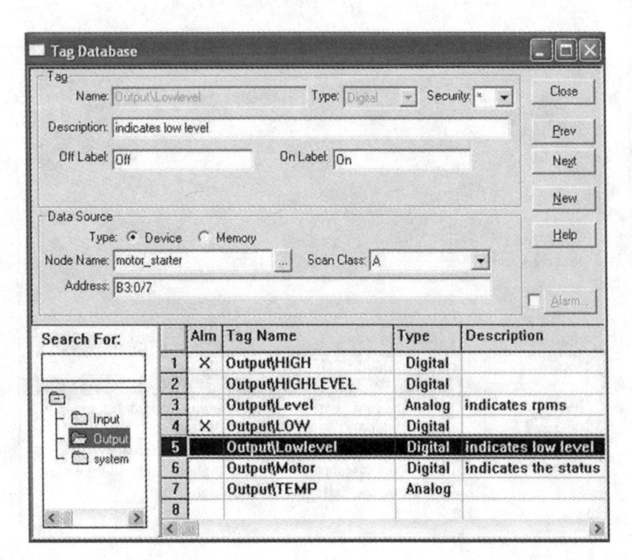

Figure 5-6-9: Enter Output Tags Screen #7
(Courtesy of Rockwell Automation, Inc.)

Tag Database

Tag — Name: Output\Motor Type: Digital Security: *

Description: indicates the status of the motor

Off Label: Off On Label: On

Data Source — Type: (•) Device ( ) Memory

Node Name: motor_starter Scan Class: A

Address: B3:0/3

Close | Prev | Next | New | Help | Alarm...

Search For:

Input / Output / system

	Alm	Tag Name	Type	Description
1	X	Output\HIGH	Digital	
2		Output\HIGHLEVEL	Digital	
3		Output\Level	Analog	indicates rpms
4	X	Output\LOW	Digital	
5		Output\Lowlevel	Digital	indicates low level
6		Output\Motor	Digital	indicates the status
7		Output\TEMP	Analog	
8				

Figure 5-6-10: Enter Output Tags Screen #8 (Courtesy of Rockwell Automation, Inc.)

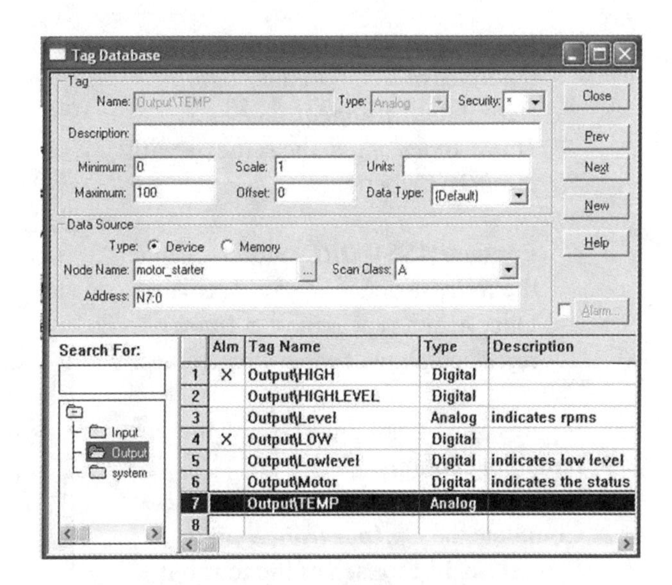

Tag Database

Tag — Name: Output\TEMP Type: Analog Security: *

Description:

Minimum: 0 Scale: 1 Units:

Maximum: 100 Offset: 0 Data Type: (Default)

Data Source — Type: (•) Device ( ) Memory

Node Name: motor_starter Scan Class: A

Address: N7:0

Close | Prev | Next | New | Help | Alarm...

Search For:

Input / Output / system

	Alm	Tag Name	Type	Description
1	X	Output\HIGH	Digital	
2		Output\HIGHLEVEL	Digital	
3		Output\Level	Analog	indicates rpms
4	X	Output\LOW	Digital	
5		Output\Lowlevel	Digital	indicates low level
6		Output\Motor	Digital	indicates the status
7		Output\TEMP	Analog	
8				

Figure 5-6-11: Enter Output Tags Screen #9 (Courtesy of Rockwell Automation, Inc.)

5.6.4 Creating the Graphic Screens

8. Figure 5-6-12 shows how the main Security Access screen should appear.

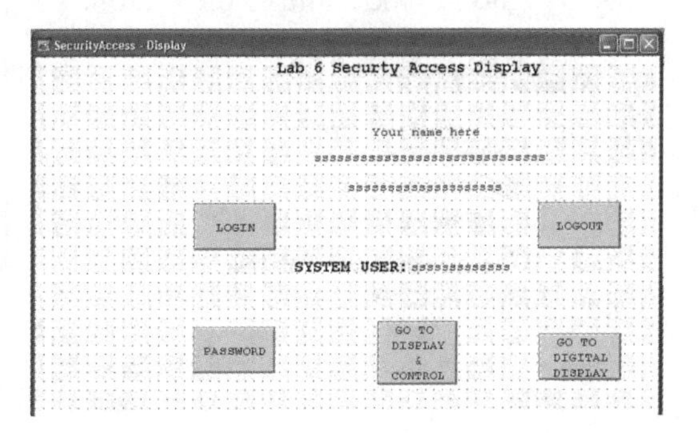

Figure 5-6-12: Main Security Access Screen (Courtesy of Rockwell Automation, Inc.)

5.6.5 *Buttons*

9. Create the five buttons shown on the ***Security Access screen*** in Figure 5-6-12 by using ***Objects > Advanced Objects > Buttons.***

10. Double click each button box — ***LOGIN, LOGOUT, and PASSWORD*** — and enter the values shown in Figure 5-6-13.

    a. Enter values shown in the ***Action*** tab for ***LOGIN***. Complete the ***Up-Appearance*** tab by entering the name of the button. Make sure that the box is checked for ***Same as Up Appearance*** in the ***Down Appearance*** Tab.

    b. For the ***LOGOUT*** button, change the ***Release*** action to the ***Logout***. Repeat the ***UpAppearance*** and ***DownAppearance***, the same as with the ***LOGIN*** button.

    c. For the ***PASSWORD*** button, change the ***Release*** action to say ***Password.*** Once again, repeat the ***UpAppearance*** and ***DownAppearance*** values.

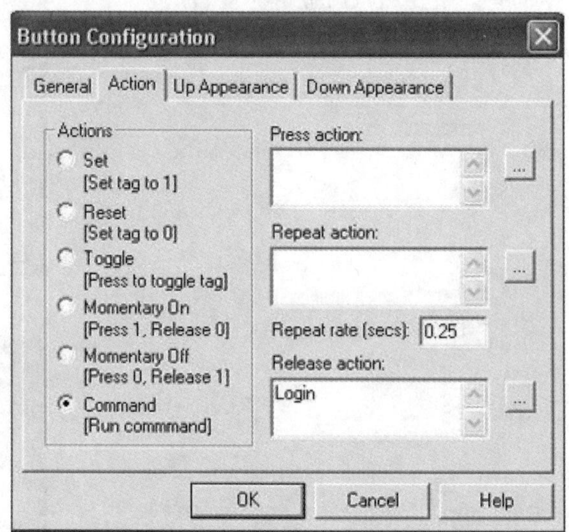

Figure 5-6-13: Buttons for LOGIN, LOGOUT, PASSWORD (Courtesy of Rockwell Automation, Inc.)

11. Figures 5-6-14 and 5-6-15 show the remaining buttons.

12. Complete the ***UpAppearance*** tab in Figure 5-6-14 for each of the four buttons by entering the name of the button. Make sure that the box is NOT checked for ***Same as Up Appearance*** in the ***Down Appearance Tab***.

13. Enter the ***name, date, and time*** in Figure 5-6-17. Create the title for this screen using the text tool.

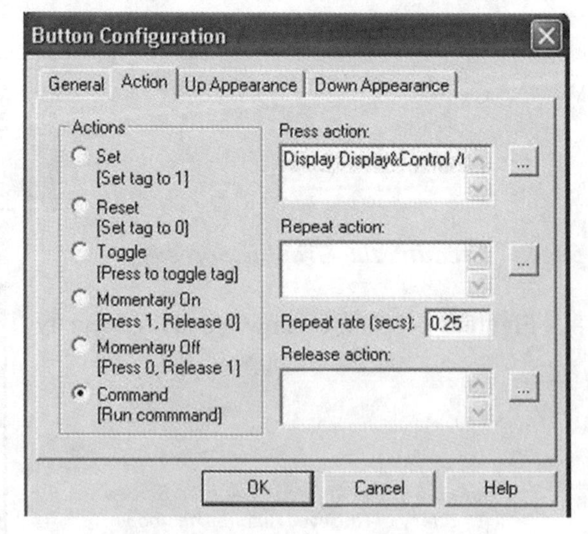

Figure 5-6-14: Go to Display&Control Button (Courtesy of Rockwell Automation, Inc.)

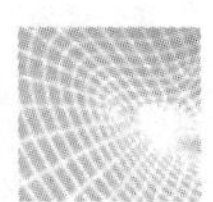

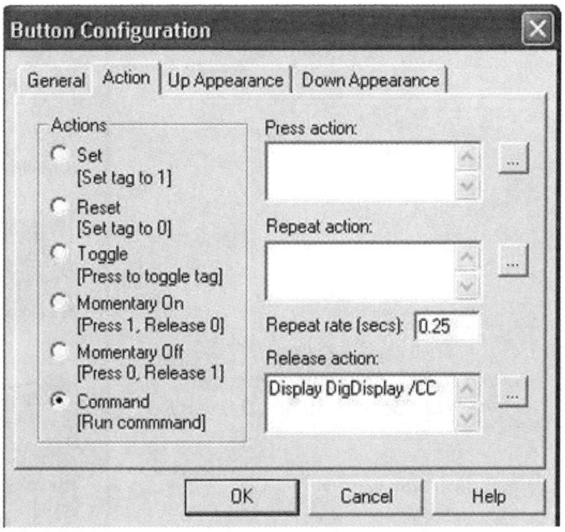

Figure 5-6-15: Go to Digital Display
(Courtesy of Rockwell Automation, Inc.)

14. Create the string display (ssssssss) on Figure 5-6-17 by clicking ***Objects*** > ***Advanced Objects*** > ***String Display***. Locate the mouse pointer in the lower right side. Select and drag a rectangle to the mouse pointer location for the string (ssss) display.

15. Double click the string display and see the ***String Display*** window. Enter the values shown in Figure 5-6-16 in the configuration windows.

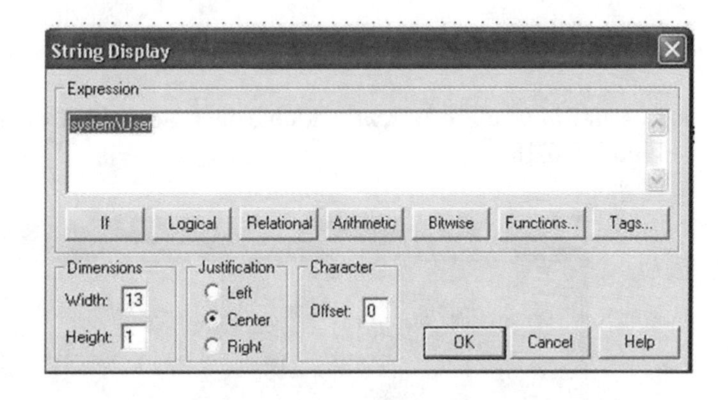

Figure 5-6-16: String Display Screen
(Courtesy of Rockwell Automation, Inc.)

5.6.6 Entering Security Settings

16. Enter the security setting through using ***Edit Mode***, then click the ***Display&Control*** screen that will appear, and then click ***Edit***. Right click in an open area of the screen to see the Display Screen Figure 5-6-17.

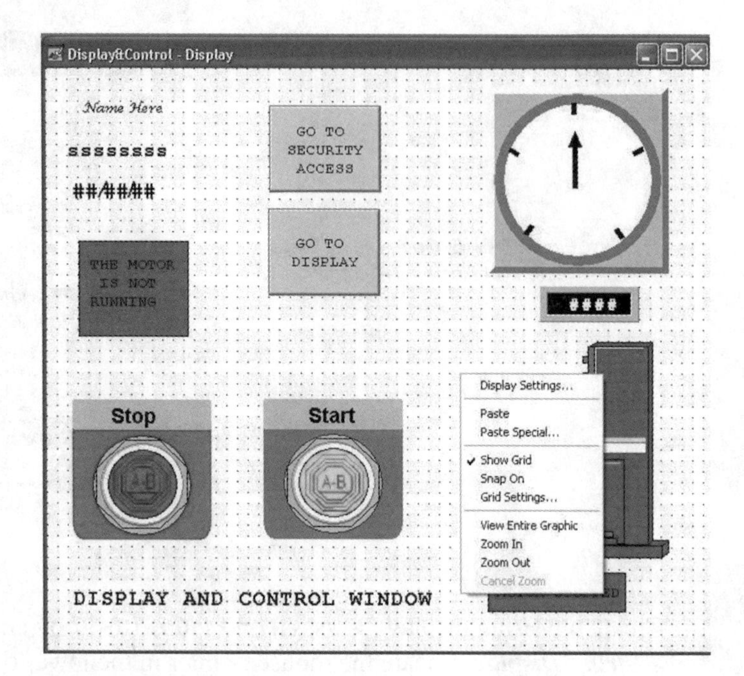

Figure 5-6-17: Main Display and Control Window (Courtesy of Rockwell Automation, Inc.)

17. Click ***Display*** and change the ***Security Code*** to ***D***.

18. Repeat this process for the ***Digital Display*** screen shown in Figure 5-6-18, except change its ***Security Code*** to ***F***.

5.6.7 Entering Security Codes

19. In ***Edit Mode***, click ***System*** > double click **Security Codes** and enter the values shown in Figure 5-6-18.

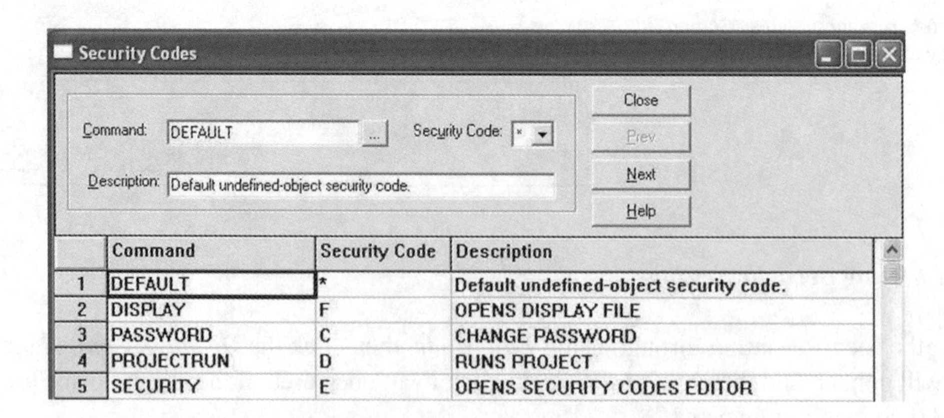

	Command	Security Code	Description
1	DEFAULT	*	Default undefined-object security code.
2	DISPLAY	F	OPENS DISPLAY FILE
3	PASSWORD	C	CHANGE PASSWORD
4	PROJECTRUN	D	RUNS PROJECT
5	SECURITY	E	OPENS SECURITY CODES EDITOR

Figure 5-6-18: Enter Security Codes (Courtesy of Rockwell Automation, Inc.)

5.6.8 Entering User Accounts

20. Click ***User Accounts*** and enter the values shown in Figure 5-6-19.

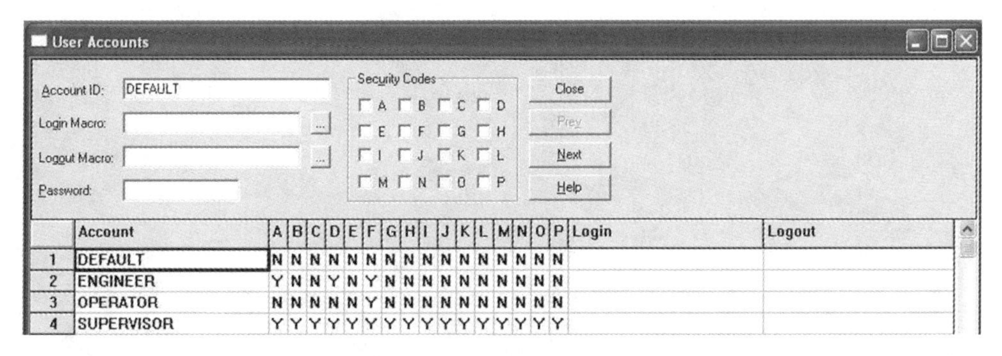

Figure 5-6-19: Enter User Accounts (Courtesy of Rockwell Automation, Inc.)

21. Click the letter boxes in the ***Security Codes*** area to set the individual codes for each account.

22. Figure 5-6-20 shows the ***passwords*** for each account.

Account	Password
DEFAULT	----
ENGINEER	Eng
OPERATOR	op1
SUPERVISOR	123

Figure 5-6-20: Passwords for Users (Courtesy of Rockwell Automation, Inc.)

23. Make sure to ***save*** the project before running it.

5.6.9 Running and Testing the Security Program

24. Start the ***SecurityAccess*** main display window; the screen in Figure 5-6-21 will appear.

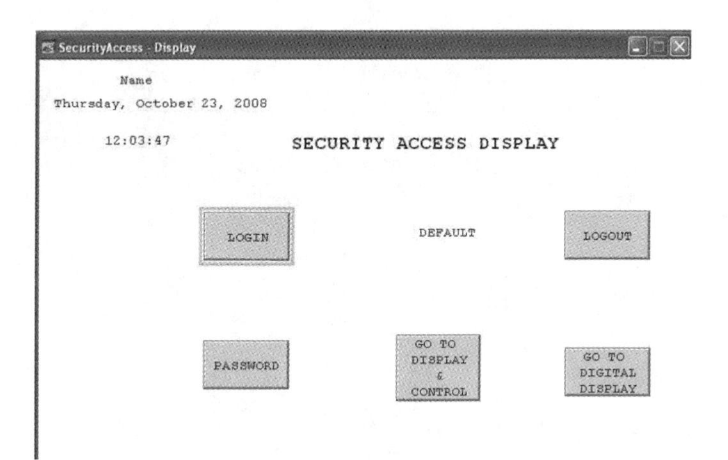

Figure 5-6-21: Security Access Display (Courtesy of Rockwell Automation, Inc.)

25. Test the program for each user and note which of the screens and buttons they can access. Be able to describe what was found about the security settings from this project.

26. This completes the lab.

RSView Lab 5-7: TrendX Charting and Data Logging

5.7.1 Introduction

In previous labs, the learning focused on how to create various types of displays, messages, and alarms. In this lab, the focus will be about creating graphical trend displays as well as creating a data log, i.e., a database.

NOTE: When instructions such *as **Edit Mode, Objects>Advanced Objects>Numeric Display and others*** appear and are not located on a figure, they are for use in RSView32 software, which is running in the background.

1. Load the PLC program from Lab 5-2 into the PLC for this lab activity (Figure 5-7-1).

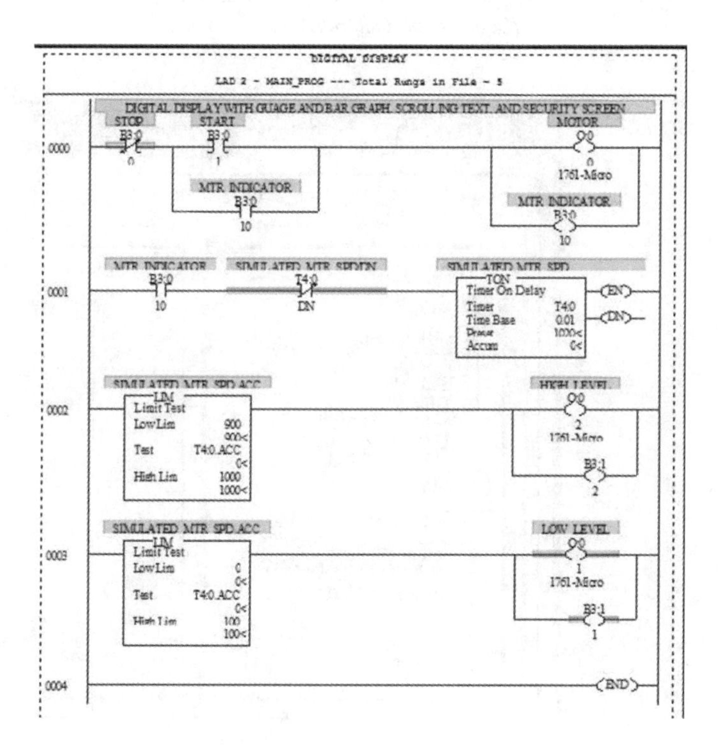

Figure 5-7-1:
Load PLC program from Lab 5-2
(Courtesy of Rockwell Automation, Inc.)

2. Make sure the PLC is in **RUN** mode.

5.7.2 Configuring RSView

3. Make sure ***RSView*** is configured under ***Edit Mode*** > ***System***. To refresh how to launch and configure ***RSView***, review Lab 5-1, which begins in Section ***5.2.2***. It provides the information on how to configure the RSView.

5.7.3 Setting Up the Tag Database

4. Enter the tags from ***RSView Lab 2*** into the ***Tag Editor***.

5.7.4 Creating the Main Screen

5. Follow Steps ***5.3.5*** to Steps ***5.3.7*** in ***Lab 5-2*** to begin creating the main screen. Move the icons if necessary to make space for the ***RSTrendX*** chart (Figure 5-7-2).

Figure 5-7-2: Main TrendX Screen (Courtesy of Rockwell Automation, Inc.)

5.7.5 Creating Data Logging, Trends, and Activating the Data Logging

6. Double click ***Data Log Setup*** under ***Edit Mode*** (Figure 5-7-3).

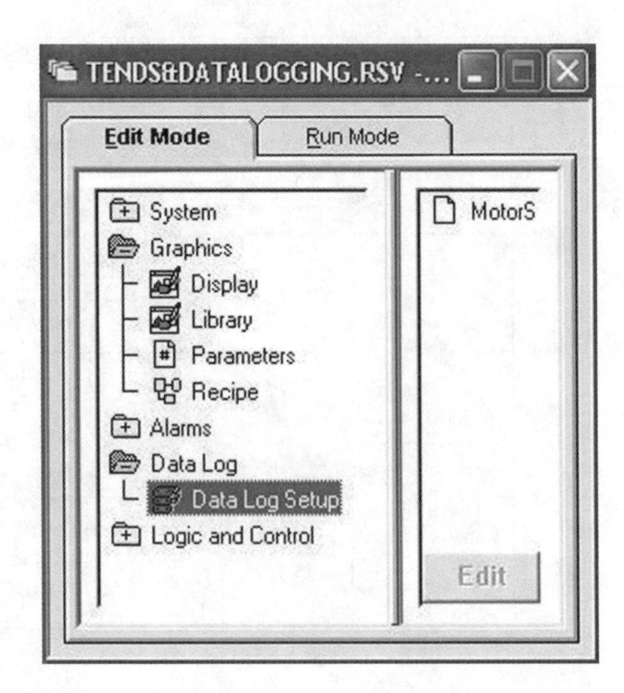

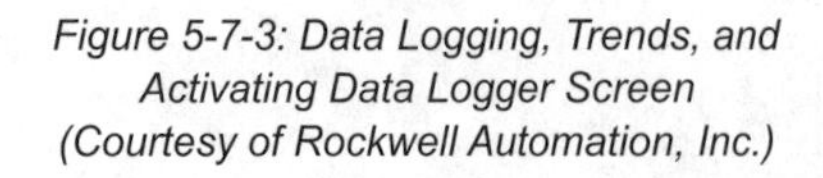
Figure 5-7-3: Data Logging, Trends, and Activating Data Logger Screen (Courtesy of Rockwell Automation, Inc.)

7. The ***Untitled – Data Log Setup*** window should appear. Enter the values shown in Figure 5-7-4; then click ***OK***.

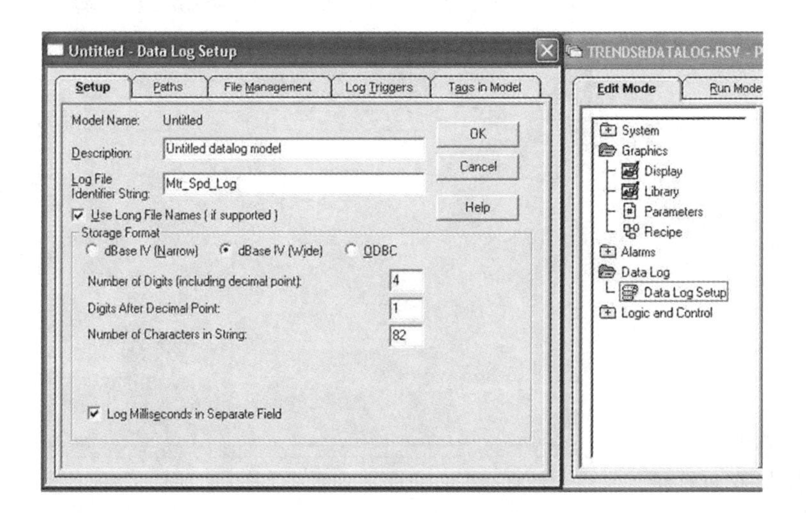

Figure 5-7-4: Data Log Setup Screen (Courtesy of Rockwell Automation, Inc.)

8. Enter values for each tab, as shown in Figure 5-7-5: ***Paths, File Management, Log Triggers, Tags in Model***.

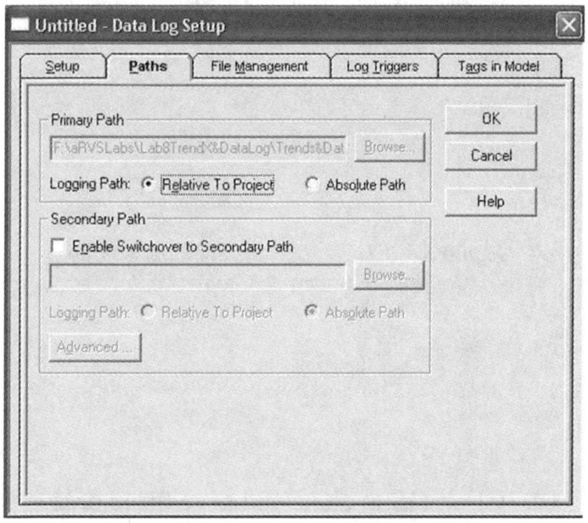

Figure 5-7-5: Data Log Path Screen (Courtesy of Rockwelll Automation, Inc.)

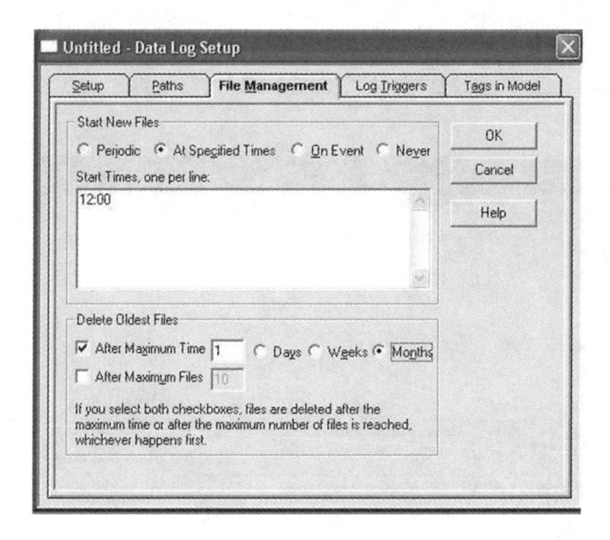

9. Under ***File Management***, click ***At Specified Times***, then ***12:00*** in the Start Times (Figure 5-7-6). This selection means that a new file will be created every day at ***12 midnight***. Once a **month**, old files will be deleted.

Figure 5-7-6: Data Log File Management Screen (Courtesy of Rockwell Automation, Inc.)

10. Enter the values for ***Log Triggers***, as shown in Figure 5-7-7.

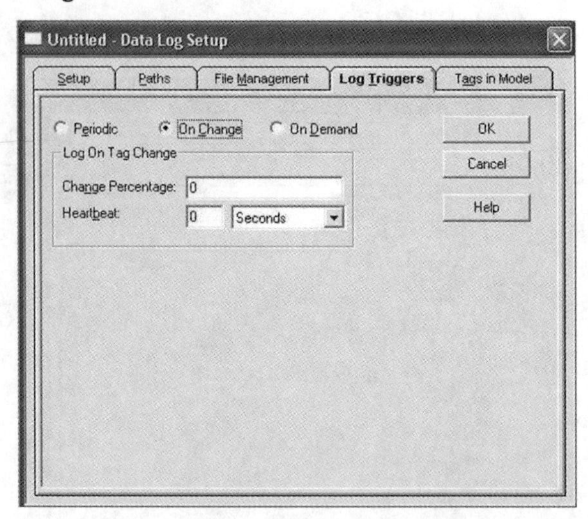

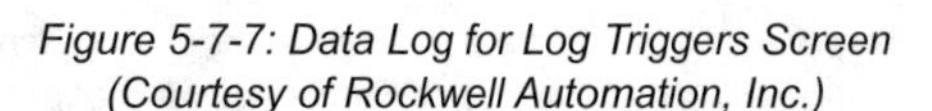

Figure 5-7-7: Data Log for Log Triggers Screen (Courtesy of Rockwell Automation, Inc.)

11. In the ***Tags in Model*** tab, click ***Tag(s) to Add:*** selection box and select ***Output\Level***. Then click the ***Add*** button to add it to the ***Tag(s) in the Model*** list, as shown in Figure 5-7-8.

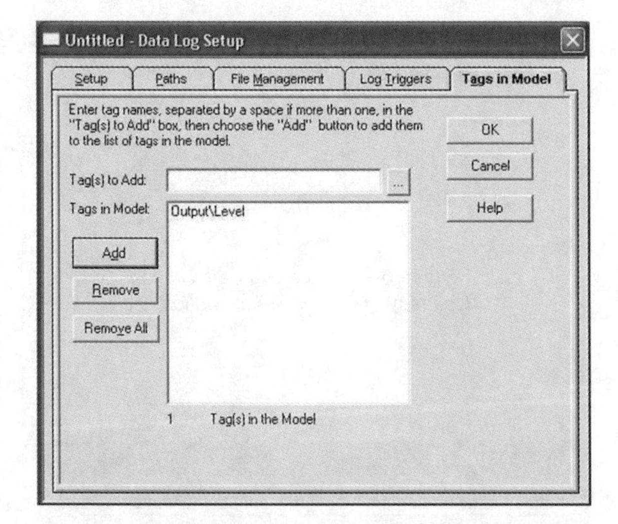

Figure 5-7-8: Data Log Tags in Model Screen (Courtesy of Rockwell Automation, Inc.)

12. Click ***OK*** to close and save these data values. A file name will need to be entered that will be the file stored in the ***DLGLOG*** folder in the HMI project. See Figure 5-7-9.

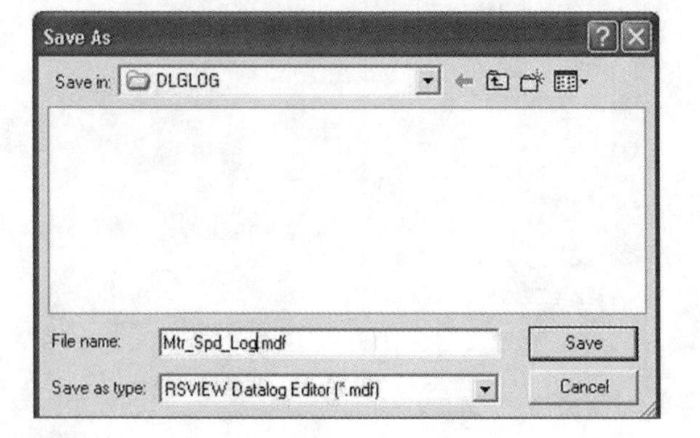

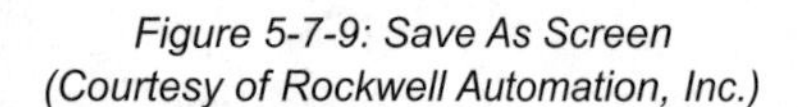

Figure 5-7-9: Save As Screen (Courtesy of Rockwell Automation, Inc.)

13. Confirm that the ***Mtr_Spd_Log.mdf*** file is visible in the ***DLGLOG*** folder (Figure 5-7-10).

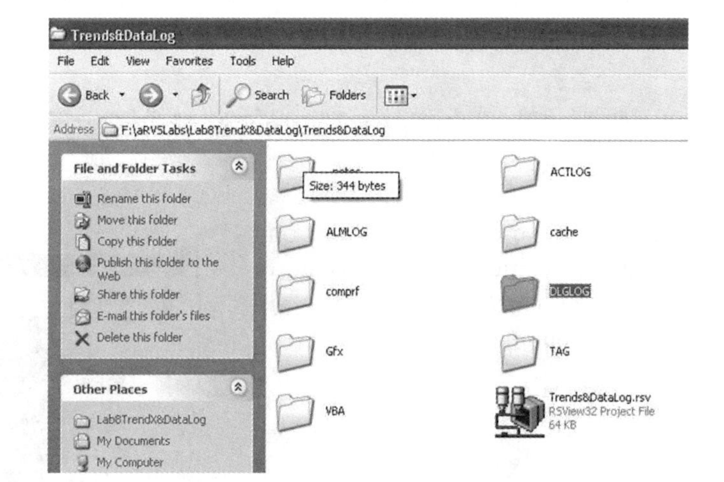

Figure 5-7-10: Trends and Data Log Screen (Courtesy of Rockwell Automation, Inc.)

14. As shown in Figure 5-7-11, click ***Save*** to store the ***MotorSpeed.mdf*** file.

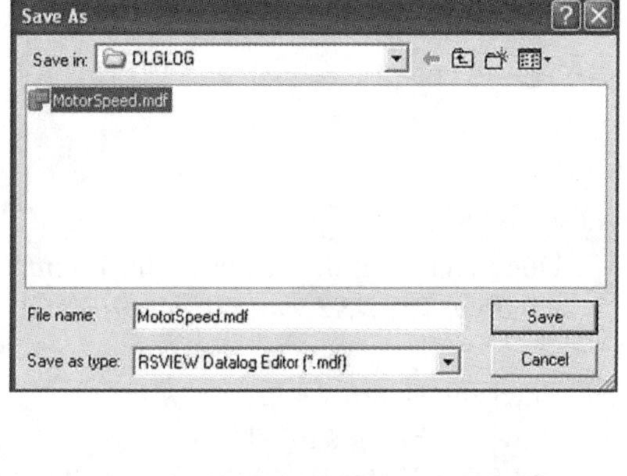

Figure 5-7-11: DLGLOG Screen (Courtesy of Rockwell Automation, Inc.)

5.7.6 *Creating TrendX Display to Show Logged Data*

15. Open a new screen by clicking in ***Edit Mode*** > ***Graphics*** > ***Display***, then double clicking ***Display*** to create a new ***Untitled-Display***. Right click anywhere in the empty graphic display and click the ***Display Settings Properties*** window. Enter the values shown in Figure 5-7-12. Skip the ***Behavior*** tab settings and save this window. Click ***OK***.

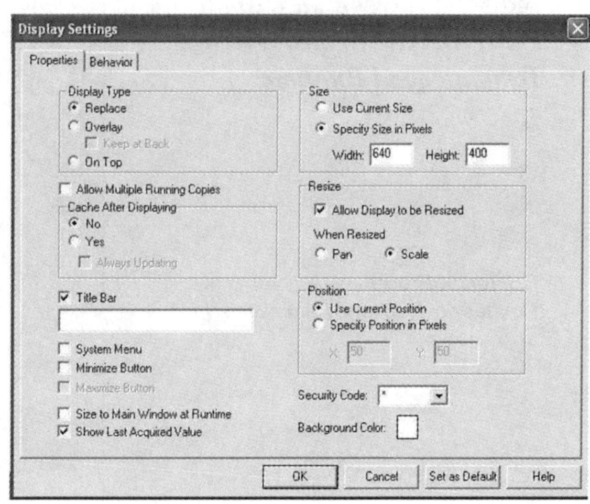

Figure 5-7-12: Display Settings Screen (Courtesy of Rockwell Automation, Inc.)

16. On the ***View*** menu, select ***ActiveX Toolbox***, if not already open. The ActiveX Toolbox will appear, as shown in Figure 5-7-13. Click the ***TrendX button*** once to cause it to be selected and indented.

Figure 5-7-13: TrendX Button (Courtesy of Rockwell Automation, Inc.)

17. Click and drag the ***RSTrendX*** object to position to fill the window display, as shown in Figure 5-7-14. Close out the ***ActiveX Toolbox***.

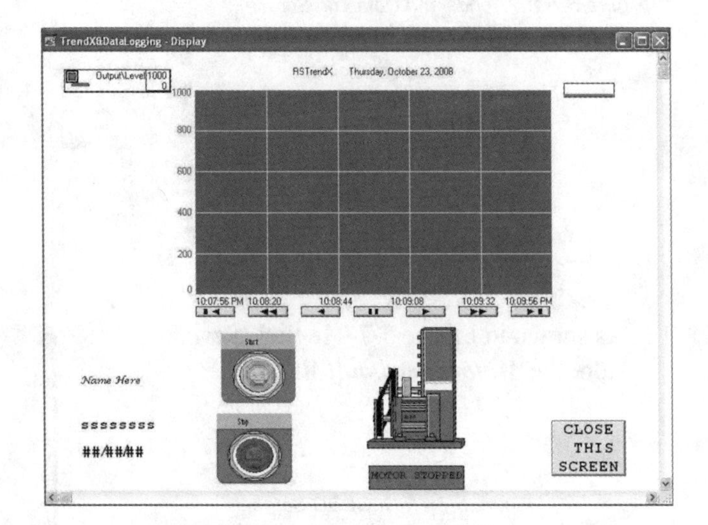

Figure 5-7-14: TrendX Data Logging Display (Courtesy of Rockwell Automation, Inc.)

18. Double click in the center of the ***TrendX*** graph display to open the ***TrendX properties*** window. The ***RSTrendX Properties*** will be displayed as a series of tabs.

19. Enter the values for each tab, as shown in Figures 5-7-15, 5-7-16, and 5-7-17: ***General, Display, Pens, X-Axis, Y-Axis, Overlays, Template, and Runtime.***

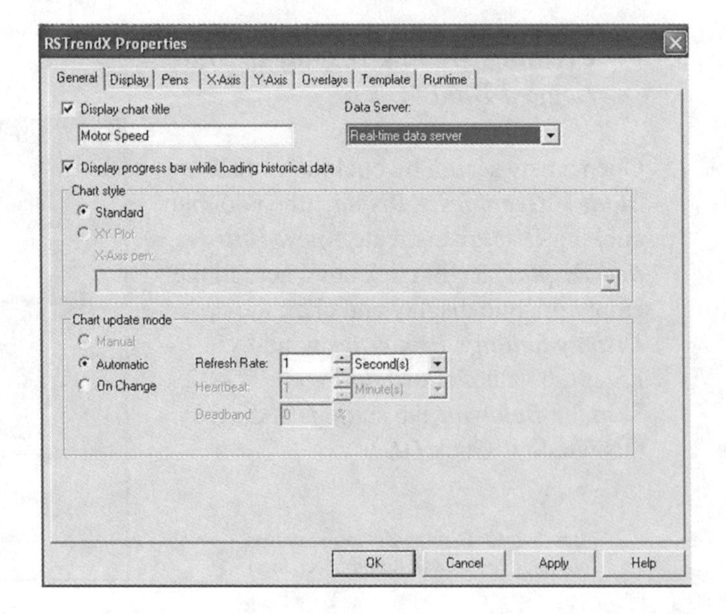

Figure 5-7-15: RSTrendX Properties General Screen (Courtesy of Rockwell Automation, Inc.)

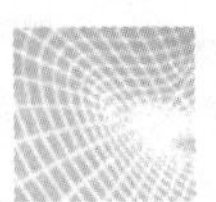

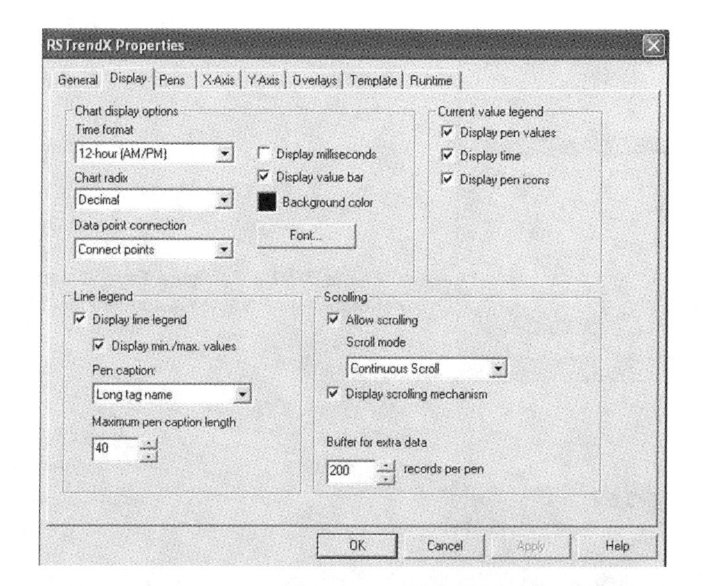

Figure 5-7-16: RSTrendX Properties Display Screen (Courtesy of Rockwell Automation, Inc.)

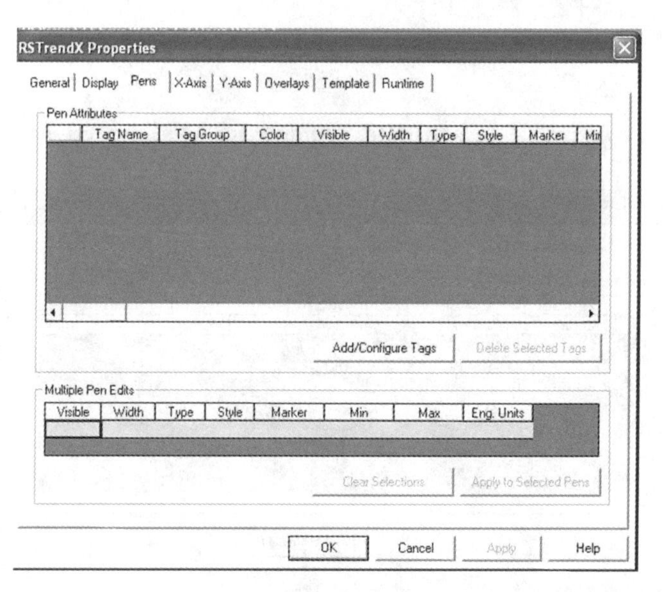

Figure 5-7-17: RSTrendX Properties Pens Screen (Courtesy of Rockwell Automation, Inc.)

20. Click the ***Add/Configure Tags*** in the just below the Pens window, as shown in Figure 5-7-17.

21. Click the ***Tag Group down arrow***, as shown at the top of Figure 5-7-18, and select ***Motor-Speed***. Enter ***Output\Level*** in the top window. Then click the ***Add All*** button and make sure the data is visible, as shown near the bottom of the window. Click ***OK***.

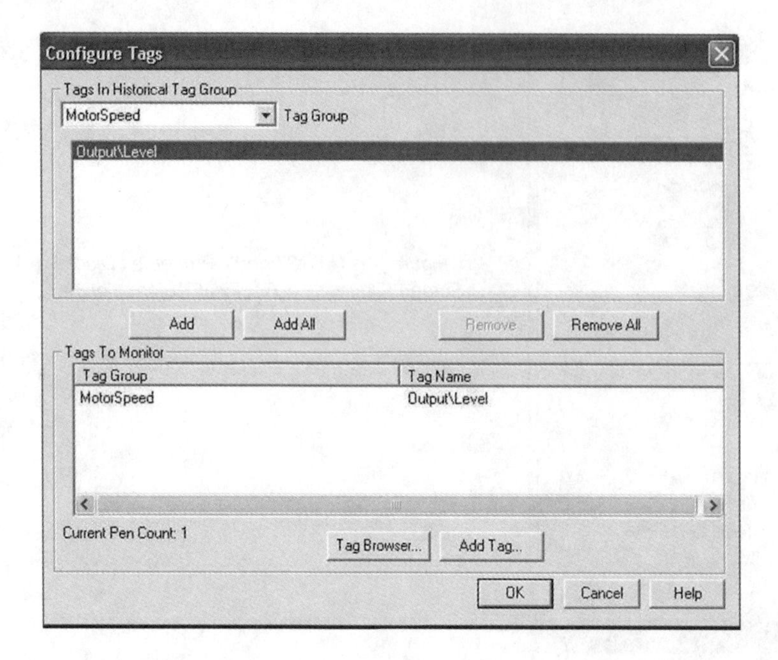

Figure 5-7-18: Configure Tags Screen (Courtesy of Rockwell Automation, Inc.)

22. A line should appear in the ***RSTrendx Properties Pen*** Attributes window for MotorSpeed, as shown in Figure 5-7-19.

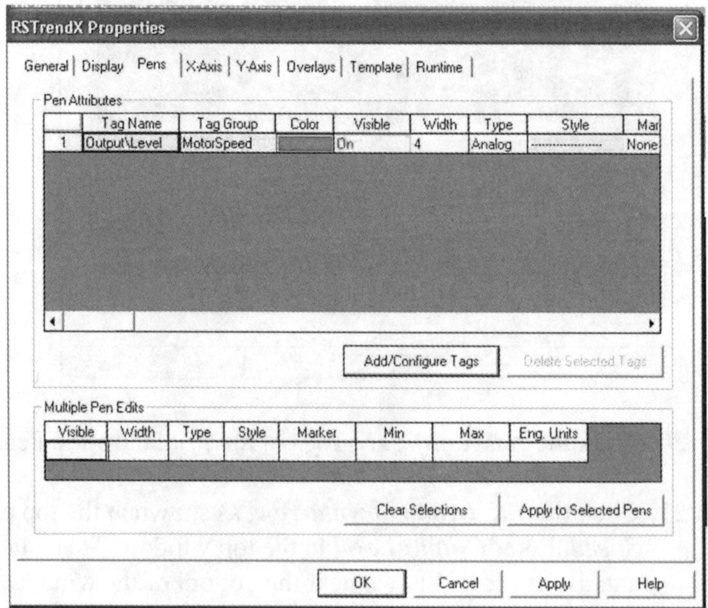

Figure 5-7-19: RSTrendX Properties for Pen Screen (Courtesy of Rockwell Automation, Inc.)

23. Change the ***Color*** column to RED by clicking to get a color palette and selecting Red. Click ***OK*** to close out the palette.

24. Double click ***Width*** and change to ***4***. Change the values in ***Type*** and ***Style***.

25. Enter values for remaining Pen Attributes as shown in the RSTrendX Properties window in Figure 5-7-20.

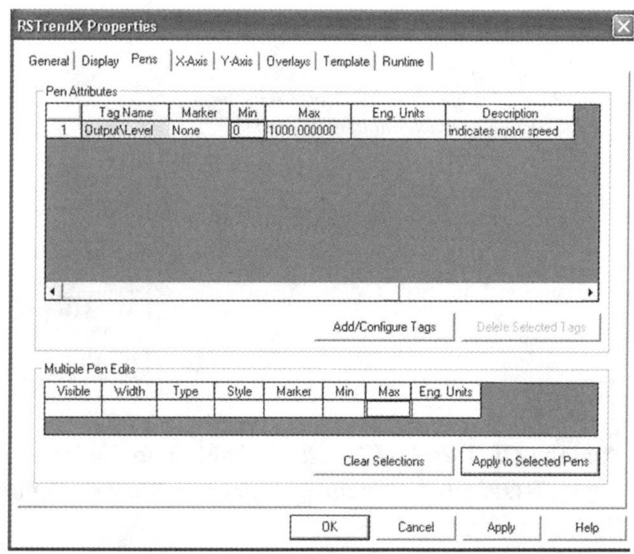

Figure 5-7-20:
RSTrendX Properties for Pen Attributes
(Courtesy of Rockwell Automation, Inc.)

26. Click ***Apply (DO NOT CLICK OK!)***

27. Click the ***X-Axis*** tab and enter the values shown in Figure 5-7-21. Click ***Apply (DO NOT CLICK OK).***

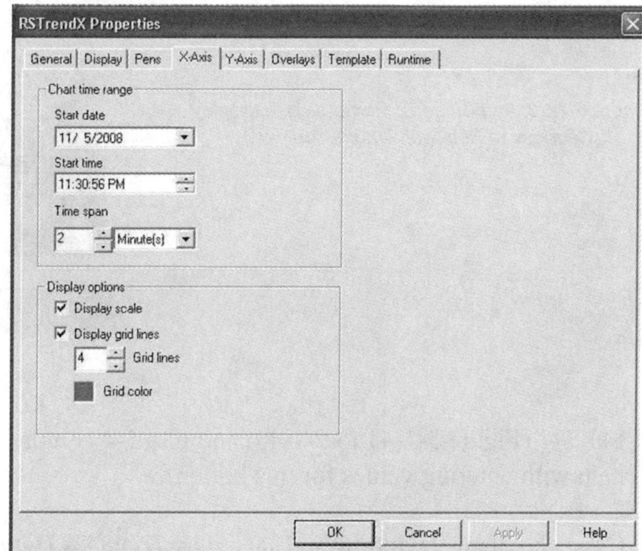

Figure 5-7-21 RSTrendX Properties for X-Axis Tag
(Courtesy of Rockwell Automation, Inc.)

28. Click the ***Y-Axis*** tab and enter the values shown in Figure 5-7-22. Click ***OK*** to save.

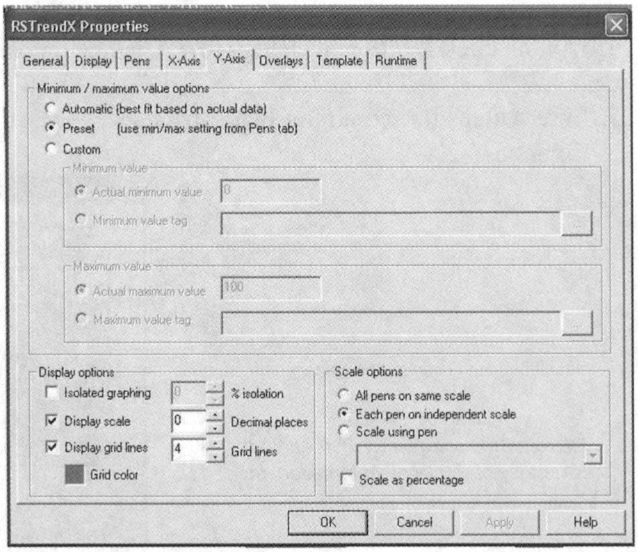

Figure 5-7-22: RSTrendX Properties for Y-Axis Tag (Courtesy of Rockwell Automation, Inc.)

29. Add the ***CLOSE THIS SCREEN*** button to the lower right corner of the graphic display by clicking ***Objects*** > ***Advanced Objects*** > ***Button*** (Figure 5-7-23).

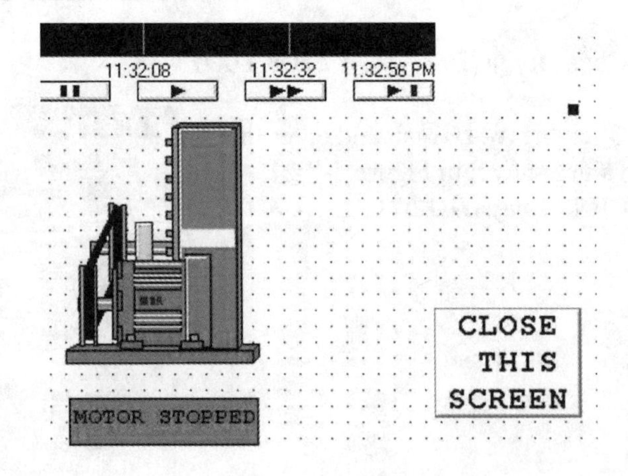

Figure 5-7-23: Add Close Screen Button to Display (Courtesy of Rockwell Automation, Inc.)

30. Lab 5-1 (Figure 5-1-41 to 5-1-43) and Lab 5-4 (Figures 5-4-39 to 5-4-42) can provide additional help with entering values for this button.

31. Close the display window and save it as ***TrendX&DataLogging.gfx.***

5.7.7 Configuring Project Startup Properties

32. In the ***Project Manager*** (right image), open ***System***. Then double click ***Startup*** to open the ***Startup editor*** (left image).

33. Enter the values shown in Figure 5-7-24 for the ***Preferences*** tab.

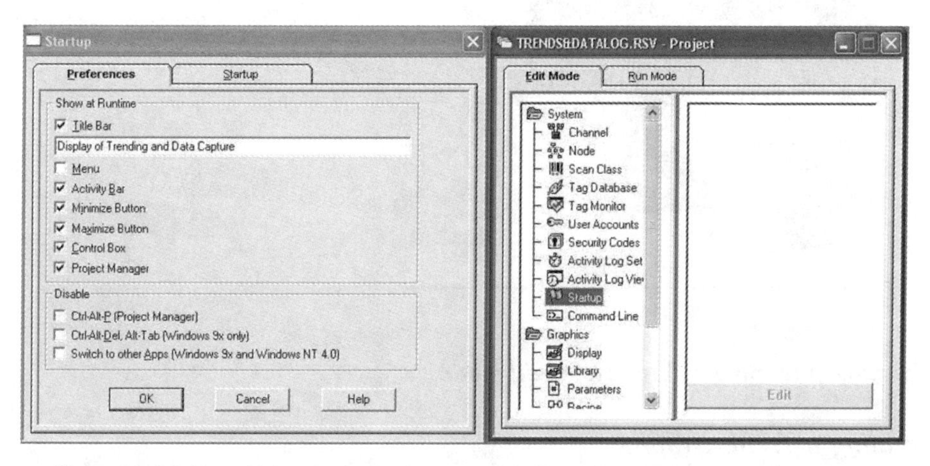

Figure 5-7-24: Enter Values in the Preferences Tab (Courtesy of Rockwell Automation, Inc.)

34. Click the ***Startup*** tab. Then click the selected check marks to choose the values to enter for ***Data Logging*** and ***Initial Graphic***. Click ***OK***.

5.7.8 Testing the Trend Graphics

35. Test the trend graphics using ***Run Mode***. Displays like the one shown in Figures 5-7-25 and then 5-7-26 should appear.

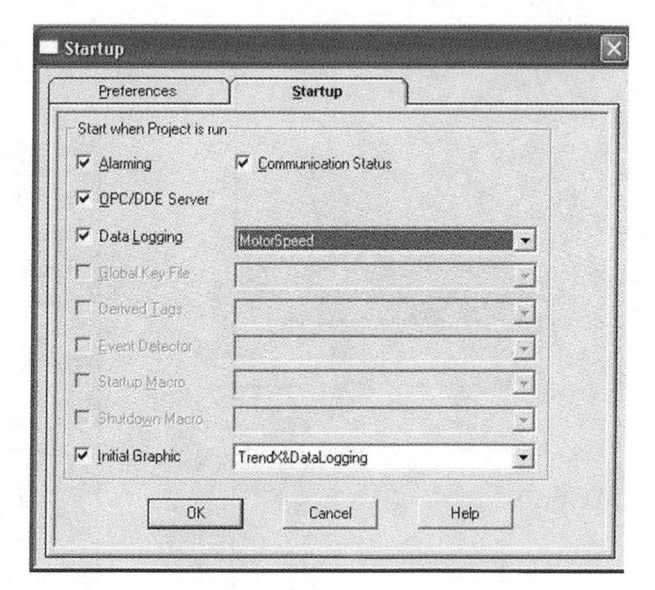

Figure 5-7-25: Test Trend Graphics (Courtesy of Rockwell Automation, Inc.)

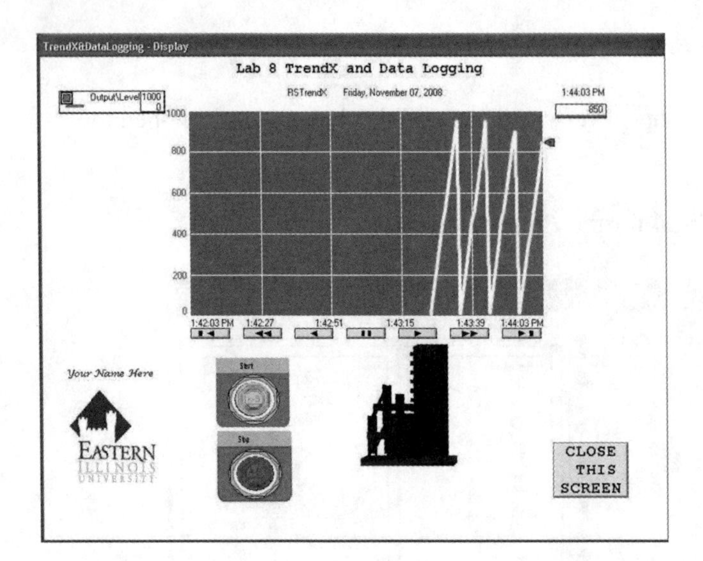

Figure 5-7-26: TrendX Graphics Screen (Courtesy of Rockwell Automation, Inc.)

36. Stop the project running and return to ***edit mode***.

5.7.9 Creating a Trend Graph

37. Click ***Graphics*** > ***Display*** then click ***Display*** again to create a new ***Untitled*** display window. Click ***Objects*** > ***Advanced Objects*** > ***Trend.*** Now drag a ***graph box*** into the bottom of the screen while leaving about 1/2 to 1-inch gap at the bottom, as shown in Figure 5-7-27.

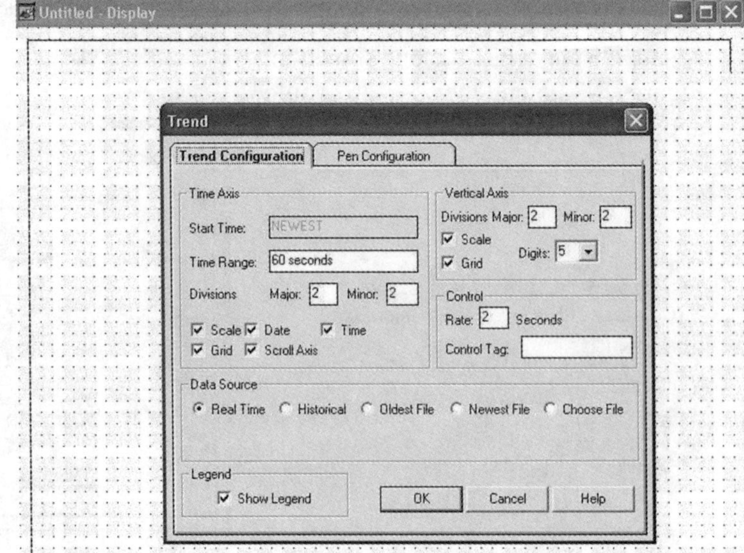

Figure 5-7-27: Trend Configuration Screen (Courtesy of Rockwell Automation, Inc.)

38. Note the ***Trend Configuration*** window opens at the same time.

39. Enter the values shown in the ***Trend Configuration*** (Figure 5-7-28) and click OK. Enter the values shown in ***Pen Configuration*** (Figure 5-7-29) and click OK.

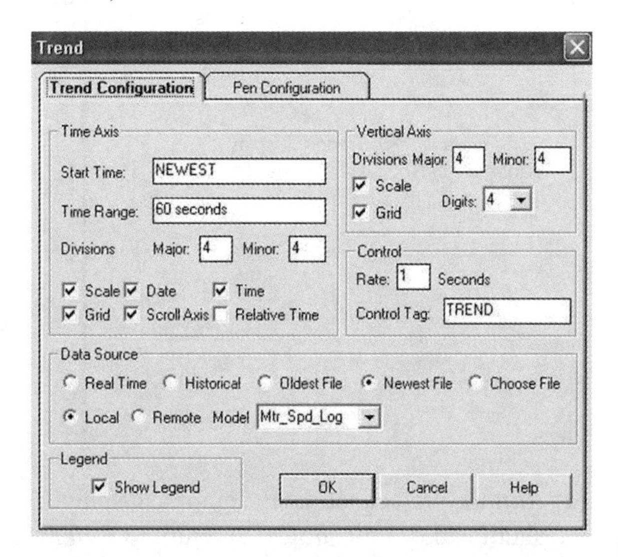

Figure 5-7-28: Values to Enter in Trend Configuration Screen (Courtesy of Rockwell Automation, Inc.)

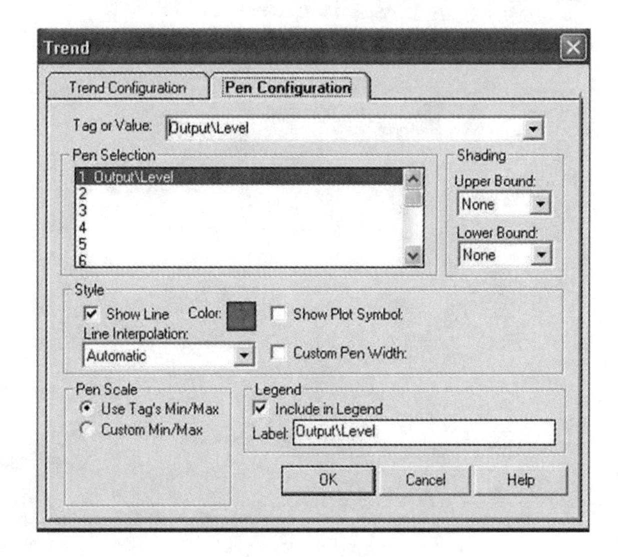

Figure 5-7-29: Pen Configuration Tag Values in Trend (Courtesy of Rockwell Automation, Inc.)

40. When the two figures above complete their operations, the trend graph screen shown in Figure 5-7-30 should appear. Add a ***CLOSE*** button at the bottom of this screen. Click the ***CLOSE*** button to close out and save this screen. Use ***Trend*** as the ***Save As*** screen name.

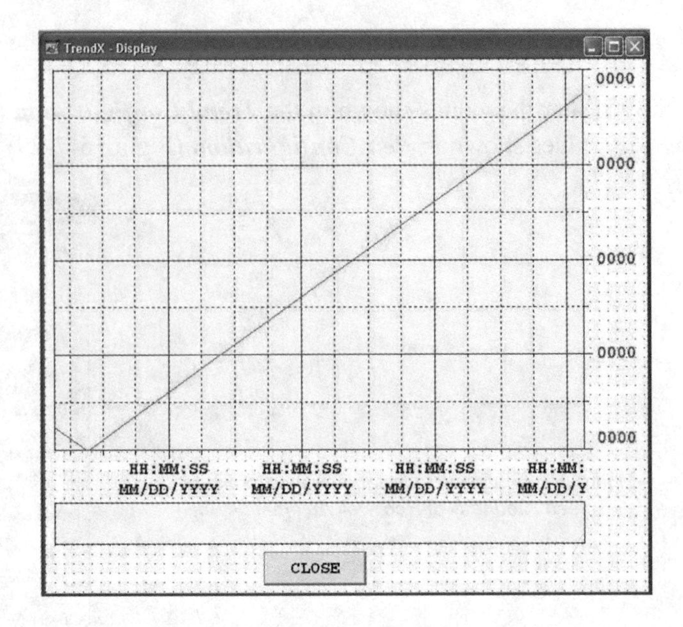

Figure 5-7-30: TrendX Display
(Courtesy of Rockwell Automation, Inc.)

41. ***Run*** the project and let the graphic display fill completely before stopping the process and shutting down the project to return to ***Edit Mode***.

RSView Lab 5-8: Moving Animation

5.8.1 Introduction

This laboratory demonstrates how RSView32 software can be used to create moving objects coupled with foreground and background objects.

The laboratory also tests the abilities to create HMI interfaces based upon previous learning experiences.

NOTE: When instructions such *as* ***Edit Mode, Objects>Advanced Objects>Numeric Display and others*** appear and are not located on a Figure, they are for use in RSView32 software, which is running in the background.

1. Enter the program shown in Figure 5-8-1 into the ***RSLogix 500.*** Then, ***Download*** it to the PLC, and put the PLC in ***Run*** mode.

Figure 5-8-1: PLC program for Lab 5-8 (Courtesy of Rockwell Automation, Inc.)

2. Configure ***RSView*** and ***RSLinx*** in a way similar to how previous laboratories were configured.

3. Create the graphic shown in Figure 5-8-2 using the ***Library*** stored icons with the exception of the roadways. Place all of the objects in the graphic display exactly as shown.

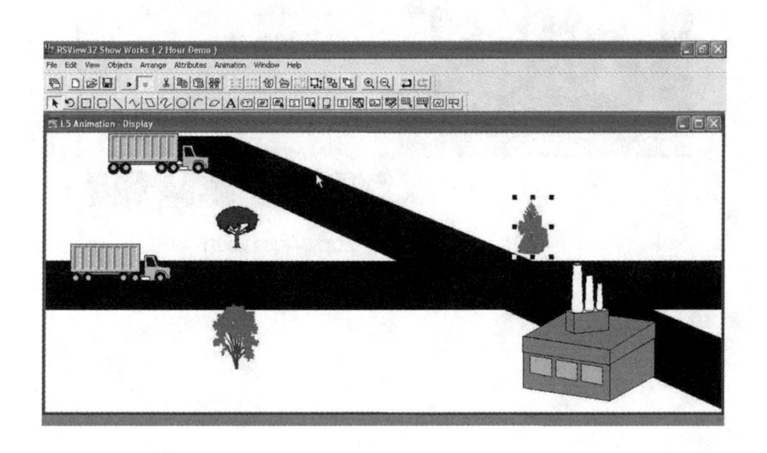

Figure 5-8-2: Picture of Roadways and Objects (Courtesy of Rockwell Automation, Inc.)

4. To create the roadways, use both the rectangle and polygon graphical creation functions in ***RSView*** (Figure 5-8-3).

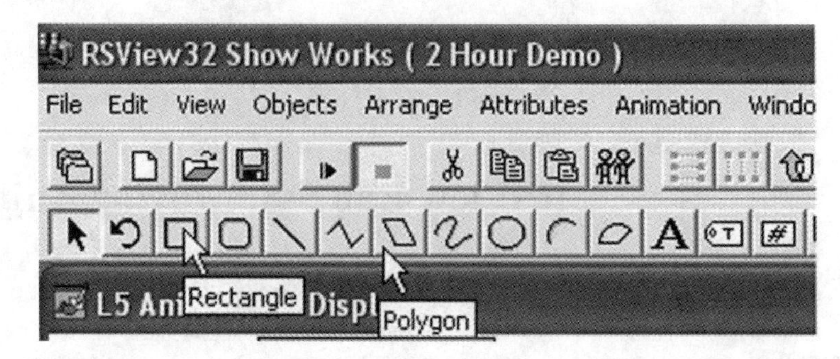

Figure 5-8-3: Shapes to create Roadways (Courtesy of Rockwell Automation, Inc.)

5. Right click objects in the display; then click ***Arrange*** and either ***Send to Back*** or ***Bring to Front*** in order to overlap icons, as needed, in front or in behind of each other.

6. To make a copy of the truck, right click and then select either ***Duplicate*** or ***Copy and Paste*** (Figure 5-8-4).

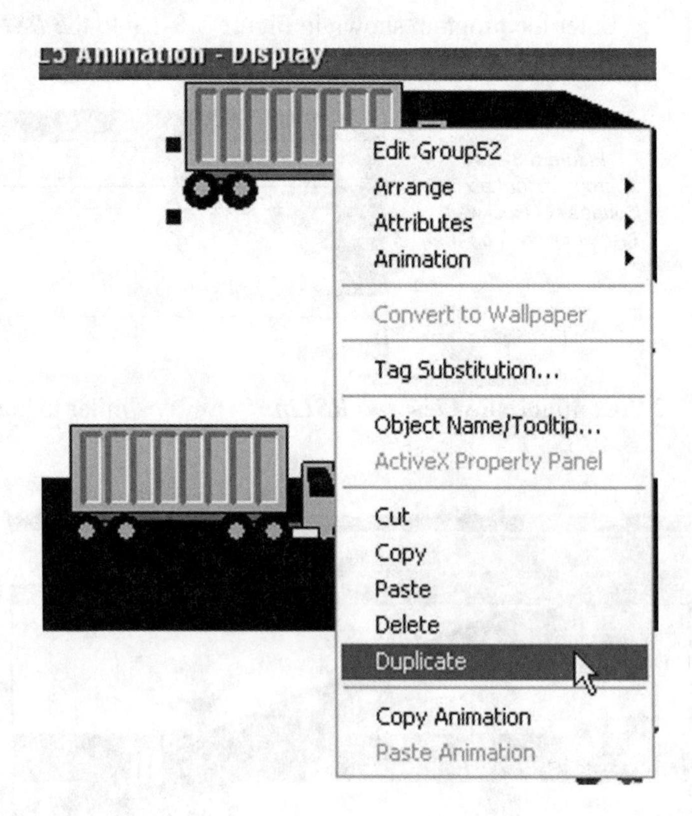

Figure 5-8-4: Adding Trucks to Roadways (Courtesy of Rockwell Automation, Inc.)

7. Click ***System*** under ***Edit Mode*** and then double click ***Tag Database***. Create an ***Outputs\ Tag*** folder and a single ***Outputs\ Tag*** for the trucks, as shown in Figure 5-8-5. This one tag will be shared between both truck icons.

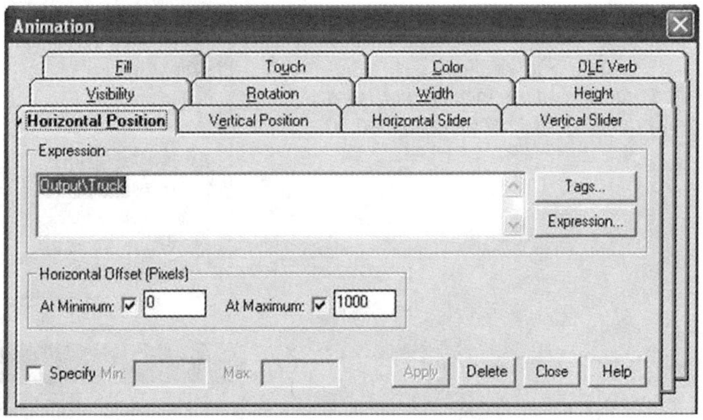

Figure 5-8-5:
Horizontal Position Animation #1 of Truck
(Courtesy of Rockwell Automtion, Inc.)

8 To animate the truck located in the middle of the display, select ***Animation** and **Horizontal Position.*** Program the screen as shown in Figure 5-8-5.

9. To animate the truck located at the top left of the display, select ***Animation, Horizontal Position, Vertical Position***, and ***Rotation*** to program these screens, as shown in Figures 5-8-6, 5-8-7, and 5-8-8.

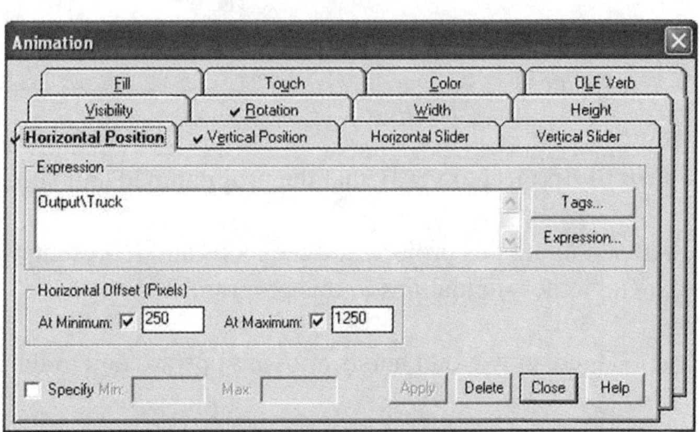

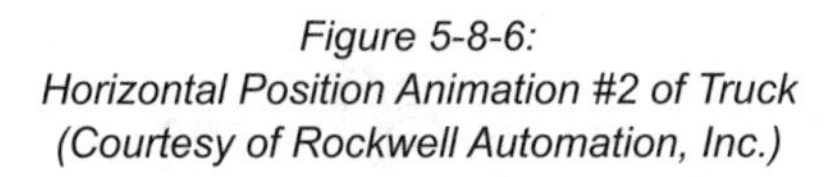
Figure 5-8-6:
Horizontal Position Animation #2 of Truck
(Courtesy of Rockwell Automation, Inc.)

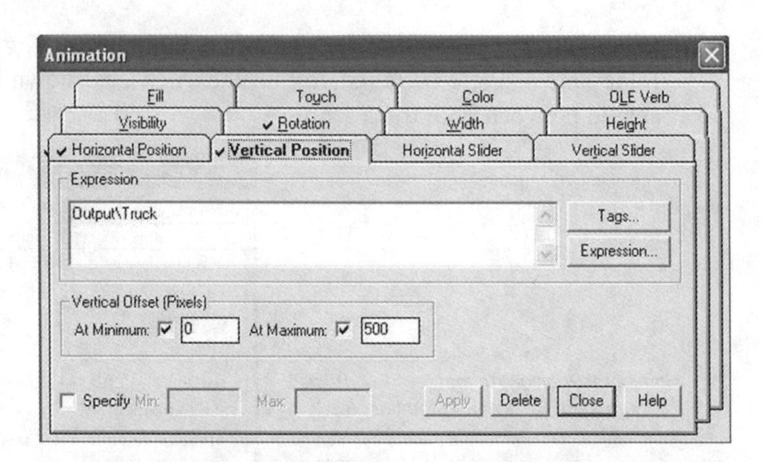

Figure 5-8-7:
Vertical Position Animation of Truck
(Courtesy of Rockwell Automation, Inc.)

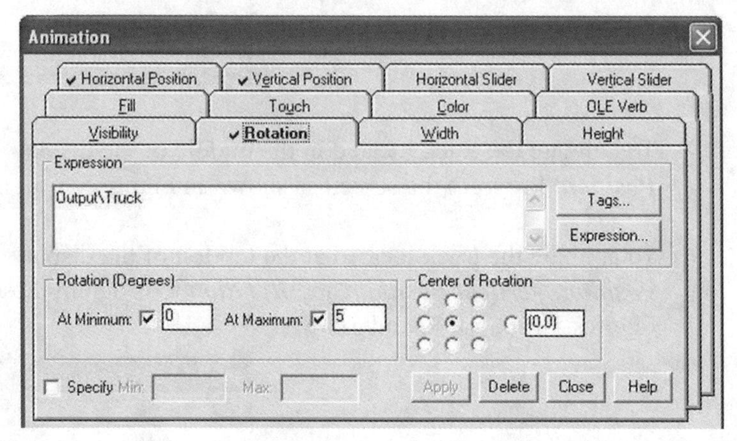

Figure 5-8-8:
Rotational Position Animation of Truck
(Courtesy of Rockwell Automation, Inc.)

10. ***Run*** this project to verify that the programmed animation works.

11. After running this project, add the icons and representations shown in Figure 5-8-9, with the following other stipulations to this program:

    a. Three, gray cloud bursts, moving upward and to the right, coming from the factory chimneys.

    b. A conveyor belt with a pallet of sacks on it, running from the factory to the left side of the screen. The pallet exits the factory ONLY after the trucks have passed the factory. A ***Timer On-Delay timer*** (TOD) will need to be added to the PLC program.

    c. An airplane lifting off at the end of the conveyor belt, flying diagonally to the right side of the screen. The pallet exits the factory ONLY after the trucks have passed the factory.

    d. Add another ***Timer On-Delay*** (TON) or ***Timer Off-Delay*** (TOD) to the PLC program for the pallet and airplane as needed, and in the sequenced order of operation.

e. It would be appropriate to work in groups of two or three to complete this portion of the laboratory.

f. Save this modified ***RSLogix***, ***RSView***, ***RSLinx*** program to the appropriate folder.

12. ***Re-Run*** this project to verify that the animation works as required, as shown in Figure 5-8-9.

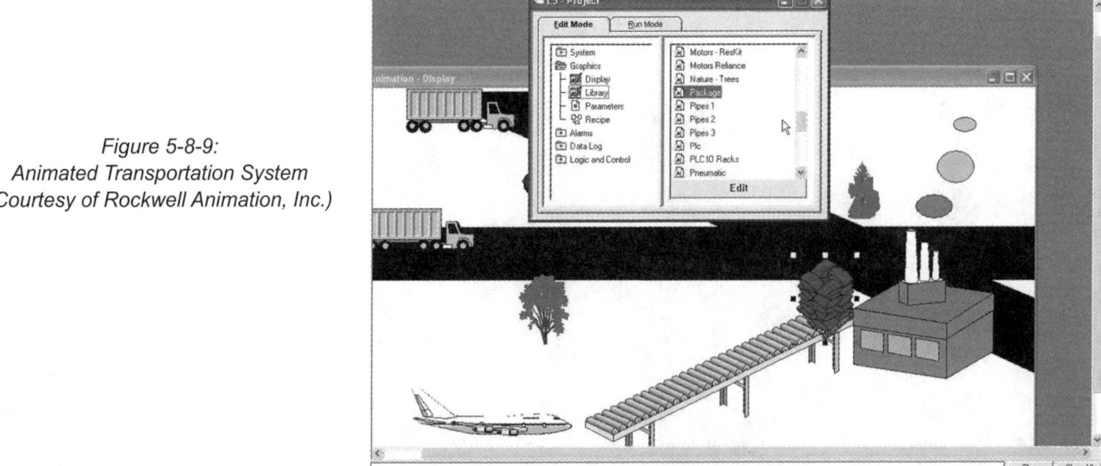

Figure 5-8-9: Animated Transportation System (Courtesy of Rockwell Animation, Inc.)

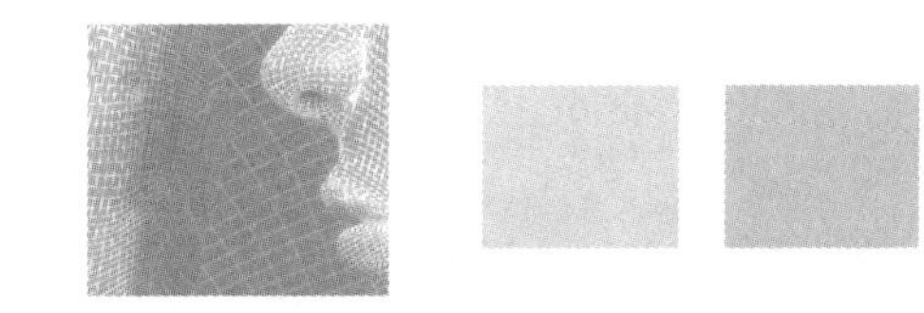

Appendix I: Hardware and Software Required for Chapters 4 and 5 Lab Activities

The lab activities require the following hardware and software:

Chapter 4 PV Labs 1–6 require the following Allen-Bradley equipment:

- MicroLogix 1200 PLC (or a MicroLogix 1100 PLC with 2 analog input channels)
- PanelView 550 graphic terminal
- 2711-NC13 PanelView programming/communication cable
- 1761-CBL-PM02 MicroLogix programming/communication cable

Chapter 4 PV Labs 1–6 require the following electromechanical devices:

- Two NC STOP pushbuttons
- Two NO START pushbuttons
- Two pilot lights
- One hobby DC electric motor

All Chapter 4 PV Labs require the following Rockwell Automation HMI software:

- RSLinx Lite or RSLinx Classic Professional
- RSLogix 500 Starter or higher
- PanelBuilder32

Chapter 4 PV Lab 4 requires a terminal emulation program. There are several available online. Here is one that works well:

- Hyperterminal Private Editions: http://www.hilgraeve.com/hyperterminal/

Chapter 5 RSView Labs 1–8 require the following:

- All Chapter 5 labs require the following Rockwell Automation HMI software:
  - RSLinx Classic Professional
  - RSLogix 500 Starter or higher
  - RSView32 (requires RSLinx Classic Professional)

Chapter 5 RSView Lab 5 requires the following Allen-Bradley equipment:

- MicroLogix 1100 PLC with two analog input channels
- Digital Temperature Sensors (see Appendix III)

Appendix II:
PLC and HMI Manuals for Software and Hardware Used in Labs

The following manuals are available for the Allen-Bradley hardware and Rockwell Automation software. We recommend that you have these manuals available for reference.

1. Allen-Bradley PanelView Standard Operating Terminals User Manual:
   2711-UM014G-EN-P.pdf
   http://literature.rockwellautomation.com/idc/groups/literature/documents/um/2711-um014_-en-p.pdf

2. RSLinx Getting Results Guide Publication:
   LINX-GR001I-EN-E.pdf
   http://literature.rockwellautomation.com/idc/groups/literature/documents/gr/linx-gr001_-en-e.pdf

3. PanelBuilder32 Quick Start Guide:
   2711-QS003-EN-P.pdf
   http://literature.rockwellautomation.com/idc/groups/literature/documents/qs/2711-qs003_-en-p.pdf

4. PanelBuilder32 Getting Results:
   2711-GR003-EN-P.pdf
   http://literature.rockwellautomation.com/idc/groups/literature/documents/gr/2711-gr003_-en-p.pdf

5. RSView32 Getting Results Guide:
   VW32-GR001-EN-E.pdf
   http://literature.rockwellautomation.com/idc/groups/literature/documents/gr/vw32-gr001_-en-e.pdf

6. RSView32 User Guide:
   VW32-UM001-EN-e.pdf
   http://literature.rockwellautomation.com/idc/groups/literature/documents/um/vw32-um001_-en-e.pdf

7. RSView32 User Manual:
   http://support.elmark.com.pl/rockwell/Literatura/2711-60.pdf

Appendix III: Constructing Digital Temperature Sensors

Both Chapter 4 Lab 5 and Chapter 5 Lab 5 require a digital temperature sensor. The commercial temperature sensors available are not designed for simple connections, such as banana plugs that are used for education level equipment and systems.

Figure A-1 shows an assembled, inexpensive Fahrenheit temperature sensor that is relativity easy to construct.

Figure A-1: Fahrenheit temperature sensor

The sensors can be scaled for Fahrenheit or Celsius temperature and are relatively accurate for educational purposes.

They consist of an integrated circuit temperature chip, resistor, connecting wires, banana plugs powered by a 9-volt battery, and a battery clip.

For details on how to build this sensor, please contact the authors.

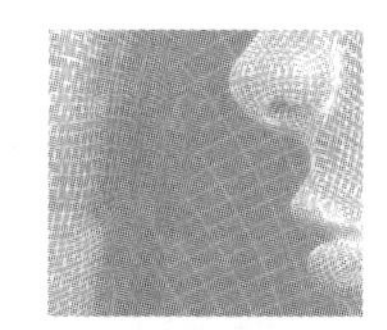

Index

A

AB Utilities 17
activate buttons 186–187
addresses 10
alarms 2, 3, 17, 73–81, 159–177
  banner screen 78–80, 87
  bar graph 166–167
  buttons 167–171, 188–189
  close button 174–177
  displays 74, 171–174
  graphic images 162–166
  message display 79
  print list 76–77
  triggers 79
Allen-Bradley
  1761-CBL-PM02, cable 33
  2711-NC13, cable 33
  website 16
analog dials 17
analog input 83–87, 179
analog meters 20
analog temperature, displays 184, 186
animation 2, 3, 17, 39, 95, 97, 104, 115–116, 141–143
  color 154
  invisibility 117–119, 143–144
  moving 215–219
  programming 116–117
  pump and tank 153–155, 157
  warning 147
annunciator 26
  buzzer 21–22
assembly 2
assembly line 28
Auto-Configure 36–38, 99–100
automated control layout 16–17, 20
Available Driver Types 36–37

B

banner screen, alarms 78–80, 87
bar graph 3, 69–70, 75–76
  alarms 166–167
  creating 61–62
  temperature 85–86
Baud Rate 43, 64
Bedford and Associates 4
binary addressing 12, 51, 97, 121
blink 17, 47, 90, 143
borders 145
brightness 16
buttons
  activate 86–187
  alarms 167–177, 188–189
  Celsius 184–185
  configuration 174–175, 185
  creating 110–112, 136–139
  icons 45–49
  security 196–197
buzzer, annunciator 21–22

C

cables, printer port 33
capactive sensors 25
Celsius 83, 85, 179, 184–186, 223
Channel 10, 102, 127
charting 3
  TrendX 201–214
circuits 6, 8
  closed 28
  latching 95
  open 28
Clear screen 78
clock, displays 139–141
Close Screen 20
closed circuit 6, 28
codes, security 198

INDEX

color
    display 15, 17
    palette 117, 142–143, 209
COMM light 11
communication driver 102
COMMUNICATION SETUP 34, 43, 56, 64
communications
    control 32
    drivers 17
    software 34–41, 98
CompactLogix PLC 4
company logo 17
computer-based HMI displays, RSView32 19–20
condition changes 21
CONFIG on power up 33–34
Config screen selector 70, 74, 80, 81, 87
configuration 33–34, 159–160
    buttons 174–175
    RSView 181, 192, 202
    startup 211
    TrendX 212–213
Configure Drivers 36–37, 98–99
contact relay 6–7, 22–23
contrast 16
control
    communications 32
    devices, physical 16
    layout 16–17
control process 2, 18
    graphic images 109–110
control relay contact 6–7, 10
control valve, directional (see DCV)
Create New Application 41–42
Create New Project 55
creating
    bar graph 61–62
    buttons 110–112, 136–139
    data logs 202–205
    folder 131
    graphic images 109–110, 135–136, 183, 195
    message display 60–61
    new project 101, 126
    programming 96–100, 121–122, 151
    scrolling message 59–60
    timer 123–124

D
data acquisition 2
Data Files 40, 97, 121
    properties 52
data logs 201–214
    setup 202
date 18, 66, 113–115, 139
DATE/TIME 34
DC power supply 32
DCOMM light 11
DCS (distributed control systems) 5
DCV (directional control valve) 22-24
digital
    analog meters 17
    clock 139–141
    display 56–58, 60
    graphic display 2
    meter, graphic images 144–145
    numeric displays 17, 20
    temperature 179, 183–184
    thermometer 83
directional control valve (see DCV)
displays
    alarms 171–177
    analog 184, 186
    color 15
    computer-based 18–19
    digital 17, 183–184
    digital clock 139–141
    graphic images 201
    loading 49–50
    message display 157–158
    monochrome 15
    motor latch 111–112, 138–139
    multiple 17
    multistate 76
    RSView32 19–20
    temperature 183–184
    TrendX 205–210
    types 15–16, 18–20
    warning 146–149
distributed control systems (DCV) 5
Download 49
drivers 36
    communications 17
    configuration 98–99

INDEX

E
electrical switch 23
electromagnets 4, 22
Elements 53, 122
Ethernet 5, 15
Excel 20

F
Factory Talk Transaction Manager 3
Fahrenheit 83, 86–87, 179, 223
failure 2, 4, 8
fans 24
FAULT light 11
field applications 97
field devices 21–29
file transfer 71
filling mode, pump and tank 154–155
folder, create 35
FORCE light 11

G
gauge display 56–58, 144–145, 149
General Motors 4
GOTO Config Screen 16–17, 33, 47–48, 50, 65, 70
Graphic Display HMI Terminal 18
graphic images 17
Graphic Terminal HMI Display 15–16, 31
graphics 48, 141–143
- alarms 162–166
- control process 109–110
- creating 135–139, 183, 195
- digital display 2
- digital meter 144–145
- pump and tank 152–153
- testing 211–212
- TrendX 212–214

H
hardware, labs 221
heating element 73
help 13
hidden color palette 117, 142–143
high level indicator 67, 152, 155–156
histograms 3, 17, 20
HMI (Human Machine Interface), introduction 1–3
HMI Diagram, creating 56, 74
Human Machine Interface (HMI), introduction 1–3

I
I/O modules 10
icons 48, 136
- creating 45–49
- library 110
- RSLogix 500 13

indicators
- gauge 57
- high level 68
- icons 45–49
- multistate 47, 68

inductive sensors 25
information 1
input modules 9, 10
interface, operator 15
internal binary input 12
Internet 5
invisibility 117–119, 143–144, 148

K
keyboard 15

L
ladder logic diagram 8, 16, 83–84, 95, 159, 179–180, 191
- introduction 5–9
- PLCs 10–11
- programming 11–14

latching circuit 8, 95
library
- icons 110
- images 17, 20
- symbols 19

library, graphic images 136
lights 19, 25–26
Limit Tests 51, 55, 121
- symbols 124–125

Local Disk (C:) 35
LOGIN 89–90
logo 17–18, 44, 48
low level indicator 152, 156

M

magnetic fields 25
Man Machine Interface (MMI) 1
manuals, labs 222
math operations 83–87, 179
memory circuit 8
message, scrolling 59–60
Message Display 60–61
messages, alarms 79, 188
meters 17
MicroLogix 1100 PLC 83–84, 179–181
MicroLogix 1200 PLC 4, 151
MicroLogix series 32–33, 95
    programming 34–41, 51–56, 63, 73
microswitches 8, 24, 26–27
MMI (Man Machine Interface) 1
modular PLC 4–5
momentary pushbuttons 1, 6, 28–29, 74–75
monitoring 2, 11
monochrome display 15, 17
Morley, Dick 4
motor graphic 17, 20
motor latch, displays 111–112, 138–139, 141
motor speed 17
motor starter 31–50, 95–119
    screen selector 70
moving animation 215–219
multiple screens 17, 51

N

Name Here 44
NC (Normally Closed) 6–7, 22–23, 29, 123
Network Nodes 43
Network Type 127
neutral 7
NO (Normally Open) 6–7, 22–23, 29
Node 103, 127–128, 134, 151
Normally Closed (see NC)
Normally Open (see NO)

O

Object Properties 61, 76
Objects 43–44
open circuit 28
operator interface 1, 15
operator panel 1–2, 15
organization name 18
output devices 16
output modules 9–10
output motor graphic 16–17

P

panel, operator 15
PanelBuilder32 16–18
    creating HMI diagram 56, 74
    motor starter 31–50
    programming 41–44
    pump and tank 63–72
PanelView
    graphic terminal display 31
    loading HMI display screen 49–50
    models 15–16
    programming 41–44
PanelView 330 (see PV300)
PanelView 550 (see PV550)
password protection 89
photo cell 25
photo sensors 27–28
photo-switches 2
pilot lights 1–2, 6–7, 10, 16, 25–26
piston 26
piston, pneumatic 22–24
PLCs (programmable logic controllers) 1, 3–5
pneumatic piston 22–24
poles 7–8
port, serial communications 32
power connections 32
POWER light 11
Print List 80–81
PRINTER configuration 34
printer port 32–33
process diagram 42
process graphics 17
Process Type 38–39
Processor Type 96
production line 2
programmable logic controllers (PLCs) 1, 3–5
programmed diagrams 40
Project Name 56
Properties 46, 76–77, 122
    configuration 211
    Data Files 52
    TrendX 206–208

pump 17–18, 63–72, 151–158
  displays 64–69
pushbuttons 1–2, 6–8
  Heater Off 74–75
  momentary 28–29
  states 46
PV300 15–16
PV550 15–17, 31–33
  back view 32
  resolution 48

R
reflectors 27–28
relay, 1–2, 4
  contact 22–23
relay ladder logic diagram 5–6, 10
Rockwell Automation, website 16
Rockwell Automation PanelBuilder32 (see PanelBuilder32)
rod 23
rotary tables 27
rotation 149
RS232
  communications 99
  connector 33
  printer port 32
  serial communications 34, 36, 38, 96
RS232 DF1, port 32
RSLinx 11, 96, 98–99
  configuration 159–160
RSLinx Classic 81
RSLinx Professional, communications 34–41
RSLinx32 17, 20
RSLogix 500 20, 32, 95–100
  configuration 159–160
  English 38, 51–52, 96, 121
  icons 13
  ladder diagrams 11–12
  programming 34–41, 101
  starter screen 39
RSView32 19–20, 96, 106
  alarms 159–177
  configuration 100–104, 151, 159–160, 181, 192, 202
  displays 121–149
  labs 95–219
  launch 100–104, 125–129
  motor starter 95–119
  moving objects 215–219
  programming 142–143
  pump and tank 151–158
RSWho 104, 128, 134
RUN 41
RUN light 11
rungs 5

S
SCADA 2
scan class 104, 128–129
SCREEN SETUP 34
scrolling text 18, 59–60
security 89–93, 191–200
  tags 192–195
sensors 21, 83
  capacitive 25
  inductive 25
  photo 27–28
  temperature 9, 29, 179, 223
serial communications 34
  RS232 96
serial connection 15
shapes 145
single line address 10
slot number 10
software
  computer-based 18–19
  labs 221
  types 15–20
solenoid 2, 7–8, 22–23
speed, motor 17
spreadsheets 3, 5, 19
sssssss string 114, 140, 197
  alarms 169
stack light units 26
START 6, 10, 16–17, 19–20, 29, 48
  properties 46
  states 47
starter program 40
states 46, 67, 77
  alarms 171, 188
  Heater Off 75
STOP 6, 10, 16–17, 19–20, 29, 48
subprogramming 13, 83–84, 180
supervisory control 2

switches 4, 6, 16, 19, 28–29
- electrical 23
- micro- 24
- momentary pushbuttons 28–29

symbols 19
- limit test 124–125

System 126–127
- folder 102

T

tags 18, 44–45, 56–57, 65, 74–75, 87, 91, 107–109, 129–135, 151–152, 160–162, 170, 173, 181–182, 202
- database 105–109
- security 192–195
- TrendX 207–208

tank 17–18, 63–72, 151–158
- displays 64–69
- graphic 67

temperature 85–87, 159
- alarms 87
- Celsius 85
- conversion 84
- digital 179, 223
- Fahrenheit 86–87
- sensors 9, 29

TEP (Termination Emulation Program) 80–81
Termination Emulation Program (TEP) 80–81
testing
- graphic images 211–212
- security 199–200

text 17, 44
- scrolling 18

thermistors 83, 179
thermocouples 179
thermometers 83
tick marks 76
time 18, 66, 139
- adding 113–115

Time Base 53
Time On-Delay (TON) 13–14, 51, 53, 63, 73, 121, 123–124, 218
- icons 54

TON (see Timer On-Delay)
touch screen 15
trainer board 8
trends 201–214
- real time 20

TrendX
- charting 201–214
- displays 205–210

triggers, alarms 79

U

user accounts 199

V

valve, directional control (see DCV)
VDC (voltage direct current) 9
Visual Basic 20
voltage direct current (VDC) 9

W

warning message 146–149, 165
- pump and tank 155–157

X

x-axis 209
X–Y graphs 3, 20

Y

y-axis 210

About the Authors

Dr. Sam Guccione is a retired Associate Professor from Eastern Illinois University's School of Technology. For 10 years, he taught undergraduate and graduate courses on topics such as Electronic Control Systems, Robots and Control Systems, Programmable Logic Controllers, Human Machine Interface (HMI), Networking, and Data Communications.

He also served as the coordinator of the School of Technology Automation Laboratory. In this role, he implemented an industrial Ethernetwork that connected Allen-Bradley PLCs, Rockwell Automation HMIs, robots, radio-frequency identification (RFID), web cameras, and machine vision cameras into classroom-based industrial processes.

For 25 years prior to his work at Eastern Illinois University, Dr. Guccione was Department Chair and Instructional Director (Assistant Dean) at Delaware Technical and Community College. He developed and taught over 30 engineering technology courses used in Associate degree programs in electrical, computer, and electro-mechanical engineering technologies.

Dr. Guccione began his career as an aerospace engineer in the National Aeronautics and Space Administration's (NASA's) Project Gemini space program and at several industrial companies. He holds Bachelor's and Master's degrees in Electrical Engineering from the University of Illinois, and a Doctorate in Education from Temple University.

He has presented papers at local, regional, and national conferences such as the Association of Technology, Management, and Applied Engineering (ATMAE) and has published in journals about his teaching techniques and research. He has been a member of ATMAE, as well as the Association of Engineering Education (ASEE).

Dr. James McKirahan is an Assistant Professor in Applied Engineering and Technology Management at Indiana State University, where he currently teaches Manufacturing Engineering Technology. He has been teaching manufacturing and industrial control systems, as well as robotics, for 17 years. He also has several years of manufacturing supervision and management experience.

Dr. McKirahan earned an Associate of Applied Science degree in Electronics Technology from Southern Illinois University–Carbondale; Bachelor's and Master's degrees in Industrial Technology and Technology, respectively, from Eastern Illinois University; and a Ph.D. in Technology Management, Manufacturing Systems, from Indiana State University.

His work includes a number of professional presentations and articles. In addition, he is a member of ATMAE; the Society of Plastics Engineers (SPE); and the Society of Manufacturing Engineers (SME).

About the Authors